T0228376

Pitman Research Notes in Mathem Series

Submission of proposals for consideration
Suggestions for publication, in the form of outlines and representative samples, are invited by the Editorial Board for assessment. Intending authors should approach one of the main editors or another member of the Editorial Board, citing the relevant AMS subject classifications. Alternatively, outlines may be sent directly to the publisher's offices. Refereeing is by members of the board and other mathematical authorities in the topic concerned, throughout the world.

Preparation of accepted manuscripts
On acceptance of a proposal, the publisher will supply full instructions for the preparation of manuscripts in a form suitable for direct photo-lithographic reproduction. Specially printed grid sheets can be provided and a contribution is offered by the publisher towards the cost of typing. Word processor output, subject to the publisher's approval, is also acceptable.

Illustrations should be prepared by the authors, ready for direct reproduction without further improvement. The use of hand-drawn symbols should be avoided wherever possible, in order to maintain maximum clarity of the text.

The publisher will be pleased to give any guidance necessary during the preparation of a typescript, and will be happy to answer any queries.

Important note
In order to avoid later retyping, intending authors are strongly urged not to begin final preparation of a typescript before receiving the publisher's guidelines. In this way it is hoped to preserve the uniform appearance of the series.

Longman Scientific & Technical
Longman House
Burnt Mill
Harlow, Essex, CM20 2JE
UK
(Telephone (0279) 426721)

Titles in this series. A full list is available from the publisher on request.

100 Optimal control of variational inequalities
V Barbu
101 Partial differential equations and dynamical systems
W E Fitzgibbon III
102 Approximation of Hilbert space operators Volume II
C Apostol, L A Fialkow, D A Herrero and D Voiculescu
103 Nondiscrete induction and iterative processes
V Ptak and F-A Potra
104 Analytic functions – growth aspects
O P Juneja and G P Kapoor
105 Theory of Tikhonov regularization for Fredholm equations of the first kind
C W Groetsch
106 Nonlinear partial differential equations and free boundaries. Volume I
J I Díaz
107 Tight and taut immersions of manifolds
T E Cecil and P J Ryan
108 A layering method for viscous, incompressible L_p flows occupying R^n
A Douglis and E B Fabes
109 Nonlinear partial differential equations and their applications: Collège de France Seminar. Volume VI
H Brezis and J L Lions
110 Finite generalized quadrangles
S E Payne and J A Thas
111 Advances in nonlinear waves. Volume II
L Debnath
112 Topics in several complex variables
E Ramírez de Arellano and D Sundararaman
113 Differential equations, flow invariance and applications
N H Pavel
114 Geometrical combinatorics
F C Holroyd and R J Wilson
115 Generators of strongly continuous semigroups
J A van Casteren
116 Growth of algebras and Gelfand–Kirillov dimension
G R Krause and T H Lenagan
117 Theory of bases and cones
P K Kamthan and M Gupta
118 Linear groups and permutations
A R Camina and E A Whelan
119 General Wiener–Hopf factorization methods
F-O Speck
120 Free boundary problems: applications and theory. Volume III
A Bossavit, A Damlamian and M Fremond
121 Free boundary problems: applications and theory. Volume IV
A Bossavit, A Damlamian and M Fremond
122 Nonlinear partial differential equations and their applications: Collège de France Seminar. Volume VII
H Brezis and J L Lions
123 Geometric methods in operator algebras
H Araki and E G Effros
124 Infinite dimensional analysis–stochastic processes
S Albeverio
125 Ennio de Giorgi Colloquium
P Krée
126 Almost-periodic functions in abstract spaces
S Zaidman
127 Nonlinear variational problems
A Marino, L Modica, S Spagnolo and M Degliovanni
128 Second-order systems of partial differential equations in the plane
L K Hua, W Lin and C-Q Wu
129 Asymptotics of high-order ordinary differential equations
R B Paris and A D Wood
130 Stochastic differential equations
R Wu
131 Differential geometry
L A Cordero
132 Nonlinear differential equations
J K Hale and P Martinez-Amores
133 Approximation theory and applications
S P Singh
134 Near-rings and their links with groups
J D P Meldrum
135 Estimating eigenvalues with *a posteriori/a priori* inequalities
J R Kuttler and V G Sigillito
136 Regular semigroups as extensions
F J Pastijn and M Petrich
137 Representations of rank one Lie groups
D H Collingwood
138 Fractional calculus
G F Roach and A C McBride
139 Hamilton's principle in continuum mechanics
A Bedford
140 Numerical analysis
D F Griffiths and G A Watson
141 Semigroups, theory and applications. Volume I
H Brezis, M G Crandall and F Kappel
142 Distribution theorems of L-functions
D Joyner
143 Recent developments in structured continua
D De Kee and P Kaloni
144 Functional analysis and two-point differential operators
J Locker
145 Numerical methods for partial differential equations
S I Hariharan and T H Moulden
146 Completely bounded maps and dilations
V I Paulsen
147 Harmonic analysis on the Heisenberg nilpotent Lie group
W Schempp
148 Contributions to modern calculus of variations
L Cesari
149 Nonlinear parabolic equations: qualitative properties of solutions
L Boccardo and A Tesei
150 From local times to global geometry, control and physics
K D Elworthy

151 A stochastic maximum principle for optimal control of diffusions
U G Haussmann
152 Semigroups, theory and applications. Volume II
H Brezis, M G Crandall and F Kappel
153 A general theory of integration in function spaces
P Muldowney
154 Oakland Conference on partial differential equations and applied mathematics
L R Bragg and J W Dettman
155 Contributions to nonlinear partial differential equations. Volume II
J I Díaz and P L Lions
156 Semigroups of linear operators: an introduction
A C McBride
157 Ordinary and partial differential equations
B D Sleeman and R J Jarvis
158 Hyperbolic equations
F Colombini and M K V Murthy
159 Linear topologies on a ring: an overview
J S Golan
160 Dynamical systems and bifurcation theory
M I Camacho, M J Pacifico and F Takens
161 Branched coverings and algebraic functions
M Namba
162 Perturbation bounds for matrix eigenvalues
R Bhatia
163 Defect minimization in operator equations: theory and applications
R Reemtsen
164 Multidimensional Brownian excursions and potential theory
K Burdzy
165 Viscosity solutions and optimal control
R J Elliott
166 Nonlinear partial differential equations and their applications: Collège de France Seminar. Volume VIII
H Brezis and J L Lions
167 Theory and applications of inverse problems
H Haario
168 Energy stability and convection
G P Galdi and B Straughan
169 Additive groups of rings. Volume II
S Feigelstock
170 Numerical analysis 1987
D F Griffiths and G A Watson
171 Surveys of some recent results in operator theory. Volume I
J B Conway and B B Morrel
172 Amenable Banach algebras
J-P Pier
173 Pseudo-orbits of contact forms
A Bahri
174 Poisson algebras and Poisson manifolds
K H Bhaskara and K Viswanath
175 Maximum principles and eigenvalue problems in partial differential equations
P W Schaefer
176 Mathematical analysis of nonlinear, dynamic processes
K U Grusa
177 Cordes' two-parameter spectral representation theory
D F McGhee and R H Picard
178 Equivariant K-theory for proper actions
N C Phillips
179 Elliptic operators, topology and asymptotic methods
J Roe
180 Nonlinear evolution equations
J K Engelbrecht, V E Fridman and E N Pelinovski
181 Nonlinear partial differential equations and their applications: Collège de France Seminar. Volume IX
H Brezis and J L Lions
182 Critical points at infinity in some variational problems
A Bahri
183 Recent developments in hyperbolic equations
L Cattabriga, F Colombini, M K V Murthy and S Spagnolo
184 Optimization and identification of systems governed by evolution equations on Banach space
N U Ahmed
185 Free boundary problems: theory and applications. Volume I
K H Hoffmann and J Sprekels
186 Free boundary problems: theory and applications. Volume II
K H Hoffmann and J Sprekels
187 An introduction to intersection homology theory
F Kirwan
188 Derivatives, nuclei and dimensions on the frame of torsion theories
J S Golan and H Simmons
189 Theory of reproducing kernels and its applications
S Saitoh
190 Volterra integrodifferential equations in Banach spaces and applications
G Da Prato and M Iannelli
191 Nest algebras
K R Davidson
192 Surveys of some recent results in operator theory. Volume II
J B Conway and B B Morrel
193 Nonlinear variational problems. Volume II
A Marino and M K V Murthy
194 Stochastic processes with multidimensional parameter
M E Dozzi
195 Prestressed bodies
D Iesan
196 Hilbert space approach to some classical transforms
R H Picard
197 Stochastic calculus in application
J R Norris
198 Radical theory
B J Gardner
199 The C^*-algebras of a class of solvable Lie groups
X Wang
200 Stochastic analysis, path integration and dynamics
K D Elworthy and J C Zambrini

201 Riemannian geometry and holonomy groups
S Salamon
202 Strong asymptotics for extremal errors and polynomials associated with Erdös type weights
D S Lubinsky
203 Optimal control of diffusion processes
V S Borkar
204 Rings, modules and radicals
B J Gardner
205 Two-parameter eigenvalue problems in ordinary differential equations
M Faierman
206 Distributions and analytic functions
R D Carmichael and D Mitrovic
207 Semicontinuity, relaxation and integral representation in the calculus of variations
G Buttazzo
208 Recent advances in nonlinear elliptic and parabolic problems
P Bénilan, M Chipot, L Evans and M Pierre
209 Model completions, ring representations and the topology of the Pierce sheaf
A Carson
210 Retarded dynamical systems
G Stepan
211 Function spaces, differential operators and nonlinear analysis
L Paivarinta
212 Analytic function theory of one complex variable
C C Yang, Y Komatu and K Niino
213 Elements of stability of visco-elastic fluids
J Dunwoody
214 Jordan decomposition of generalized vector measures
K D Schmidt
215 A mathematical analysis of bending of plates with transverse shear deformation
C Constanda
216 Ordinary and partial differential equations. Volume II
B D Sleeman and R J Jarvis
217 Hilbert modules over function algebras
R G Douglas and V I Paulsen
218 Graph colourings
R Wilson and R Nelson
219 Hardy-type inequalities
A Kufner and B Opic
220 Nonlinear partial differential equations and their applications: Collège de France Seminar. Volume X
H Brezis and J L Lions
221 Workshop on dynamical systems
E Shiels and Z Coelho
222 Geometry and analysis in nonlinear dynamics
H W Broer and F Takens
223 Fluid dynamical aspects of combustion theory
M Onofri and A Tesei
224 Approximation of Hilbert space operators. Volume I. 2nd edition
D Herrero
225 Operator theory: proceedings of the 1988 GPOTS–Wabash conference
J B Conway and B B Morrel
226 Local cohomology and localization
J L Bueso Montero, B Torrecillas Jover and A Verschoren
227 Nonlinear waves and dissipative effects
D Fusco and A Jeffrey
228 Numerical analysis 1989
D F Griffiths and G A Watson
229 Recent developments in structured continua. Volume II
D De Kee and P Kaloni
230 Boolean methods in interpolation and approximation
F J Delvos and W Schempp
231 Further advances in twistor theory. Volume I
L J Mason and L P Hughston
232 Further advances in twistor theory. Volume II
L J Mason and L P Hughston
233 Geometry in the neighborhood of invariant manifolds of maps and flows and linearization
U Kirchgraber and K Palmer
234 Quantales and their applications
K I Rosenthal
235 Integral equations and inverse problems
V Petkov and R Lazarov
236 Pseudo-differential operators
S R Simanca
237 A functional analytic approach to statistical experiments
I M Bomze
238 Quantum mechanics, algebras and distributions
D Dubin and M Hennings
239 Hamilton flows and evolution semigroups
J Gzyl
240 Topics in controlled Markov chains
V S Borkar
241 Invariant manifold theory for hydrodynamic transition
S Sritharan
242 Lectures on the spectrum of $L^2(\Gamma\backslash G)$
F L Williams
243 Progress in variational methods in Hamiltonian systems and elliptic equations
M Girardi, M Matzeu and F Pacella
244 Optimization and nonlinear analysis
A Ioffe, M Marcus and S Reich
245 Inverse problems and imaging
G F Roach
246 Semigroup theory with applications to systems and control
N U Ahmed
247 Periodic-parabolic boundary value problems and positivity
P Hess
248 Distributions and pseudo-differential operators
S Zaidman
249 Progress in partial differential equations: the Metz surveys
M Chipot and J Saint Jean Paulin
250 Differential equations and control theory
V Barbu

251 Stability of stochastic differential equations with respect to semimartingales
X Mao
252 Fixed point theory and applications
J Baillon and M Théra
253 Nonlinear hyperbolic equations and field theory
M K V Murthy and S Spagnolo
254 Ordinary and partial differential equations. Volume III
B D Sleeman and R J Jarvis
255 Harmonic maps into homogeneous spaces
M Black
256 Boundary value and initial value problems in complex analysis: studies in complex analysis and its applications to PDEs 1
R Kühnau and W Tutschke
257 Geometric function theory and applications of complex analysis in mechanics: studies in complex analysis and its applications to PDEs 2
R Kühnau and W Tutschke
258 The development of statistics: recent contributions from China
X R Chen, K T Fang and C C Yang
259 Multiplication of distributions and applications to partial differential equations
M Oberguggenberger
260 Numerical analysis 1991
D F Griffiths and G A Watson
261 Schur's algorithm and several applications
M Bakonyi and T Constantinescu
262 Partial differential equations with complex analysis
H Begehr and A Jeffrey
263 Partial differential equations with real analysis
H Begehr and A Jeffrey
264 Solvability and bifurcations of nonlinear equations
P Drábek
265 Orientational averaging in mechanics of solids
A Lagzdins, V Tamuzs, G Teters and A Kregers
266 Progress in partial differential equations: elliptic and parabolic problems
C Bandle, J Bemelmans, M Chipot, M Grüter and J Saint Jean Paulin
267 Progress in partial differential equations: calculus of variations, applications
C Bandle, J Bemelmans, M Chipot, M Grüter and J Saint Jean Paulin
268 Stochastic partial differential equations and applications
G Da Prato and L Tubaro
269 Partial differential equations and related subjects
M Miranda
270 Operator algebras and topology
W B Arveson, A S Mishchenko, M Putinar, M A Rieffel and S Stratila
271 Operator algebras and operator theory
W B Arveson, A S Mishchenko, M Putinar, M A Rieffel and S Stratila
272 Ordinary and delay differential equations
J Wiener and J K Hale
273 Partial differential equations
J Wiener and J K Hale
274 Mathematical topics in fluid mechanics
J F Rodrigues and A Sequeira
275 Green functions for second order parabolic integro-differential problems
M G Garroni and J F Menaldi
276 Riemann waves and their applications
M W Kalinowski
277 Banach C(K)-modules and operators preserving disjointness
Y A Abramovich, E L Arenson and A K Kitover
278 Limit algebras: an introduction to subalgebras of C*-algebras
S C Power
279 Abstract evolution equations, periodic problems and applications
D Daners and P Koch Medina
280 Emerging applications in free boundary problems
J Chadam and H Rasmussen
281 Free boundary problems involving solids
J Chadam and H Rasmussen
282 Free boundary problems in fluid flow with applications
J Chadam and H Rasmussen
283 Asymptotic problems in probability theory: stochastic models and diffusions on fractals
K D Elworthy and N Ikeda
284 Asymptotic problems in probability theory: Wiener functionals and asymptotics
K D Elworthy and N Ikeda
285 Dynamical systems
R Bamon, R Labarca, J Lewowicz and J Palis
286 Models of hysteresis
A Visintin
287 Moments in probability and approximation theory
G A Anastassiou
288 Mathematical aspects of penetrative convection
B Straughan
289 Ordinary and partial differential equations. Volume IV
B D Sleeman and R J Jarvis
290 *K*-theory for real *C**-algebras
H Schröder
291 Recent developments in theoretical fluid mechanics
G P Galdi and J Necas
292 Propagation of a curved shock and nonlinear ray theory
P Prasad
293 Non-classical elastic solids
M Ciarletta and D Ieşan
294 Multigrid methods
J Bramble
295 Entropy and partial differential equations
W A Day
296 Progress in partial differential equations: the Metz surveys 2
M Chipot
297 Nonstandard methods in the calculus of variations
C Tuckey
298 Barrelledness, Baire-like- and (LF)-spaces
M Kunzinger
299 Nonlinear partial differential equations and their applications. Collège de France Seminar. Volume XI
H Brezis and J L Lions
300 Introduction to operator theory
T Yoshino

301 Generalized fractional calculus and applications
V Kiryakova
302 Nonlinear partial differential equations and their applications. Collège de France Seminar Volume XII
303 Numerical analysis 1993
D F Griffiths and G A Watson
304 Topics in abstract differential equations
S Zaidman
305 Complex analysis and its applications
C C Yang, G C Wen, K Y Li and Y M Chiang
306 Computational methods for fluid-structure interaction
J M Crolet and R Ohayon
307 Random geometrically graph directed self-similar multifractals
L Olsen

Lars Olsen
University of North Texas, USA

Random geometrically graph directed self-similar multifractals

CRC Press
Taylor & Francis Group
Boca Raton London New York

CRC Press is an imprint of the
Taylor & Francis Group, an **informa** business

First published 1994 by Longman Group Limited

Published 2019 by CRC Press
Taylor & Francis Group
6000 Broken Sound Parkway NW, Suite 300
Boca Raton, FL 33487-2742

First issued in paperback 2019

ISBN 13: 978-0-367-44948-3 (pbk)
ISBN 13: 978-0-582-25381-0 (hbk)

Visit the Taylor & Francis Web site at
http://www.taylorandfrancis.com

and the CRC Press Web site at
http://www.crcpress.com

Copublished in the United States with
John Wiley & Sons Inc.

ISSN 0269-3674

British Library Cataloguing in Publication Data

A catalogue record for this book is
available from the British Library

Library of Congress Cataloging-in-Publication Data

Olsen, Lars.
Random geometrically graph directed self-similar multifractals /
Lars Olsen.
p. cm. -- (Pitman research notes in mathematics series ;)
Includes bibliographical references.
1. Fractals. 2. Random measure. I. Title. II. Series.
QA614.86.047 1994
514'.74--dc20 94-10256
CIP

To My
Parents

"Multifractality" is a fancy word for a relatively simple concept.

—Thomas C. Halsey

Multifractality, Scaling, and Diffusive Growth, Fractals' Physical Origin and Properties, Proceedings of the Special Seminar on Fractals at the Ettore Majorana Centre for Scientific Culture, Erice (Trapani), Italy, October, 1988 (editor L. Pietronero), pp. 205–216, Plenum Press, New York, 1989.

The attractiveness of multifractals may to some extent be due to mystery.

—Benoit Mandelbrot

An introduction to multifractal distribution functions, Proceedings of the NATO Advanced Study Institute on Random Fluctuations and Pattern Growth: Experiments and Models, Cargése, Corsica, France, July 18–31, 1988 (editors H. E. Stanley & N. Ostrowsky), pp. 279–291, NATO ASI Series, Series E: Applied Sciences, Vol. 157, Kluwer Academic Press, 1988.

Contents

Acknowledgements

This research was supported by grants 11-9421-1 PD and 11-9421-2 PD from the Danish Natural Science Research Council.

Chapter 1

Introduction

The so-called multifractal theory was introduced by the theoretical physicists Frisch & Parisi [Fr] and Halsey et al. [Ha] in 1986. Recently much physics literature has been devoted to the study of multifractals, cf. e.g. [Bo,Co1,Gr1,Gr2,He,Pa, Te1,Te2]. The first rigorous results on self-similar multifractals were obtained by Cawley & Mauldin [Ca], Edgar & Mauldin [Ed] and Falconer [Fa4] during the period 1991 through 1992. Multifractals have subsequently been studied by a large number of mathematicians [Av,Bo,Br,Co1,Co2,Fen,Ho,Kah5,Ki1,Ki2,Lo1,Lo2,Lo3, Ol,Pe1,Pe2,Pe3,Pey3,Ra,Str]. The purpose of this exposition is to present a rigorous foundation for the multifractal structure of random geometrically graph directed self-similar measures along the lines introduced and developed by Olsen in [Ol]. The random graph directed self-similar measures that we study are natural measure-theoretical extensions of the random self-similar sets that appear in Graf [Gra1], and the results that we obtain are natural multifractal extensions of the main results in Graf [Gra1].

If X is a topological space, let $\mathcal{P}(X)$ denote the family of Borel probability measures on X. Now, let X be a metric space. If $x \in X$ and $r > 0$ then $B(x,r)$ will denote the closed ball with center x and radius $r > 0$. Now fix $\mu \in \mathcal{P}(X)$. The upper resp. lower local dimension of μ at a problem $x \in X$ is defined by

$$\overline{\alpha}_\mu(x) = \limsup_{r \searrow 0} \frac{\log \mu B(x,r)}{\log r} \tag{1.0.1}$$

resp.

$$\underline{\alpha}_\mu(x) = \liminf_{r \searrow 0} \frac{\log \mu B(x,r)}{\log r}. \tag{1.0.2}$$

If $\overline{\alpha}_\mu(x)$ and $\underline{\alpha}_\mu(x)$ agree we refer to the common value as the local dimension of μ at x and denote it by $\alpha_\mu(x)$. Upper and lower local dimensions have been investigated by a large number of authors, cf. e.g. [Bi1,Bi2 p. 141,Cu1,Cu2,Fro,Haa3,Sh,Yo].

For $\alpha \geq 0$ write

$$\begin{aligned}
\overline{\Delta}^{\alpha}(\mu) &= \{x \in \operatorname{supp}\mu \mid \overline{\alpha}_\mu(x) \leq \alpha\}, \\
\overline{\Delta}_{\alpha}(\mu) &= \{x \in \operatorname{supp}\mu \mid \alpha \leq \overline{\alpha}_\mu(x)\}, \\
\underline{\Delta}^{\alpha}(\mu) &= \{x \in \operatorname{supp}\mu \mid \underline{\alpha}_\mu(x) \leq \alpha\}, \\
\underline{\Delta}_{\alpha}(\mu) &= \{x \in \operatorname{supp}\mu \mid \alpha \leq \underline{\alpha}_\mu(x)\}
\end{aligned}$$

and

$$\Delta_\mu(\alpha) = \underline{\Delta}_{\alpha}(\mu) \bigcap \overline{\Delta}^{\alpha}(\mu)$$

where $\operatorname{supp}\mu$ denotes the topological support of μ. One should think of the family $\{\Delta_\mu(\alpha) \mid \alpha \geq 0\}$ as a multifractal decomposition of the support of μ – i.e. we have decomposed the (perhaps fractal) set $\operatorname{supp}\mu$ into a family $\{\Delta_\mu(\alpha) \mid \alpha \geq 0\}$ of subfractals according to the measure μ and indexed by $\alpha \in \mathbb{R}_+$.

The main problem in multifractal theory is to estimate the size of $\Delta_\mu(\alpha)$. This is done by introducing the functions f_μ and F_μ defined by

$$\begin{aligned}
f_\mu(\alpha) &= \dim\{x \in \operatorname{supp}\mu \mid \alpha_\mu(x) = \alpha\} = \dim\Delta_\mu(\alpha), \\
F_\mu(\alpha) &= \operatorname{Dim}\{x \in \operatorname{supp}\mu \mid \alpha_\mu(x) = \alpha\} = \operatorname{Dim}\Delta_\mu(\alpha)
\end{aligned}$$

for $\alpha \geq 0$, and where dim and Dim denotes the Hausdorff dimension and packing dimension respectively. These and similar functions are generically known as "the multifractal spectrum of μ", "the singularity spectrum of μ", "the spectrum of scaling indices" or simply "the $f(\alpha)$-spectrum". The function $f(\alpha) = f_\mu(\alpha)$ was first explicitly defined by the physicists Halsey et al. in 1986 in their seminal paper [Ha]. There are (apart from trivial cases) so far only five types of measures μ for which the f_μ function has been rigorously determined, namely

1) graph directed self-similar measures in $\mathbb{R}^d$ with totally disconnected support, cf. Cawley & Mauldin [Ca] and Edgar & Mauldin [Ed] (see also Olsen [Ol]); cf. Feng [Fen], King [Ki1] and King & Geronimo [Ki2] for a multifractal analysis of some technical extensions of the graph directed self-similar measures considered by [Ca] and [Ed].
2) "Cookie-Cutters" (i.e. Gibbs states on 0-dimensional hyperbolic attractors in $\mathbb{R}$), cf. Bohr & Rand [Bo], Rand [Ra] and Collet et al. [Col]; the multifractal structure of some Gibbs state like measures in $\mathbb{R}$ has also been studied by Brown, Michon & Peyrière [Br] and Peyrière [Pey3].
3) invariant measures of maximal entropy for rational maps of the complex plane, cf. Lopes [Lo1,Lo2].
4) random statistically self-similar measures, cf. Falconer [Fa4]. (We came to know Falconer's results only after having finished almost all of the following investigations. However, our results are more general than the results

obtained by Falconer since we treat the general case of graph directed self-similar measures rather than confine ourselves to the subclass of self-similar measures. We also note that our results are sharper than Falconer's, i.e. sharper versions of Falconer's results are obtained as corollaries to our results by considering graphs with only one vertex. Moreover, Falconer does not study the generalized multifractal Hausdorff measures and packing measures introduced in [Ol].)

5) random measures generated by "strongly bounded random cascades", c.f. Holley and Waymire [Ho].

In all five cases it turns out that there exist numbers $\underline{a} \leq \overline{a}$ such that $f_\mu(\alpha) = 0$ for $\alpha \in [0, \infty[\setminus [\underline{a}, \overline{a}]$ and f_μ is concave and smooth on $]\underline{a}, \overline{a}[$. The proofs in [Br, Ca,Ed,Ki1,Pey3] are based on the ergodic theorem and some combinatoric geometric arguments whereas the proofs in [Bo,Co1,Ki2,Lo1,Lo2,Ra] are based on the thermodynamic formalism developed by Bowen [Bow] and Ruelle [Ru].

The purpose of this exposition is to investigate the multifractal structure of random graph directed self-similar measures using the formalism introduced in [Ol]. Our results are, due to the fact that we make use of the generalized multifractal Hausdorff and packing measures introduced by Olsen in [Ol], natural multifractal generalizations of Graf's and Tsujii's main theorems on random geometrically self-similar sets, in particular Graf [Gra1, Theorem 7.4, Theorem 7.6, Theorem 7.7, Theorem 7.8] and Tsujii [Tsu2, Theorem 3.7, Theorem 3.11, Theorem 4.1, Theorem 4.2]. In fact, the results in [Gra1] and [Tsu2] are obtained as corollaries to our results be setting the "multifractal parameter" q equal to 0. We will now give a brief and informal description of non-random self-similar sets and measures, random self-similar sets and random self-similar measures in order to 1) illustrate the nature of our multifractal generalizations of Graf's and Tsujii's results and 2) motivate our definitions.

1.1 Non-random Self-Similar Sets and Measures.

A theory for (non-random) self similar sets and measures has been developed by Hutchinson [Hu] in 1981 (based on some result due to Moran [Mo].) This theory was subsequently extended to so-called (non-random) graph directed self-similar sets (and measures) by Barnsley et al. [Bar], Bandt [Ban], Mauldin & Williams [Mau2] and others during the period 1985-1988. Graph directed self-similar sets and measures are defined and constructed as follows. Let (V, E) be a directed multigraph; here V is the set of vertices and E is the set of edges. If $u, v \in V$ are vertices and $e \in E$ is an edge, then we denote the set of edges with initial vertex u by E_u, the set of edges from u to v by E_{uv}, and the terminal vertex of e by $\tau(e)$. A list

$$\big((V,E),(X_u)_{u\in V},(S_e)_{e\in E},(p_e)_{e\in E}\big)$$

where

1) (V, E) is a directed multigraph;
2) X_u is a compact metric space;
3) $S_e : X_v \to X_u$ is a contraction for all $u, v \in V$ and all edges $e \in E_{uv}$;
4) $p_e \in [0,1]$ with $\sum_{e\in E_u} p_e = 1$ for each $u \in V$;

is called a Mauldin-Williams graph (MW-graph). If $G = \big((V,E),(X_u)_{u\in V},(S_e)_{e\in E},(p_e)_{e\in E}\big)$ is a MW-graph, then there exist (cf. [Hu] and [Mau2]) a unique list $(K_u)_u$, where K_u is a non-empty compact subset of X_u, satisfying

$$K_u = \bigcup_{e\in E_u} S_e(K_{\tau(e)}) \quad \text{for all } u \in V,$$

and a unique list $(\mu_u)_u \in \prod_{u\in V} \mathcal{P}(X_u)$ of probability measures satisfying

$$\mu_u = \sum_{e\in E_u} p_e \mu_{\tau(e)} \circ S_e^{-1} \quad \text{for all } u \in V.$$

In fact,

$$K_u = \operatorname{supp} \mu_u \,.$$

Moreover, the sets $(K_u)_u$ and measures $(\mu_u)_u$ can according to [Hu] be constructed by the following recursive process:
For each $u \in V$ choose a non-empty compact subset C_u of X_u and a probability measure $\nu_u \in \mathcal{P}(X_u)$ on X_u and construct recursively for each $u \in V$ a sequence $(K_{u,n})_{n\in\mathbb{N}}$ of non-empty compact subsets of X_u and a sequence $(\mu_{u,n})_{n\in\mathbb{N}}$ of probability measures on X_u as follows.

1. Let

$$\mu_{u,1} = \sum_{e_1\in E_u} p_{e_1} \nu_{\tau(e_1)} \circ S_{e_1}^{-1}$$

$$K_{u,1} = \bigcup_{e_1\in E_u} S_{e_1}(C_{\tau(e_1)})$$

2. Let

$$\begin{aligned}\mu_{u,2} &= \sum_{e_1\in E_u} p_{e_1}\mu_{\tau(e_1),1}\circ S_{e_1}^{-1}\\ &= \sum_{e_1\in E_u}\sum_{e_2\in E_{\tau(e_1)}} p_{e_1}p_{e_2}\nu_{\tau(e_2)}\circ(S_{e_1}\circ S_{e_2})^{-1}\\ K_{u,2} &= \bigcup_{e_1\in E_u} S_{e_1}(K_{\tau(e_1),1})\\ &= \bigcup_{e_1\in E_u}\bigcup_{e_2\in E_{\tau(e_1)}}(S_{e_1}\circ S_{e_2})(C_{\tau(e_2)})\end{aligned}$$

3. Let

$$\begin{aligned}\mu_{u,3} &= \sum_{e_1\in E_u} p_{e_1}\mu_{\tau(e_2),1}\circ S_{e_1}^{-1}\\ &= \sum_{e_1\in E_u}\sum_{e_2\in E_{\tau(e_1)}}\sum_{e_3\in E_{\tau(e_2)}} p_{e_1}p_{e_2}p_{e_3}\nu_{\tau(e_3)}\circ(S_{e_1}\circ S_{e_2}\circ S_{e_3})^{-1}\\ K_{u,3} &= \bigcup_{e_1\in E_u} S_{e_1}(K_{\tau(e_2),1})\\ &= \bigcup_{e_1\in E_u}\bigcup_{e_2\in E_{\tau(e_1)}}\bigcup_{e_3\in E_{\tau(e_2)}}(S_{e_1}\circ S_{e_2}\circ S_{e_3})(C_{\tau(e_3)})\end{aligned}$$

⋮

Continue this process

⋮

We then obtain a sequence of non-empty compact sets

$$\left(K_{u,n} := \bigcup_{e_1\in E_u}\bigcup_{e_2\in E_{\tau(e_1)}}\cdots\bigcup_{e_n\in E_{\tau(e_{n-1})}}(S_{e_1}\circ\cdots\circ S_{e_n})(C_{\tau(e_n)})\right)_{n\in\mathbb{N}},$$

and a sequence of probability measures

$$\left(\mu_{u,n} := \sum_{e_1\in E_u}\sum_{e_2\in E_{\tau(e_1)}}\cdots\sum_{e_n\in E_{\tau(e_{n-1})}} p_{e_1}\cdot\ldots\cdot p_{e_n}\nu_{\tau(e_n)}\circ(S_{e_1}\circ\cdots\circ S_{e_n})^{-1}\right)_{n\in\mathbb{N}}$$

It follows from a slight extension of the results in Hutchinson [Hu] (cf. also Mauldin & Williams [Mau2]) that

$$K_{u,n} \to K_u \quad \text{w.r.t. Hausdorff metric}$$

and

$$\mu_{u,n} \xrightarrow{\text{w}} \mu_u$$

where $\xrightarrow{\text{w}}$ denotes weak convergence. The sets $(K_u)_{u\in V}$ are called the graph directed self-similar sets associated with G, and the measures $(\mu_u)_{u\in V}$ are called the graph directed self-similar measures associated with G. In the original case studied by Hutchinson [Hu], the graph (V, E) is required only to have one vertex, $*$ say – in this case the set $K := K_*$ and the measure $\mu := \mu_*$ are called self-similar. The sets K_u and the measures μ_u are called graph directed self-affine if $X_u \subseteq \mathbb{R}^d$ for all u and all the maps S_e are affine. In the case where the graph (V, E) only has one vertex, $*$ say, the set $K_* := K$ and the measure $\mu_* := \mu$ are called self-affine if $X_* \subseteq \mathbb{R}^d$ and all the maps S_e are affine.

Self-similar and graph directed self-similar sets and measures have been investigated by a large number of authors. However, in order to obtain formulas for the Hausdorff and packing dimensions of self-similar sets and graph directed self-similar sets one needs to impose certain technical conditions on the maps S_e guaranteeing that the overlaps $S_e(X_{\tau(e)}) \bigcap S_\varepsilon(X_{\tau(\varepsilon)})$ are small for all $u \in V$ and $e, \varepsilon \in E_u$ with $e \neq \varepsilon$. If $(S_e(X_{\tau(e)}))_{e\in E_{uv}}$ is a pairwise disjoint family for each u, then clearly $S_e(X_{\tau(e)}) \bigcap S_\varepsilon(X_{\tau(\varepsilon)}) = \varnothing$ for all $u \in V$ and $e, \varepsilon \in E_u$ with $e \neq \varepsilon$, and the overlaps $S_e(X_{\tau(e)}) \bigcap S_\varepsilon(X_{\tau(\varepsilon)})$ are thus (very) small in every sense. However, it is not too difficult to see that if $(S_e(X_{\tau(e)}))_{e\in E_{uv}}$ is a pairwise disjoint family for each u, then the recursively constructed limit sets K_u are totally disconnected for all $u \in V$, and the theory would therefore not include e.g. Sierpinski triangles, Menger spronges and von Koch curves. This is clearly very unsatisfactory and it would be desirable with a less restrictive non-overlapping condition. The following so-called "open set condition" (which is satisfied by Sierpinski triangles, Menger spronges and von Koch curves) was introduced by Hutchinson [Hu] in 1981,

> The Open Set Condition:
> there exists a list $(U_u)_{u\in V}$ of sets, where U_u is a non-empty, open and bounded subset of X_u, satisfying
> $$S_e(U_{\tau(e)}) \bigcap S_\varepsilon(U_{\tau(\varepsilon)}) = \varnothing$$
> for all $u \in V$ and $e, \varepsilon \in E_u$ with $e \neq \varepsilon$. (1.1.1)

The open set condition is on the one hand strong enough to ensure good theoretical results, whereas it on the other hand is weak enough to include a large number of interesting examples. The reader is referred to Bandt & Graf [BanG] and Schief [Sc] for a thorough discussion of the open set condition and some equivalent statements.

Hutchinson [Hu] computes the Hausdorff dimension of a self-similar set in $\mathbb{R}^d$ in the case where the open set condition is satisfied and all the maps S_e are similarities (cf. also the textbooks [Fa1,Fa2] by Falconer and the textbook [Edg] by Edgar). Mauldin & Williams [Mau2] and later Stella [Ste], Pesin & Weiss [Pes] and Strichartz [Str] have calculated the Hausdorff dimension α (and investigated the positivity and the finiteness of the Hausdorff measure) of graph directed self-similar sets in $\mathbb{R}^d$ in the case where all the maps S_e are similarities and the open set condition is satisfied. If in addition the graph (V, E) is strongly connected then the Hausdorff dimension α is determined by

$$r\left(\left(\sum_{e\in E_{uv}} \operatorname{Lip}(S_e)^\alpha\right)_{u,v\in V}\right) = 1 \tag{1.1.2}$$

where $r(\cdot)$ denotes the spectral radius, and Lip denotes the Lipschitz constant. McMullen [Mc] and Bedford & Urbanski [Bed] compute the Hausdorff and box dimension of some self-affine sets in $\mathbb{R}^2$ satisfying the open set condition, Falconer [Fa5,Fa6] and Peres [Per] compute the Hausdorff and box dimension of some self-affine sets in $\mathbb{R}^d$ satisfying the open set condition, and Keynon & Peres [Ke] compute the Hausdorff dimension of some graph directed self-affine sets in $\mathbb{R}^2$ in the case where a suitable disjointness condition is satisfied.

Strichartz [Str] and Deliu et al. [Del] have computed the Hausdorff dimiension of graph directed self-similar measures in $\mathbb{R}^d$ under the assumption that all the maps S_e are similarities and a strong disjointness condition is satisfied; in fact, [Str] and [Del] show that such measures are dimensional exact in the sense of Cutler [Cu1,Cu2], cf. also [Haa3].

1.2 Random Self-Similar Sets

Random geometrically self-similar sets have been studied by several authors, e.g. Arbeiter [Ar], Falconer [Fa3], Graf [Gra1], Graf et al. [Gra2], Mauldin & Williams [Mau], Patzschke & Zähle [Pat], Tohiki & Tsujii [To] and Tsujii [Tsu1,Tsu2]. The basic idea in [Ar,Fa3,Gra1,Gra2,Mau1,Pat,To,Tsu1,Tsu2] is to randomize each step in the above mentioned deterministic construction of the sets $(K_u)_u$:
For each $u \in V$ let ν_u a probability distribution on the set

$$\{(S_s)_{e \in E_u} \mid S_e : X_{\tau(e)} \to X_u \text{ is a contraction}\}.$$

For each finite string $e_1 \dots e_n$ of vertices let $C_{e_1 \dots e_n}$ be a non-emty compact subset of $X_{\tau(e_n)}$. Fix an initial vertex u and construct a sequence of random non-empty compact subsets of X_u as follows.

1. First choose a list $(S_{e_1})_{e_1 \in E_u}$ of contractions $S_{e_1} : X_{\tau(e_1)} \to X_u$ according to ν_u, and put
$$K_{u,1} = \bigcup_{e_1 \in E_u} S_{e_1}(C_{e_1}).$$
2. For each $e_1 \in E_u$ choose a list $(S_{e_1 e_2})_{e_2 \in E_{\tau(e_1)}}$ of contractions $S_{e_1 e_2} : X_{\tau(e_2)} \to X_{\tau(e_1)}$ according to $\nu_{\tau(e_1)}$, and put
$$K_{u,2} = \bigcup_{e_1 \in E_u} \bigcup_{e_2 \in E_{\tau(e_1)}} S_{e_1} \circ S_{e_1 e_2}(C_{e_1 e_2}).$$
3. For each $e_1 \in E_u$ and $e_2 \in E_{\tau(e_1)}$ choose a list $(S_{e_1 e_2 e_3})_{e_3 \in E_{\tau(e_2)}}$ of contractions $S_{e_1 e_2 e_3} : X_{\tau(e_3)} \to X_{\tau(e_2)}$ according to $\nu_{\tau(e_2)}$, and put
$$K_{u,3} = \bigcup_{e_1 \in E_u} \bigcup_{e_2 \in E_{\tau(e_1)}} \bigcup_{e_3 \in E_{\tau(e_2)}} S_{e_1} \circ S_{e_1 e_2} \circ S_{e_1 e_2 e_3}(C_{e_1 e_2 e_3}).$$

⋮

Continue this process

⋮

We then obtain a sequence

$$\left(K_{u,n} = \bigcup_{e_1 \in E_u} \bigcup_{e_2 \in E_{\tau(e_1)}} \cdots \bigcup_{e_n \in E_{\tau(e_{n-1})}} S_{e_1} \circ S_{e_1 e_2} \circ \cdots \circ S_{e_1 \dots e_n}(C_{e_1 \dots e_n})\right)_{n \in \mathrm{N}} \tag{1.2.1}$$

of random non-empty compact sets. Now consider the self-similar case, i.e. the case in which the graph (V, E) is required to have only one vertex, $*$ say. Write $K_{*,n} = K_n$ for $n \in \mathbb{N}$. The sequence $(K_{*,n})_n = (K_n)_n$ is thus given by

$$\left(K_n = \bigcup_{e_1 \in E} \bigcup_{e_2 \in E} \cdots \bigcup_{e_n \in E} S_{e_1} \circ S_{e_1 e_2} \circ \cdots \circ S_{e_1 \ldots e_n}(C_{e_1 \ldots e_n}) \right)_{n \in \mathbb{N}} \tag{1.2.2}$$

It follows from [Ar,Fa3,Gra1] that the sequence in (1.2.2) converges almost surely w.r.t. the Hausdorff metric to a random non-empty compact limit set

$$K := \lim_n K_n \,. \tag{1.2.3}$$

We prove in section 2.2 that the same result holds in the random graph directed self-similar case, i.e. for each vertex $u \in V$ then the sequence $(K_{u,n})_n$ in (1.2.1) converges almost surely w.r.t. the Hausdorff metric to a random non-empty compact limit set

$$K_u := \lim_n K_{u,n} \,.$$

The Hausdorff dimension and the Hausdorff measure of the limit set $K = \lim_n K_n$ in (1.2.3) have been studied intensively by e.g. Arbeiter [Ar], Falconer [Fa3], Graf [Gra1], Graf et al. [Gra2] and Mauldin & Williams [Mau1]. In fact, two different approaches to random self-similar sets have appeared. The construction described above, obtained by randomizing each step of Hutchinson's deterministic construction was introduced in 1986 by Falconer [Fa3], Mauldin & Williams [Mau] and Graf [Gra1]. However, at the same time and independently of Falconer, Graf and Mauldin & Williams, Zähle in a series of papers [Zä1,Zä2,Zä3] introduced an axiomatic notion of statistically self-similarity (for random measures rather than random sets) based on Palm distributions (cf. Kallenberg [Kal] for a discussion of Palm distributions), and studied the quality of that notion by means of the connection between a suitable similarity index and the Hausdorff dimension. In 1990 Patzschke & Zähle [Pat] proved that the random recursive construction model of Falconer, Graf and Mauldin & Williams fitted into the notion introduced by Zähle in [Zä1] if a certain (strong) disjointness condition [Pat, Condition (1) in Theorem 4.1] was satisfied. Falconer [Fa3] and Mauldin & Williams [Mau1] computed the almost sure Hausdorff dimension of a random self-similar set in $\mathbb{R}^d$ in the case where the open set condition is satisfied and all the maps $S_{e_1 \ldots e_n}$ are similarities whose Lipschitz constants are almost surely uniformly bounded from below – in fact, Mauldin & Williams considered the far more general case in which the graph $(V := \{*\}, E)$ has countable many edges (recall that (V, E) only has one vertex since Graf [Gra2] and Mauldin & Williams [Mau1] considered the self-similar case.) However, Mauldin & Williams

[Mau1] left some problems unsolved, in particular questions concerning the almost sure positivity and finiteness of the Hausdorff measure at the almost sure Hausdorff dimension. These questions were later taken up and answered by Graf [Gra1] and Graf et al. [Gra2]. Graf gives in [Gra1] a sufficient condition [Gra1, the hypothesis in Theorem 7.7] guaranteeing that the Hausdorff measure $\mathcal{H}^\alpha(K)$ of K, at the almost sure Hausdorff dimension α, is zero almost surely; and a necessary and sufficient condition guaranteeing that the Hausdorff measure $\mathcal{H}^\alpha(K)$ of K, at the almost sure Hausdorff dimension α, is positive almost surely. In the case where [Gra1, the hypothesis in Theorem 7.7] is satisfied, and consequently (by [Gra1]), the Hausdorff measure is zero almost surely, Graf et al. [Gra2] compute the almost sure exact Hausdorff dimension function. Later Tsujii [Tsu1,Tsu2] and Tohoki & Tsujii [To] generalized all the results in [Gra1] to the graph directed self-similar case. Tsujii [Tsu1,Tsu2] (and in the self-similar case Graf [Gra1]) introduces for each vertex u a measure P_u on the family of graph directed self-similar subsets of X_u, which in a natural way respects the self-similarity. If certain technical conditions are satisfied Tsujii [Tsu1,Tsu2] (and in the self-similar case Graf [Gra1]) proves that

$$\dim K_u = \alpha \quad \text{for } P_u\text{-a.a. } K_u \tag{1.2.4}$$

where α is determined by

$$r\left(\left(\mathrm{E}_{\nu_u}\left[\sum_{e\in E_{uv}} \mathrm{Lip}(S_e)^\alpha\right]\right)_{u,v\in V}\right) = 1 \tag{1.2.5}$$

(here E_{ν_u} denotes expectation w.r.t. ν_u.) Equation (1.2.4) and (1.2.5) are natural random analogues of (1.1.2). Furthermore Tsujii [Tsu1,Tsu2] (and in the self-similar case Graf [Gra1]) shows that if certain technical conditions are satisfied then the following statements are equivalent.

1.2.i) For all u then

$$\rho_u^{-1}\sum_v \sum_{e\in E_{uv}} \mathrm{Lip}(S_e)^\alpha \rho_v = 1 \quad \text{for } \nu_u\text{-a.a. } (S_e)_{e\in E_u}$$

where $(\rho_u)_{u\in V}$ is a certain "Perron-Frobenius eigenvector".

1.2.ii) For all u

$$P_u(\mathcal{H}^\alpha(K_u) > 0) > 0\,.$$

1.2.iii) For all u

$$P_u(\mathcal{H}^\alpha(K_u) > 0) = 1\,.$$

1.2.iv) There exists a u such that

$$P_u(\mathcal{H}^\alpha(K_u) > 0) > 0\,.$$

1.2.v) There exists a u such that

$$P_u(\mathcal{H}^\alpha(K_u) > 0) = 1\,.$$

Moreover, Tsujii [Tsu1,Tsu2] (and in the self-similar case Graf [Gra1]) also proves that if certain technical conditions are satisfied then the following implication holds,

$$\left.\begin{array}{l} \text{If there exists } u \in V \text{ such that} \\ \quad \nu_u\left(\left\{(S_e)_{e\in E_u} \,\middle|\, \sum_{e\in E_u} \mathrm{Lip}(S_e)^\alpha \neq 1\right\}\right) > 0 \\ \text{then} \\ \qquad \mathcal{H}^\alpha(K_v) = 0 \quad \text{for } P_v\text{-a.a. } K_v \\ \text{for all } v \in V. \end{array}\right\} \tag{1.2.6}$$

Finally Tsujii [Tsu1,Tsu2] proves that if certain conditions are satisfied then the Hausdorff measure $\mathcal{H}^\alpha(K_u)$ of K_u is constant almost surely, i.e. there exists a constant c_u only depending on the vertex $u \in V$ such that

$$\mathcal{H}^\alpha(K_u) = c_u \quad \text{for } P_u\text{-a.a. } K_u\,. \tag{1.2.7}$$

We also remark that some of the results due to Falconer [Fa3], Graf [Gra1] and Mauldin & Williams [Mau1] have been extended to the non-compact case by Arbeiter [Ar].

1.3 Random Self-Similar Measures

The random geometrically self-similar measures that we study are also obtained by randomizing each step in the above mentioned deterministic construction – and are thus natural analogues of the random self-similar sets investigated by [Ar,Fa3, Gra1,Gra2,Mau1,Pat,To,Tsu1,Tsu2].

For each $u \in V$ let λ_u be a probability distribution on the set

$$\{(S_s)_{e\in E_u} \mid S_e : X_{\tau(e)} \to X_u \text{ is a contraction}\}\times \{(p_e)_{e\in E_u} \mid p_e \in [0,1], \sum_{e\in E_u} p_e = 1\}.$$

For each finite string $e_1 \dots e_n$ of vertices let $\nu_{e_1\dots e_n}$ be a Borel probability measure on $X_{\tau(e_n)}$. Fix an initial vertex u and construct a sequence of random Borel probability measures on X_u as follows,

1. First choose an ordered pair $\big((S_{e_1})_{e_1\in E_u}, (p_{e_1})_{e_1\in E_u}\big)$ according to λ_u consisting of contractions $S_{e_1} : X_{\tau(e_1)} \to X_u$ and a probability vector $(p_{e_1})_{e_1\in E_u}$, i.e. $p_{e_1} \geq 0$ and $\sum_{e_1} p_{e_1} = 1$. Now put

$$\mu_{u,1} = \sum_{e_1\in E_u} p_{e_1}\nu_{e_1} \circ S_{e_1}^{-1}.$$

2. For each $e_1 \in E_u$ choose an ordered pair $\big((S_{e_1e_2})_{e_2\in E_{\tau(e_1)}}, (p_{e_1e_2})_{e_2\in E_{\tau(e_1)}}\big)$ according to $\lambda_{\tau(e_1)}$ consisting of contractions $S_{e_1e_2} : X_{\tau(e_2)} \to X_{\tau(e_1)}$ and a probability vector $(p_{e_1e_2})_{e_2\in E_{\tau(e_1)}}$, i.e. $p_{e_1e_2} \geq 0$ and $\sum_{e_2} p_{e_1e_2} = 1$. Now put

$$\mu_{u,2} = \sum_{e_1\in E_u} \sum_{e_2\in E_{\tau(e_1)}} p_{e_1}p_{e_1e_2}\nu_{e_1e_2} \circ (S_{e_1} \circ S_{e_1e_2})^{-1}.$$

3. For each $e_1 \in E_u$ and $e_2 \in E_{\tau(e_1)}$ choose an ordered pair $\big((S_{e_1e_2e_3})_{e_3\in E_{\tau(e_2)}}, (p_{e_1e_2e_3})_{e_3\in E_{\tau(e_2)}}\big)$ according to $\lambda_{\tau(e_2)}$ consisting of contractions $S_{e_1e_2e_3}$: $X_{\tau(e_3)} \to X_{\tau(e_2)}$ and a probability vector $(p_{e_1e_2e_3})_{e_3\in E_{\tau(e_2)}}$, i.e. $p_{e_1e_2e_3} \geq 0$ and $\sum_{e_3} p_{e_1e_2e_3} = 1$. Now put

$$\mu_{u,3} = \sum_{e_1\in E_u} \sum_{e_2\in E_{\tau(e_1)}} \sum_{e_3\in E_{\tau(e_2)}} p_{e_1}p_{e_1e_2}p_{e_1e_2e_3}\cdot \nu_{e_1e_2e_3} \circ (S_{e_1} \circ S_{e_1e_2} \circ S_{e_1e_2e_3})^{-1}.$$

$$\vdots$$

Continue this process

$$\vdots$$

We then obtain a sequence

$$\left(\mu_{u,n} = \sum_{e_1 \in E_u} \sum_{e_2 \in E_{\tau(e_1)}} \cdots \sum_{e_n \in E_{\tau(e_{n-1})}} p_{e_1} p_{e_1 e_2} \cdot \dots \cdot p_{e_1 \dots e_n} \cdot \nu_{e_1 \dots e_n} \circ (S_{e_1} \circ S_{e_1 e_2} \circ \dots \circ S_{e_1 \dots e_n})^{-1}\right)_{n \in \mathbb{N}}$$

of random Borel probability measures. A special case of a somewhat similar construction appears implicitly in Arbeiter [Ar]. In section 2.3 we investigate under which conditions the sequence $(\mu_{u,n})_n$ converges weakly almost surely to a random probability measure μ_u – moreover we show that the limit measure is independent of the start measures $\nu_{e_1 \dots e_n}$. The main purpose of this exposition is to study the multifractal structure of the limit measure μ_u. We do that by introducing an auxiliary function $\beta : \mathbb{R} \to \mathbb{R}$ defined by the requirement

$$r\left(\left(\mathbf{E}_{\lambda_u}\left[\sum_{e \in E_{u,v}} p_e^q \operatorname{Lip}(S_e)^{\beta(q)}\right]\right)_{u,v \in V}\right) = 1 \tag{1.3.1}$$

(details will be given in section 2.5.) Next we introduce for each vertex $u \in V$, a probability measure P_u on the set of graph directed self-similar measures on X_u. The definition of the measure P_u ensures that it in a natural way respects the self-similar structure of μ_u. We will always in the following discussion assume that the technical conditions (I), (III), (IV) and (V) stated in section 2.5 are satisfied. It is then shown that there exists numbers $0 \leq a_{\min} \leq a_{\max}$ such that if $\alpha \geq 0$, P_u-a.a. $\mu \in \mathcal{P}(X_u)$ satisfy,

$$f_\mu(\alpha) = F_\mu(\alpha) = \begin{cases} \inf_q(\alpha q + \beta(q)) := \beta^*(\alpha) & \text{for } \alpha \in]a_{\min}, a_{\max}[\\ 0 & \text{for } \alpha \notin [a_{\min}, a_{\max}] \end{cases} \tag{1.3.2}$$

(we prove, in fact, a stronger result; details will be given in section 2.5.) One of the main technical tools used in the proof of (1.3.2) is the generalized multifractal Hausdorff measure $\mathcal{H}_\mu^{q,t}$ and the generalized multifractal packing measure $\mathcal{P}_\mu^{q,t}$ introduced by Olsen in [Ol] (here $q, t \in \mathbb{R}$ and μ is a Borel probability measure on a metric space; details will be given in section 2.1.) As a by product we obtain a proof of the fact that there exist two extended real valued numbers $q_{\min}, q_{\max} \in [-\infty, \infty]$ with $-\infty \leq q_{\min} < 0 < 1 < q_{\max} \leq \infty$, such that for each $q \in]q_{\min}, q_{\max}[$, then the following conditions are equivalent.

1.3.i) For all u then

$$\rho_u(q)^{-1}\sum_{v}\sum_{e\in E_{uv}} p_e^q \operatorname{Lip}(S_e)^{\beta(q)}\rho_u(q)=1 \quad \text{for } \lambda_u\text{-a.a. } \left((S_e)_{e\in E_u},(p_e)_{e\in E_u}\right)$$

where $(\rho_u(q))_{u\in V}$ is a certain "Perron-Frobenius eigenvector" depending on q.

1.3.ii) For all u,

$$P_u(\mathcal{H}_\mu^{q,\beta(q)}(\operatorname{supp}\mu)>0)>0\,.$$

1.3.iii) For all u,

$$P_u(\mathcal{H}_\mu^{q,\beta(q)}(\operatorname{supp}\mu)>0)=1\,.$$

1.3.iv) There exists a u such that

$$P_u(\mathcal{H}_\mu^{q,\beta(q)}(\operatorname{supp}\mu)>0)>0\,.$$

1.3.v) There exists a u such that

$$P_u(\mathcal{H}_\mu^{q,\beta(q)}(\operatorname{supp}\mu)>0)=1\,.$$

The equivalence between 1.3.i) through 1.3.v) is a multifractal extension of the Graf-Tsujii equivalence between 1.2.i) through 1.2.v) – in fact, the equivalence between 1.2.i) through 1.2.v) is obtained as a corollary to the equivalence between 1.3.i) through 1.3.v) by setting $q=0$.

We also prove that for a fixed $q\in\mathbb{R}$, then the following implication holds,

$$\left.\begin{array}{l}\text{If there exists } u\in V \text{ such that}\\[4pt] \lambda_u\left(\left\{\left((S_e)_{e\in E_u},(p_e)_{e\in E_u}\right)\,\middle|\,\sum\limits_{e\in E_u} p_e^q\operatorname{Lip}(S_e)^{\beta(q)}\neq 1\right\}\right)>0\\[4pt] \text{then}\\[4pt] \qquad \mathcal{H}_\mu^{q,\beta(q)}(\operatorname{supp}\mu)=0 \quad \text{for } P_v\text{-a.a. } \mu\\[4pt] \text{for all } v\in V.\end{array}\right\} \tag{1.3.3}$$

The result in (1.3.3) is a natural multifractal extension of the Graf-Tsujii implication in (1.2.6) – in fact, implication (1.2.6) is obtained as a corollary to implication (1.3.3) by setting $q=0$.

Finally we prove that for a fixed $q\in]q_{\min},q_{\max}[$, then the multifractal Hausdorff measure $\mathcal{H}_\mu^{q,\beta(q)}(\operatorname{supp}\mu)$ is constant for P_u-a.a. μ, i.e. there exists a constant $c_{u,q}$ only depending on q and the vertex u such that

$$\mathcal{H}_\mu^{q,\beta(q)}(\operatorname{supp}\mu)=c_{u,q} \quad \text{for } P_u\text{-a.a. } \mu \tag{1.3.4}$$

(we do, in fact, prove a stronger result; details will be given in section 2.8.) The result in (1.3.4) is a natural multifractal extension of Tsujii's result in equation (1.2.7) – in fact, equation (1.2.7) is obtained as a corollary to (1.3.4) by setting $q=0$.

We will now give a brief description of the organization of the exposition.

Chapter 2. Chapter 2 contains the basic definitions and states the main results. All proofs will be given in Chapter 4 through Chapter 7. Chapter 2 is divided into 9 sections.

Section 2.1. This section recalls the definition of the Hausdorff and packing measure and the corresponding dimensions. Section 2.1 also recalls the definition of the multifractal Hausdorff measure $\mathcal{H}_{\mu}^{q,\beta(q)}$ and the multifractal packing measure $\mathcal{P}_{\mu}^{q,\beta(q)}$ introduced in [Ol]. Finally, section 2.1 states the main theorems from [Ol] describing the relation between $\mathcal{H}_{\mu}^{q,\beta(q)}$ and $\mathcal{P}_{\mu}^{q,\beta(q)}$ and the multifractal spectra functions f_{μ} and F_{μ}. These theorems will be one of our main technical tools in the study of the multifractal structure of random graph directed self-similar measures.

Section 2.2. This section defines random graph directed self-similar sets, and states an existence and uniqueness theorem for random graph directed self-similar sets.

Section 2.3. This section defines random graph directed self-similar measures, and states an existence and uniqueness theorem for random graph directed self-similar measures.

Section 2.4. In this section we introduce the probability measure P_u on the family of Borel probability measures $\mathcal{P}(X_u)$ on X_u w.r.t. which we, among other things, will compute the almost sure multifractal spectrum. We also state a theorem which asserts that the measure P_u in a certain sense is unique.

Section 2.5. In order to find an expression for the P_u almost sure multifractal spectra functions f_{μ} and F_{μ} we need an auxiliary function β. Section 2.5 defines and investigates the properties of β.

Section 2.6 In this section we state the main results expressing the P_u almost sure multifractal spectra functions f_{μ} and F_{μ} in terms of β. We also find P_u almost sure expressions for the generalized multifractal Hausdorff dimension function b_{μ} and the generalized multifractal packing dimension function B_{μ} (introduced in [Ol]) in terms of β.

Section 2.7 This section investigates the P_u almost sure positivity and finiteness of the generalized multifractal Hausdorff measures $\mathcal{H}_{\mu}^{q,\beta(q)}(\operatorname{supp}\mu)$ and $\mathcal{H}_{\mu}^{q,\beta(q)}(\Delta_{\mu}(\alpha(q)))$ where $\alpha := -\beta'$ (it will be shown that β is differentiable, in fact, real analytic.) These investigations are natural multifractal extension of the investigations in Graf [Gra1] and Tsujii [Tsu1,Tsu2], and the conclusions are natural multifractal generalizations and extensions of the results in Graf [Gra1] and Tsujii [Tsu1,Tsu2] – in fact, Graf's and Tsujii's results are (under some mild restrictions) obtained as corollaries to our multifractal results by setting $q = 0$.

Section 2.8 This section investigates the P_u almost sure constancy of the

following two generalized multifractal Hausdorff measures $\mathcal{H}_\mu^{q,\beta(q)}(\operatorname{supp}\mu)$ and $\mathcal{H}_\mu^{q,\beta(q)}(\Delta_\mu(\alpha(q)))$ where $\alpha := -\beta'$ (it will be shown that β is differentiable, in fact, real analytic.) These investigations are natural multifractal extension of the investigations in Graf [Gra1] and Tsujii [Tsu1,Tsu2], and the conclusions are natural multifractal generalizations and extensions of the results in Graf [Gra1] and Tsujii [Tsu1,Tsu2] – in fact, Graf's and Tsujii's results are (under some mild restrictions) obtained as corollaries to our multifractal results by setting $q = 0$.
Section 2.9 Section 2.9 gives a thermodynamical interpretation of our main results using the concept of phase transitions.

Chapter 3. This chapter contains some examples which illustrate our results.

Chapter 4. This chapter contains the proofs of the results in section 2.2 through 2.5.

Chapter 5. In order to prove the main theorems in section 2.6 through section 2.9 we need an auxiliary random variable $X_{u,q}$. The random variable $X_{u,q}$ is defined in Chapter 5 and the positive and negative moments of $X_{u,q}$ are studied.

Chapter 6. It is necessary to establish a connection between the random variable $X_{u,q}$ and the measures μ_u in order to make full use of the former in the proofs of the theorems in section 2.6 through section 2.9. Chapter 6 introduces auxiliary (random) measures $\mathcal{M}_{u,q}$ which establish a connection between the random variable $X_{u,q}$ and the graph directed self-similar measures μ_u. We remark that the measures $\mathcal{M}_{u,q}$ are multifractal generalizations of the so-called "random construction measure" ν_ω in Mauldin & Williams [Mau1] and Graf et al. [Gra2].

Chapter 7. This chapter contains the proofs of the main theorems in section 2.6 through section 2.9.

Chapter 2

Definitions and Statements of Results

This chapter contains the basic definitions and states the main results. The proofs will be given in Chapter 4 through Chapter 7.

2.1 The Multifractal Spectrum and the Multifractal Measures $\mathcal{H}_\mu^{q,t}$ and $\mathcal{P}_\mu^{q,t}$

This section has two purposes. The first purpose is to define the (centered) Hausdorff measure and the Hausdorff dimension, and the packing measure and the packing dimension. The second purpose is to define the multifractal generalizations of the Hausdorff and packing measure introduced by Olsen [Ol] and recall the main results from [Ol].

We first recall the definition of the Hausdorff measure, the centered Hausdorff measure and the packing measure. Let X be a metric space, $E \subseteq X$ and $\delta > 0$. A countable family $\mathcal{B} = (B(x_i, r_i))_i$ of closed balls in X is called a centered δ-covering of E if $E \subseteq \cup_i B(x_i, r_i)$, $x_i \in E$ and $0 < r_i < \delta$ for all i. The family $\mathcal{B}$ is called a centered δ-packing of E if $x_i \in E$, $0 < r_i < \delta$ and $B(x_i, r_i) \cap B(x_j, r_j) = \varnothing$ for all $i \neq j$. Let $E \subseteq X$, $s \geq 0$ and $\delta > 0$. Now put

$$\mathcal{H}_\delta^s(E) = \inf\{\sum_i \operatorname{diam}(E_i)^s \mid E \subseteq \bigcup_{i=1}^\infty E_i\,,\ \operatorname{diam} E_i < \delta\}\,.$$

The s-dimensional Hausdorff measure $\mathcal{H}^s(E)$ of E is defined by

$$\mathcal{H}^s(E) = \sup_{\delta > 0} \mathcal{H}_\delta^s(E)\,.$$

The reader is referred to [Fal] for more information on $\mathcal{H}^s$. Next we define the centered Hausdorff measure introduced by Raymond & Tricot in [Ray]. Put

$$\overline{\mathcal{C}}_\delta^s(E) = \inf\left\{\sum_{i=1}^\infty (2r_i)^s \mid (B(x_i, r_i))_i \text{ is a centered } \delta\text{-covering of } E\right\}\,.$$

The s-dimensional centered pre-Hausdorff measure $\overline{\mathcal{C}}^s(E)$ of E is defined by

$$\overline{\mathcal{C}}^s(E) = \sup_{\delta>0} \overline{\mathcal{C}}^s_\delta(E).$$

The set function $\overline{\mathcal{C}}^s$ is not necessarily monotone, and hence not necessarily an outer measure, c.f. [Ray, pp. 137-138]. But $\overline{\mathcal{C}}^s$ gives rise to a Borel measure, called the s-dimensional centered Hausdorff measure $\mathcal{C}^s(E)$ of E, as follows

$$\mathcal{C}^s(E) = \sup_{F\subseteq E} \overline{\mathcal{C}}^s(F).$$

It is easily seen (c.f. [Ray,Lemma 3.3]) that

$$2^{-s}\mathcal{C}^s \leq \mathcal{H}^s \leq \mathcal{C}^s$$

We will now define the packing measure. Write

$$\overline{\mathcal{P}}^s_\delta(E) = \sup\left\{\sum_{i=1}^{\infty}(2r_i)^s \mid (B(x_i, r_i))_i \text{ is a centered } \delta\text{-packing of } E\right\}.$$

The s-dimensional prepacking measure $\overline{\mathcal{P}}^s(E)$ of E is defined by

$$\overline{\mathcal{P}}^s(E) = \inf_{\delta>0} \overline{\mathcal{P}}^s_\delta(E).$$

The set function $\overline{\mathcal{P}}^s$ is not necessarily countable subadditive, and hence not necessarily an outer measure, c.f. [Ta] or [Fa2]. But $\overline{\mathcal{P}}^s$ give rise to a Borel measure, namely the s-dimensional packing measure $\mathcal{P}^s(E)$ of E, as follows

$$\mathcal{P}^s(E) = \inf_{E\subseteq\cup_{i=1}^\infty E_i} \sum_{i=1}^{\infty} \overline{\mathcal{P}}^s(E_i).$$

The packing measure was introduced by Taylor and Tricot in [Ta] using centered δ-packings of open balls, and by Raymond and Tricot in [Ray] using centered δ-packings of closed balls.

Also recall that the Hausdorff dimension $\dim(E)$, the packing dimension $\mathrm{Dim}(E)$ and the logarithmic index $\Delta(E)$ of E is defined by

$$\begin{aligned} \dim(E) &= \sup\{s \geq 0 \mid \mathcal{H}^s(E) = \infty\} \\ \mathrm{Dim}(E) &= \sup\{s \geq 0 \mid \mathcal{P}^s(E) = \infty\} \\ \Delta(E) &= \sup\{s \geq 0 \mid \overline{\mathcal{P}}^s(E) = \infty\}. \end{aligned}$$

We refer the reader to [Tr] and [Ray] for more information on the centered Hausdorff measure, the packing measure and the packing dimension.

Olsen [Ol] suggested that some multifractal generalizations $\mathcal{H}_\mu^{q,t}$ and $\mathcal{P}_\mu^{q,t}$ of the centered Hausdorff measure and the packing measure might be useful in the study of multifractals. We will now define the multifractal measures $\mathcal{H}_\mu^{q,t}$ and $\mathcal{P}_\mu^{q,t}$. For $q \in \mathbb{R}$ define $\varphi_q : [0,\infty[\to \overline{\mathbb{R}}_+ = [0,\infty]$ by

$$
\begin{aligned}
\varphi_q(x) &= \begin{cases} \infty & \text{for } x = 0 \\ x^q & \text{for } 0 < x \end{cases} && \text{for } q < 0 \\
\varphi_q(x) &= 1 \qquad\qquad \text{for } q = 0 && \\
\varphi_q(x) &= \begin{cases} 0 & \text{for } x = 0 \\ x^q & \text{for } 0 < x \end{cases} && \text{for } 0 < q
\end{aligned}
$$

For $\mu \in \mathcal{P}(X)$, $E \subseteq X$, $q,t \in \mathbb{R}$ and $\delta > 0$ write

$$
\begin{aligned}
\overline{\mathcal{H}}_{\mu,\delta}^{q,t}(E) &= \inf\{\sum_i \varphi_q(\mu(B(x_i,r_i)))(2r_i)^t \mid (B(x_i,r_i))_i \text{ is a centered} \\
&\qquad\qquad \delta\text{-covering of } E\}\,, \quad E \neq \varnothing \\
\overline{\mathcal{H}}_{\mu,\delta}^{q,t}(\varnothing) &= 0 \\
\overline{\mathcal{H}}_{\mu}^{q,t}(E) &= \sup_{\delta>0} \overline{\mathcal{H}}_{\mu,\delta}^{q,t}(E) \\
\mathcal{H}_{\mu}^{q,t}(E) &= \sup_{F\subseteq E} \overline{\mathcal{H}}_{\mu}^{q,t}(F)\,.
\end{aligned}
$$

We also make the dual definitions

$$
\begin{aligned}
\overline{\mathcal{P}}_{\mu,\delta}^{q,t}(E) &= \sup\{\sum_i \varphi_q(\mu(B(x_i,r_i)))(2r_i)^t \mid (B(x_i,r_i))_i \text{ is a centered} \\
&\qquad\qquad \delta\text{-packing of } E\}\,, \quad E \neq \varnothing \\
\overline{\mathcal{P}}_{\mu,\delta}^{q,t}(\varnothing) &= 0 \\
\overline{\mathcal{P}}_{\mu}^{q,t}(E) &= \inf_{\delta>0} \overline{\mathcal{P}}_{\mu,\delta}^{q,t}(E) \\
\mathcal{P}_{\mu}^{q,t}(E) &= \inf_{E\subseteq\cup_i E_i} \sum_i \overline{\mathcal{P}}_{\mu}^{q,t}(E_i)\,.
\end{aligned}
$$

It is proved in [Ol] that $\mathcal{H}_\mu^{q,t}$ and $\mathcal{P}_\mu^{q,t}$ are metric outer measures and thus Borel measures on X. The measure $\mathcal{H}_\mu^{q,t}$ is of course a multifractal generalization of the centered Hausdorff measure, whereas $\mathcal{P}_\mu^{q,t}$ is a multifractal generalization of the packing measure.

The next result shows that the measures $\mathcal{H}_\mu^{q,t}$ and $\mathcal{P}_\mu^{q,t}$ and the pre-measure $\overline{\mathcal{P}}_\mu^{q,t}$ in the usual way assign a dimension to each subset E of X.

Proposition 2.1.1.

i) *There exists a unique number* $\Delta_\mu^q(E)$ *in* $[-\infty,\infty]$ *such that*

$$\overline{\mathcal{P}}_\mu^{q,t}(E)=\begin{cases}\infty & \text{for } t<\Delta_\mu^q(E)\\ 0 & \text{for } \Delta_\mu^q(E)<t\end{cases}$$

ii) *There exists a unique number* $\mathrm{Dim}_\mu^q(E)$ *in* $[-\infty,\infty]$ *such that*

$$\mathcal{P}_\mu^{q,t}(E)=\begin{cases}\infty & \text{for } t<\mathrm{Dim}_\mu^q(E)\\ 0 & \text{for } \mathrm{Dim}_\mu^q(E)<t\end{cases}$$

iii) *There exists a unique number* $\dim_\mu^q(E)$ *in* $[-\infty,\infty]$ *such that*

$$\mathcal{H}_\mu^{q,t}(E)=\begin{cases}\infty & \text{for } t<\dim_\mu^q(E)\\ 0 & \text{for } \dim_\mu^q(E)<t\end{cases}$$

Proof. See [Ol,Proposition 1.1]. □

The number $\dim_\mu^q(E)$ is an obvious multifractal analogue of the Hausdorff dimension $\dim(E)$ of E whereas $\mathrm{Dim}_\mu^q(E)$ and $\Delta_\mu^q(E)$ are obvious multifractal analogues of the packing dimension $\mathrm{Dim}(E)$ and the logarithmic index $\Delta(E)$ of E respectively. In fact, it follows immediately from the definitions that

$$\begin{aligned}\dim(E)=&\ \dim_\mu^0(E)\\ \mathrm{Dim}(E)=&\ \mathrm{Dim}_\mu^0(E)\\ \Delta(E)=&\ \Delta_\mu^0(E)\,.\end{aligned}\tag{2.1.1}$$

It is also readily seen that

$$\begin{gathered}0\le\dim_\mu^q(E)\ \text{ for }\ q\le 1\ \text{ and }\ \mu(E)>0\\ \Delta_\mu^q(E)\le 0\ \text{ for }\ 1\le q\end{gathered}\tag{2.1.2}$$

Since $\dim_\mu^q$ and Dim_μ^q are defined in terms of outer measures the following hold

1) $\dim_\mu^q$, Dim_μ^q are monotone, i.e.

$$\begin{aligned}\dim_\mu^q(E)\le\dim_\mu^q(F)\quad&\text{for}\quad E\subseteq F\\ \mathrm{Dim}_\mu^q(E)\le\mathrm{Dim}_\mu^q(F)\quad&\text{for}\quad E\subseteq F\,.\end{aligned}$$

2) $\dim_\mu^q$, Dim_μ^q are σ-stable, i.e.

$$\begin{aligned}\dim_\mu^q(\cup_{n\in\mathbb{N}}E_n)=&\ \sup_{n\in\mathbb{N}}\dim_\mu^q(E_n)\\ \mathrm{Dim}_\mu^q(\cup_{n\in\mathbb{N}}E_n)=&\ \sup_{n\in\mathbb{N}}\mathrm{Dim}_\mu^q(E_n)\,.\end{aligned}$$

These properties will be used tactically throughout the exposition.

As in the non multifractal case, the Hausdorff dimension $\dim_\mu^q$ is majorized by the packing dimension Dim_μ^q.

Proposition 2.1.2. *Let $\mathcal{B}(\mathbb{R}^d)$ denote the Borel algebra in $\mathbb{R}^d$ and let $\mu \in \mathcal{P}(\mathbb{R}^d)$. Then the following hold for $q, t \in \mathbb{R}$,*

i) $\mathcal{P}_\mu^{q,t} \leq \overline{\mathcal{P}}_\mu^{q,t}$.
ii) $\mathcal{H}_\mu^{q,t} \leq \mathcal{P}_\mu^{q,t}$ *on* $\mathcal{B}(\mathbb{R}^d)$ *for* $q \leq 0$.
iii) *There exists an integer* $\zeta \in \mathbb{N}$ *such that* $\mathcal{H}_\mu^{q,t} \leq \zeta \mathcal{P}_\mu^{q,t}$.

In particular

$$\dim_\mu^q \leq \mathrm{Dim}_\mu^q \leq \Delta_\mu^q .$$

Proof. See [Ol]. □

The main point in [Ol] is that the functions

$$b_\mu : q \to \dim_\mu^q(\operatorname{supp} \mu)$$
$$B_\mu : q \to \mathrm{Dim}_\mu^q(\operatorname{supp} \mu)$$

are related to the multifractal spectrum of μ, whereas the function

$$\Lambda_\mu : q \to \Delta_\mu^q(\operatorname{supp} \mu)$$

is related to the generalized Rényi dimensions of μ. This fact is implicitly contained in the inequalities in Proposition 2.1.3 through Proposition 2.1.6. Let X be a metric space and $\mu \in \mathcal{P}(X)$. Fix $\alpha \geq 0$, q, $t \in \mathbb{R}$ and $\delta > 0$ with

$$0 < \alpha q + t$$

Then the following inequalities hold.

Proposition 2.1.3.

i) $\mathcal{H}^{\alpha q+t+\delta}(\overline{\Delta}^\alpha(\mu)) \leq 2^{\alpha q+\delta}\mathcal{H}_\mu^{q,t}(\overline{\Delta}^\alpha(\mu))$ *for* $0 \leq q$.
ii) $\mathcal{H}^{\alpha q+t+\delta}(\underline{\Delta}_\alpha(\mu)) \leq 2^{\alpha q+\delta}\mathcal{H}_\mu^{q,t}(\underline{\Delta}_\alpha(\mu))$ *for* $q \leq 0$.
iii) *If* $0 \leq \alpha q + b_\mu(q)$ *and* $X = \mathbb{R}^d$ *then*

$$\dim(\overline{\Delta}^\alpha(\mu)) \leq \alpha q + b_\mu(q) \qquad \text{for } 0 \leq q$$
$$\dim(\underline{\Delta}_\alpha(\mu)) \leq \alpha q + b_\mu(q) \qquad \text{for } q \leq 0.$$

iv) *If* $0 \leq \alpha q + B_\mu(q)$ *then*

$$\dim(\overline{\Delta}_\alpha(\mu)) \leq \alpha q + B_\mu(q) \qquad \text{for } q \leq 0$$
$$\dim(\underline{\Delta}^\alpha(\mu)) \leq \alpha q + B_\mu(q) \qquad \text{for } 0 \leq q .$$

Proposition 2.1.4.

i) $\mathcal{P}^{\alpha q+t+\delta}(\overline{\Delta}^{\alpha}(\mu)) \leq 2^{\alpha q+\delta}\mathcal{P}^{q,t}_{\mu}(\overline{\Delta}^{\alpha}(\mu))$ *for* $0 \leq q$.

ii) $\mathcal{P}^{\alpha q+t+\delta}(\underline{\Delta}_{\alpha}(\mu)) \leq 2^{\alpha q+\delta}\mathcal{P}^{q,t}_{\mu}(\underline{\Delta}_{\alpha}(\mu))$ *for* $q \leq 0$.

iii) *If* $0 \leq \alpha q + B_\mu(q)$ *then*

$$\begin{aligned} \operatorname{Dim}(\overline{\Delta}^{\alpha}(\mu)) &\leq \alpha q + B_\mu(q) \qquad \text{for } 0 \leq q\,, \\ \operatorname{Dim}(\underline{\Delta}_{\alpha}(\mu)) &\leq \alpha q + B_\mu(q) \qquad \text{for } q \leq 0\,. \end{aligned}$$

Proposition 2.1.5.

i) *If* $A \subseteq \overline{\Delta}^{\alpha}(\mu)$ *is Borel then* $\mathcal{H}^{q,t}_{\mu}(A) \leq 2^t\mathcal{H}^{\alpha q+t-\delta}(A)$ *for* $q \leq 0$.

ii) *If* $A \subseteq \underline{\Delta}_{\alpha}(\mu)$ *is Borel then* $\mathcal{H}^{q,t}_{\mu}(A) \leq 2^t\mathcal{H}^{\alpha q+t-\delta}(A)$ *for* $0 \leq q$.

Proposition 2.1.6.

i) *If* $A \subseteq \overline{\Delta}^{\alpha}(\mu)$ *is Borel then* $\mathcal{P}^{q,t}_{\mu}(A) \leq 2^{-\alpha q+\delta}\mathcal{P}^{\alpha q+t-\delta}(A)$ *for* $q \leq 0$.

ii) *If* $A \subseteq \underline{\Delta}_{\alpha}(\mu)$ *is Borel then* $\mathcal{P}^{q,t}_{\mu}(A) \leq 2^{-\alpha q+\delta}\mathcal{P}^{\alpha q+t-\delta}(A)$ *for* $0 \leq q$.

The reader is referred to [Ol] for the proofs of Proposition 2.1.3 through Proposition 2.1.6. The previous four propositions constitute the basic technical results used in the proof of one of the main results in [Ol] giving upper bounds of f_μ and F_μ in terms of b_μ and B_μ. However before we can state the result we need some definitions. Write

$$\begin{aligned} \underline{a}_\mu &= \underline{a} := \sup_{0<q} -\frac{b_\mu(q)}{q} \qquad & \overline{a}_\mu &= \overline{a} := \inf_{q<0} -\frac{b_\mu(q)}{q} \\ \underline{A}_\mu &= \underline{A} := \sup_{0<q} -\frac{B_\mu(q)}{q} \qquad & \overline{A}_\mu &= \overline{A} := \inf_{q<0} -\frac{B_\mu(q)}{q} \end{aligned}$$

If $f : \mathbb{R} \to \mathbb{R}$ is a real valued function, let $f^* : \mathbb{R} \to [-\infty, \infty[$ denote the following Legendre transform of f,

$$f^*(x) = \inf_y(xy + f(y))\,, \quad x \in \mathbb{R}$$

We are now ready to state the main result relating b_μ and f_μ (and B_μ and F_μ).

Theorem 2.1.7 (Upper bound estimate). *Let X be a metric space, $\mu \in \mathcal{P}(X)$ and $\alpha \geq 0$. Then the following assertions hold.*

i) $\underline{a} \leq \inf_x \overline{\alpha}_\mu(x) \leq \sup_x \overline{\alpha}_\mu(x) \leq \overline{A}$,
$\underline{A} \leq \inf_x \underline{\alpha}_\mu(x) \leq \sup_x \underline{\alpha}_\mu(x) \leq \overline{a}$.

ii)

$$f_\mu(\alpha) := \dim(\Delta_\mu(\alpha)) = \begin{cases} \leq b^*(\alpha) & \alpha \in \,]\underline{a}, \overline{a}[\\ 0 & \alpha \in \mathbb{R}_+ \setminus [\underline{a}, \overline{a}] \end{cases}$$

iii)

$$F_\mu(\alpha) := \mathrm{Dim}(\Delta_\mu(\alpha)) = \begin{cases} \leq B^*(\alpha) & \alpha \in]\underline{a}, \overline{a}[\\ 0 & \alpha \in \mathbb{R}_+ \setminus [\underline{a}, \overline{a}] \end{cases}$$

Proof. See [Ol,Theorem 2.17]. □

We remark that [Ol] also contains a (somewhat more technical) result giving lower bounds of f_μ and F_μ in terms b_μ and B_μ.

We now recall some of the main results from [Ol] concerning the behavior of the functions b_μ and B_μ. We first define two subfamilies $\mathcal{P}_0(X)$ and $\mathcal{P}_1(X)$ of $\mathcal{P}(X)$. For $\mu \in \mathcal{P}(X)$, $E \subseteq \operatorname{supp}\mu$ and $a > 1$ write

$$T_a(E) = \limsup_{r \searrow 0} \left(\sup_{x \in E} \frac{\mu B(x, ar)}{\mu B(x, r)} \right)$$

We will write $T_a(x) = T_a(\{x\})$ for $x \in \operatorname{supp}\mu$.

For $E \subseteq \operatorname{supp}\mu$ put

$$\begin{aligned} \mathcal{P}_0(X, E) = &\ \{\mu \in \mathcal{P}(X) \mid \exists a > 1 : \forall x \in E : T_a(x) < \infty\} \\ \mathcal{P}_1(X, E) = &\ \{\mu \in \mathcal{P}(X) \mid \exists a > 1 : T_a(E) < \infty\}, \end{aligned}$$

and write $\mathcal{P}_0(X, \operatorname{supp}\mu) = \mathcal{P}_0(X)$ and $\mathcal{P}_1(X, \operatorname{supp}\mu) = \mathcal{P}_1(X)$. It is easily seen that $T_a(E) < \infty$ for some $a > 1$ if and only if $T_a(E) < \infty$ for all $a > 1$ (c.f. [Ol,Lemma 2.1]), and the subfamilies $\mathcal{P}_0(X, E)$ and $\mathcal{P}_1(X, E)$ are thus well defined (i.e. independent of the number $a > 1$ that appears in the definition). Some differentiation results for measures μ satisfying $T_5(x) < \infty$ for $x \in \operatorname{supp}\mu$ (and consequently $T_a(x) < \infty$ for all $a > 1$) appear in [Fed, pp.160–163] and [Mat]. Furthermore, some global density theorems for measures μ satisfying $T_3(x) < \infty$ for μ-a.a. x appear in Sato [Sa]. Measures satisfying $T_3(x) < \infty$ for μ-a.a. x are called Federer measures by Sato [Sa].

The next propositions summarizes the properties of b_μ, B_μ and Λ_μ.

Proposition 2.1.8. *The following statements hold.*

i) Λ_μ *is decreasing and convex.*
ii) B_μ *is decreasing and convex.*
iii) b_μ *is decreasing.*

The map b_μ need not be convex for $\mu \in \mathcal{P}_1(X)$. Olsen [Ol] constructs a measure $\mu \in \mathcal{P}_1(\mathbb{R})$ such that

$$b_\mu(q) = d(1 - q) \wedge D(1 - q),$$

where $0 < d < D < 1$. However, the next proposition shows that if $\mu \in \mathcal{P}_0(\mathbb{R}^d)$, then b_μ satisfies a "weak" form of convexity: instead of $b_\mu(\alpha p + (1-\alpha)q) \leq \alpha b_\mu(p) + (1-\alpha) b_\mu(q)$ then the following inequality holds $b_\mu(\alpha p + (1-\alpha)q) \leq \alpha B_\mu(p) + (1-\alpha) b_\mu(q)$; i.e. we have replaced the smaller number $b_\mu(p)$ with the (perhaps) somewhat larger number $B_\mu(p)$ (here $p, q \in \mathbb{R}$ and $\alpha \in [0, 1]$.)

Proposition 2.1.9. *Let $\mu \in \mathcal{P}(\mathbb{R}^d)$, $E \subseteq \mathbb{R}^d$, $p, q \in \mathbb{R}$ and $\alpha \in [0,1]$.*

i) *If $\alpha p + (1-\alpha)q \leq 0$ then*

$$b_\mu(\alpha p + (1-\alpha)q) \leq \alpha B_\mu(p) + (1-\alpha) b_\mu(q)$$

ii) *If $0 < \alpha p + (1-\alpha)q$ and in addition $\mu \in \mathcal{P}_0(\mathbb{R}^d)$ then*

$$b_\mu(\alpha p + (1-\alpha)q) \leq \alpha B_\mu(p) + (1-\alpha) b_\mu(q)$$

The reader is referred to [Ol] for the proofs of Proposition 2.1.8 and Proposition 2.1.9.

The next proposition investigates the behavior of $B_\mu(q)$ when $|q|$ is large. The proposition shows that B_μ have affine asymptotes.

Proposition 2.1.10. *The following statements hold.*

i) *There exists a number $\underline{E} \geq 0$ such that*

$$B_\mu(q) + \underline{A} q \searrow \underline{E} \quad \text{as} \quad q \to \infty .$$

i) *There exists a number $\overline{E} \geq 0$ such that*

$$B_\mu(q) + \overline{A} q \searrow \overline{E} \quad \text{as} \quad q \to -\infty .$$

Proof. See [Ol,Proposition 2.13]. □

We close this section with some results on multifractal box dimensions and generalized Rényi dimensions. We begin by recalling the definition of the upper and lower box-dimension. Let $E \subseteq \mathbb{R}^d$ and $N_\delta(E)$ denote the largest number of disjoint balls of radius δ with centres in E. Then the lower respectively upper box-dimension of E are defined as

$$\underline{C}(E) = \liminf_{\delta \searrow 0} \frac{\log N_\delta(E)}{-\log \delta}$$
$$\overline{C}(E) = \limsup_{\delta \searrow 0} \frac{\log N_\delta(E)}{-\log \delta} .$$

If $\overline{C}(E) = \underline{C}(E)$ we refer to the common value as the box-dimension and denote it by $C(E)$. The reader is referred to [Fa2] for more information about box-dimensions.

We will now define multifractal box-dimensions. Let $\mu \in \mathcal{P}(\mathbb{R}^d)$ and $q \in \mathbb{R}$. For $E \subseteq \mathbb{R}^d$ and $\delta > 0$ write

$$S^q_{\mu,\delta}(E) = \sup \left\{ \sum_i \mu(B(x_i,\delta))^q \mid (B(x_i,\delta))_{i \in \mathbb{N}} \text{ is a centered packing of } E \right\} .$$

The upper respectively lower multifractal q-box dimension $\overline{C}^q_\mu(E)$ and $\underline{C}^q_\mu(E)$ of E (with respect to the measure μ) are defined by

$$\overline{C}^q_\mu(E) = \limsup_{\delta \searrow 0} \frac{\log S^q_{\mu,\delta}(E)}{-\log \delta}$$

$$\underline{C}^q_\mu(E) = \liminf_{\delta \searrow 0} \frac{\log S^q_{\mu,\delta}(E)}{-\log \delta}$$

cf. Olsen [Ol]. If $\overline{C}^q_\mu(E) = \underline{C}^q_\mu(E)$ we refer to the common value as the q-box dimension of E (with respect to the measure μ) and denote it by $C^q_\mu(E)$. A somewhat similar definition appears in [Fa2, p. 225] and [Str]. Also observe that

$$\underline{C}^0_\mu(E) = \underline{C}(E), \quad \overline{C}^0_\mu(E) = \overline{C}(E).$$

Hentschel & Procaccia [He], Grassberger & Procaccia [Gr1] and Grassberger [Gr2] proposed in 1983 a multifractal formalism parallel to (but independent of) the $f(\alpha)$ formalism introduced by Halsey et al. [Ha]. Hentschel & Procaccia [He] and Grassberger & Procaccia [Gr1] introduced a one-parameter family of numbers $(D_q)_{q\in\mathbb{R}}$ based on some generalized entropies due to Rényi [Re1,Re2]. Let $\mu \in \mathcal{P}(\mathbb{R}^d)$. For $q \in \mathbb{R}$ and $\delta > 0$ write

$$h^q_\delta(\mu) = \frac{1}{1-q} \log(S^q_{\mu,\delta}(\operatorname{supp}\mu)) \quad \text{for } q \neq 1\,,$$

and

$$h^1_\delta(\mu) = \inf\left\{ -\sum_i \mu(E_i)\log\mu(E_i) \,\middle|\, (E_i)_i \text{ is countable Borel partition of } \operatorname{supp}\mu, \ \operatorname{diam} E_i \le \delta \quad \text{for all } i \right\}.$$

Following Hentschel & Procaccia [He, formula (3.13)] we define the q Rényi dimensions d^q_μ and D^q_μ of μ by

$$d^q_\mu = \liminf_{\delta\searrow 0} \frac{h^q_\delta(\mu)}{-\log\delta} \quad \text{for } q \le 1$$

$$d^q_\mu = \limsup_{\delta\searrow 0} \frac{h^q_\delta(\mu)}{-\log\delta} \quad \text{for } 1 < q$$

$$D^q_\mu = \limsup_{\delta\searrow 0} \frac{h^q_\delta(\mu)}{-\log\delta} \quad \text{for } q < 1$$

$$D^q_\mu = \liminf_{\delta\searrow 0} \frac{h^q_\delta(\mu)}{-\log\delta} \quad \text{for } 1 \le q$$

(in [He] all limits are assumed to exist and Hentschel et al. therefore only consider $D_\mu^q = \lim_{\delta \searrow 0} \frac{h_\delta^q(\mu)}{-\log \delta}$). Observe that

$$(1-q)d_\mu^q \leq (1-q)D_\mu^q .$$

A parallel development of q Rényi dimensions using integrals was also suggested in [He, formula (3.14)]. For $r > 0$ and $q \in \mathbb{R} \setminus \{0\}$ write

$$I_{\mu,r}^q = \frac{1}{q} \log \left(\int_{\operatorname{supp} \mu} \mu(B(x,r))^q d\mu(x) \right)$$

and

$$\overline{I}_\mu^q = \limsup_{r \searrow 0} \frac{I_{\mu,r}^q}{-\log r}$$

$$\underline{I}_\mu^q = \liminf_{r \searrow 0} \frac{I_{\mu,r}^q}{-\log r} .$$

Our final propositions recall some results from [Ol] on the relation between the dimension functions b_μ, B_μ and Λ_μ, the multifractal box dimensions and the generalized Rényi dimensions.

Theorem 2.1.11.

i) *If $\mu \in \mathcal{P}(\mathbb{R}^d)$ and $q \leq 0$ then*

$$b_\mu(q) \leq \underline{C}_\mu^q(\operatorname{supp} \mu) \leq \overline{C}_\mu^q(\operatorname{supp} \mu) \leq \Lambda_\mu(q) .$$

ii) *If $\mu \in \mathcal{P}_1(\mathbb{R}^d)$ and $0 < q$ then*

$$b_\mu(q) \leq \underline{C}_\mu^q(\operatorname{supp} \mu) \leq \overline{C}_\mu^q(\operatorname{supp} \mu) \leq \Lambda_\mu(q) .$$

Proof. See [Ol,Proposition 2.19 and Proposition 2.22]. □

Theorem 2.1.12. *If $\mu \in \mathcal{P}_1(\mathbb{R}^d)$ and $q \in \mathbb{R}$ then*

$$(q-1)I_\mu^{q-1} = \Lambda_\mu(q) = (1-q)D_\mu^q .$$

Proof. See [Ol,Proposition 2.24]. □

We remark that Olsen [Ol] also defines and studies other types of multifractal box dimensions.

2.2 Random Geometrically Graph Directed Self-Similar Sets

The purpose of this section is to introduce the notion of a random graph directed self-similar set.

Let (V, E) be a finite directed multigraph. The set V is the set of vertices and E is the set of edges. For $u, v \in V$ let E_{uv} denote the set of edges from u to v write $E_u = \cup_{v \in V} E_{uv}$. A path in the graph is a finite string $e_1 e_2 \ldots e_n$ of edges such that the terminal vertex of the edge e_i is the initial vertex of the next edge e_{i+1} and an infinite path in the graph is an infinite string $e_1 e_2 \ldots$ of edges such that $e_1 \ldots e_n$ is a path for all $n \in \mathbb{N}$. For $e \in E$ let $\iota(e)$ and $\tau(e)$ denote the initial and terminal vertex of e respectively. For $u, v \in V$ and $n \in \mathbb{N}$ write

$$E_{uv}^{(n)} = \{e_1 \ldots e_n \text{ is a path such that } \iota(e_1) = u \text{ and } \tau(e_n) = v\}$$
$$E_{uv}^{(*)} = \bigcup_{n \in \mathbb{N}} E_{uv}^{(n)}$$

$$E_u^{(n)} = \bigcup_{v \in V} E_{uv}^{(n)}, \qquad E_u^{(*)} = \bigcup_{v \in V} E_{uv}^{(*)}$$
$$E^{(n)} = \bigcup_{u \in V} E_u^{(n)}, \qquad E^{(*)} = \bigcup_{u \in V} E_u^{(*)}$$

$$E_u^{\mathbb{N}} = \{e_1 e_2 \ldots \text{ is an infinite path such that } \iota(e_1) = u\}$$
$$E^{\mathbb{N}} = \bigcup_{u \in V} E_u^{\mathbb{N}}$$

If $\alpha = \alpha_1 \ldots \alpha_n$, $\beta = \beta_1 \ldots \beta_m \in E^{(*)}$ are paths and the terminal vertex $\tau(\alpha_n)$ of α_n is equal to the initial vertex $\iota(\beta_1)$ of β_1 then we write $\alpha\beta = \alpha_1 \ldots \alpha_n \beta_1 \ldots \beta_m$. If $\alpha = \alpha_1 \ldots \alpha_n \in E^{(n)}$ and $k \in \{1, \ldots, n\}$ then we write $\alpha|k = \alpha_1 \ldots \alpha_k$. Similarly, if $\alpha = \alpha_1 \ldots \alpha_n \in E^{(*)}$ is a path, $\omega = \omega_1 \omega_2 \cdots \in E^{\mathbb{N}}$ is an infinite path with $\tau(\alpha_n) = \iota(\omega_1)$ and $m \in \mathbb{N}$ is an integer then write $\alpha\omega = \alpha_1 \ldots \alpha_n \omega_1 \omega_2 \ldots$ and $\omega|m = \omega_1 \ldots \omega_m$. For $\alpha = \alpha_1 \ldots \alpha_n \in E^{(n)}$ put $|\alpha| = n$. If $\alpha = \alpha_1 \ldots \alpha_n \in E^{(n)}$, $\beta = \beta_1 \ldots \beta_m \in E^{(m)}$ with $|\alpha| \leq |\beta|$ and $\alpha_1 = \beta_1, \ldots, \alpha_n = \beta_n$ then we write $\alpha \prec \beta$. Similarly, if $\alpha = \alpha_1 \ldots \alpha_n \in E^{(*)}$ is a path and $\omega = \omega_1 \omega_2 \cdots \in E^{\mathbb{N}}$ is an infinite path with $\alpha_1 = \omega_1, \ldots, \alpha_n = \omega_n$ we write $\alpha \prec \omega$. For $\alpha \in E^{(*)}$ we define the cylinder $[\alpha]$ generated by α by

$$[\alpha] := \{\omega \in E^{\mathbb{N}} \mid \alpha \prec \omega\}.$$

Finally, if $\alpha \in E^{(n)}$ and $\omega \in E^{\mathbb{N}}$ then we will always write $\alpha = \alpha_1 \ldots \alpha_n$ or $\alpha = \alpha(1) \ldots \alpha(n)$ and $\omega = \omega_1 \omega_2 \ldots$ or $\omega = \omega(1)\omega(2) \ldots$.

For metric spaces X, Y write

$$\mathrm{Con}(X, Y) = \{f : X \to Y \mid f \text{ is a contraction}\}.$$

If (X, d_X) and (Y, d_Y) are metric spaces and $f \in \mathrm{Con}(X, Y)$, $\mathrm{Lip}(f)$ denotes the Lipschitz constant of f, i.e. $\mathrm{Lip}(f) = \sup_{x,y \in X, x \neq y} \frac{d_Y(f(x), f(y))}{d_X(x,y)}$.

For each $u \in V$ let X_u be a complete metric space. For fixed $u \in V$ write

$$\Delta_u = \{(p_e)_{e \in E_u} \mid p_e \in [0,1],\ \sum_{e \in E_u} p_e = 1\}$$
$$\Gamma_u = \prod_{e \in E_u} \mathrm{Con}(X_{\tau(e)}, X_u)$$
$$\Xi_u = \Gamma_u \times \Delta_u$$
$$\Omega_u = \Xi_u \times \prod_{\alpha \in E_u^{(*)} \setminus \{\varnothing\}} \Xi_{\tau(\alpha_{|\alpha|})}$$
$$\Omega = \prod_{u \in V} \Omega_u$$

For $u \in V$, $\alpha \in E_u^{(*)}$ and $e \in E_{\tau(\alpha_{|\alpha|})}$ define projections

$$\pi_{u,\alpha} : \Omega_u = \Xi_u \times \prod_{\gamma \in E_u^{(*)} \setminus \{\varnothing\}} \Xi_{\tau(\gamma_{|\gamma|})} \to \begin{cases} \Xi_u & \text{for } \alpha = \varnothing \\ \Xi_{\tau(\alpha_{|\alpha|})} & \text{for } \alpha \neq \varnothing \end{cases} \tag{2.2.1}$$

$$S_{\alpha e} : \Omega_u = \Xi_u \times \prod_{\gamma \in E_u^{(*)} \setminus \{\varnothing\}} \Xi_{\tau(\gamma_{|\gamma|})} \to \begin{cases} \mathrm{Con}(X_{\tau(e)}, X_u) & \text{for } \alpha = \varnothing \\ \mathrm{Con}(X_{\tau(\alpha e_{|\alpha e|})}, X_{\tau(\alpha_{|\alpha|})}) & \text{for } \alpha \neq \varnothing \end{cases} \tag{2.2.2}$$

$$p_{\alpha e} : \Omega_u = \Xi_u \times \prod_{\gamma \in E_u^{(*)} \setminus \{\varnothing\}} \Xi_{\tau(\gamma_{|\gamma|})} \to [0,1] \tag{2.2.3}$$

by the requirement: if $\omega \in \Omega_u$ then

$$\omega = \left(\pi_{u,\alpha}(\omega)\right)_{\alpha \in E_u^{(*)}}$$

and

$$\pi_{u,\varnothing}(\omega) = \left((S_e(\omega))_{e \in E_u}, (p_e(\omega))_{e \in E_u}\right) \in \Xi_u \qquad \text{for } \alpha = \varnothing$$
$$\pi_{u,\alpha}(\omega) = \left((S_{\alpha e}(\omega))_{e \in E_{\tau(\alpha_{|\alpha|})}}, (p_{\alpha e}(\omega))_{e \in E_{\tau(\alpha_{|\alpha|})}}\right) \in \Xi_{\tau(\alpha_{|\alpha|})} \qquad \text{for } \alpha \neq \varnothing$$

For $u \in V$ and $e \in E_u$ we also define projections

$$S_e : \Xi_u \to \mathrm{Con}(X_{\tau(e)}, X_u) \tag{2.2.4}$$
$$p_e : \Xi_u \to [0,1] \tag{2.2.5}$$

by the requirement: if $\xi \in \Xi_u$ then

$$\xi = \left((S_e(\xi))_{e\in E_u}, (p_e(\xi))_{e\in E_u}\right).$$

We remark that random variables

$$S_e : \Omega_u \to \mathrm{Con}(X_{\tau(e)}, X_u), \quad p_e : \Omega_u \to [0,1]$$

have already been defined in (2.2.2) and (2.2.3); however, no confusion will arise since it will always be clear from the context whether S_e and p_e denote the random variables defined in (2.2.2) and (2.2.3) or the random variables defined in (2.2.4) and (2.2.5). Put

$$(\Omega_u)_0 = \{\omega \in \Omega_u \mid \forall \sigma \in E_u^{\mathbb{N}} : \prod_{n=1}^{\infty} \mathrm{Lip}(S_{\sigma|n}(\omega)) = 0\}$$

$$\Omega_0 = \prod_{u\in V} (\Omega_u)_0 .$$

Finally define for $\alpha \in E_u^{(*)} \setminus \{\varnothing\}$ the "shift" map $\mathcal{S}_\alpha : \Omega_u \to \Omega_{\tau(\alpha_{|\alpha|})}$ by

$$\mathcal{S}_\alpha(\omega) = \sigma$$

where

$$\pi_{\tau(\alpha_{|\alpha|}),\gamma}(\sigma) = \pi_{u,\alpha\gamma}(\omega) \quad \text{for } \gamma \in E^{(*)}_{\tau(\alpha_{|\alpha|})} .$$

If X is a metric space then $\mathcal{C}(X)$ will denote the family of all non-empty, compact subsets of X, $\mathcal{F}(X)$ will denote the family of all non-empty, closed subsets of X and $\mathcal{F}_b(X)$ will denote the family of all non-empty, closed and bounded subsets of X. Also, D_X will denote the Hausdorff metric on $\mathcal{F}_b(X)$.

For $u \in V$, $n \in \mathbb{N}$, $\gamma \in E_u^{(*)}$ and $\mathbf{K} = \left(K_\varnothing, (K_\alpha)_{\alpha\in E_u^{(*)}\setminus\{\varnothing\}}\right) \in \mathcal{C}(X_u) \times \prod_{\alpha\in E_u^{(*)}\setminus\{\varnothing\}} \mathcal{C}(X_{\tau(\alpha_{|\alpha|})})$ define maps

$$C_{u,n}^{\mathbf{K}} : \Omega_u \to \mathcal{C}(X_u)$$
$$C_{u,\gamma} : \Omega_u \to \mathcal{F}(X_u)$$
$$C_u : \Omega_u \to \mathcal{F}(X_u)$$

by

$$C_{u,n}^{\mathbf{K}} = \bigcup_{\alpha\in E_u^{(n)}} S_{\alpha|1} \circ \cdots \circ S_{\alpha|n}(K_\alpha)$$
$$C_{u,\gamma} = \overline{S_{\gamma|1} \circ \cdots \circ S_{\gamma|n}(X_{\tau(\gamma_{|\gamma|})})}$$
$$C_u = \bigcap_{n\in\mathbb{N}} \bigcup_{\alpha\in E_u^{(n)}} \overline{S_{\alpha|1} \circ \cdots \circ S_{\alpha|n}(X_{\tau(\alpha_n)})}$$

For $n \in \mathbb{N}$ and $\mathbf{K} = (\mathbf{K}_u)_{u \in V} \in \prod_{u \in V} \left(\mathcal{C}(X_u) \times \prod_{\alpha \in E_u^{(*)} \setminus \{\varnothing\}} \mathcal{C}(X_{\tau(\alpha_{|\alpha|})}) \right)$ define maps

$$C_n^{\mathbf{K}} : \Omega \to \prod_{u \in V} \mathcal{C}(X_u)$$
$$C : \Omega \to \prod_{u \in V} \mathcal{F}(X_u)$$

by

$$C_n^{\mathbf{K}}(\omega) = \left(C_{u,n}^{\mathbf{K}_u}(\omega_u) \right)_{u \in V}$$
$$C(\omega) = \left(C_u(\omega_u) \right)_{u \in V}$$

for $\omega = (\omega_u)_{u \in V} \in \prod_{u \in V} \Omega_u = \Omega$.

The next theorem defines the notion of a random graph directed self-similar set.

Theorem 2.2.1. *Let X_v be bounded for all $v \in V$. Let $\omega = (\omega_u)_u \in \Omega_0$ and $\mathbf{K} = (\mathbf{K}_u)_{u \in V} \in \prod_{u \in V} \left(\mathcal{C}(X_u) \times \prod_{\alpha \in E_u^{(*)} \setminus \{\varnothing\}} \mathcal{C}(X_{\tau(\alpha_{|\alpha|})}) \right)$. Then the following statements hold.*

i) *The sequence*

$$\left(C_{u,n}^{\mathbf{K}_u}(\omega_u) \right)_n$$

is convergent in $(\mathcal{C}(X_u), D_{X_u})$; in particular the sequence

$$\left(C_n^{\mathbf{K}}(\omega) \right)_n$$

is convergent in $(\prod_{u \in V} \mathcal{C}(X_u), D)$, where D denotes the maximum metric in $\prod_{u \in V} \mathcal{C}(X_u)$, i.e.

$$D((K_u)_u, (L_u)_u) = \sup_u D_{X_u}(K_u, L_u)$$

for $(K_u)_u, (L_u)_u \in \prod_{u \in V} \mathcal{C}(X_u)$.

ii) *We have*

$$\lim_n C_{u,n}^{\mathbf{K}_u}(\omega_u) = C_u(\omega_u)$$

in particular

$$\lim_n C_n^{\mathbf{K}}(\omega) = C(\omega)$$

whence $C(\omega) \in \prod_{u \in V} \mathcal{C}(X_u)$.

The maps $C_u = \lim_n C_{u,n}^{\mathbf{K}_u}$ are called random graph directed self-similar sets. Theorem 2.2.1 is a graph directed extension of [Gra1, Theorem 2.2].

Let $\omega \in (\Omega_u)_0$. It is easily seen that

$$\begin{aligned} C_u(\omega) &= \bigcap_{n \in \mathbb{N}} \bigcup_{\alpha \in E_u^{(n)}} \overline{S_{\alpha|1}(\omega) \circ \cdots \circ S_{\alpha|n}(\omega)(X_{\tau(\alpha_{|\alpha|})})} \\ &= \bigcup_{\sigma \in E_u^{\mathbb{N}}} \bigcap_{n \in \mathbb{N}} \overline{S_{\sigma|1}(\omega) \circ \cdots \circ S_{\sigma|n}(\omega)(X_{\tau(\sigma_n)})}. \end{aligned}$$

Since for each $\sigma \in E_u^{\mathbb{N}}$, $\left(\overline{S_{\sigma|1}(\omega) \circ \cdots \circ S_{\sigma|n}(\omega)(X_{\tau(\sigma_n)})}\right)_n$ is a decreasing sequence of non-empty compact sets whose diameter tends to zero as n tends to infinity, and since X_u is complete, then $\bigcap_n \overline{S_{\sigma|1}(\omega) \circ \cdots \circ S_{\sigma|n}(\omega)(X_{\tau(\sigma_n)})}$ is a singleton. Now define

$$\pi_u(\omega) : E_u^{\mathbb{N}} \to C_u(\omega)$$

by

$$\{\pi_u(\omega)(\sigma)\} = \bigcap_n \overline{S_{\sigma|1}(\omega) \circ \cdots \circ S_{\sigma|n}(\omega)(X_{\tau(\sigma_n)})}.$$

Then $\pi_u(\omega)$ is continuous and

$$\pi_u(\omega)(E_u^{\mathbb{N}}) = C_u(\omega).$$

2.3 Random Geometrically Graph Directed Self-Similar Measures

In this section we define the notion of random graph directed self-similar measures.

We first recall the definition of a random measure. If X is a locally compact, second countable Hausdorff topological space then $\mathcal{M}(X)$ denotes the space of positive Radon measures on X, and $\mathrm{va}(\mathcal{M}(X))$ denotes the vague topology on $\mathcal{M}(X)$. Let $\mathrm{w}(\mathcal{P}(X))$ denote the weak topology on the set $\mathcal{P}(X)$ of Borel probability measures on X. If (Ω, Σ, P) is a probability space and $\Phi : \Omega \to \mathcal{M}(X)$ is a map from Ω to $\mathcal{M}(X)$ then Φ is called a random measure if and only if Φ is Σ–$\mathrm{va}(\mathcal{M}(X))$ measurable (cf. Kallenberg [Kal, p. 13].)

We will now define the notion of graph directed self-similar measures. For $u \in V$, $n \in \mathbb{N}$ and $\gamma = (\gamma_\varnothing, (\gamma_\alpha)_{\alpha \in E_u^{(*)} \setminus \{\varnothing\}}) \in \mathcal{P}(X_u) \times \prod_{\alpha \in E_u^{(*)} \setminus \{\varnothing\}} \mathcal{P}(X_{\tau(\alpha_{|\alpha|})})$ define

$$\mu_{u,n}^{\gamma} : \Omega_u \to \mathcal{P}(X_u)$$

by

$$\mu_{u,n}^{\gamma} = \sum_{\alpha \in E_u^{(n)}} p_{\alpha|1} \cdots p_{\alpha|n} \gamma_\alpha \circ (S_{\alpha|1} \circ \cdots \circ S_{\alpha|n})^{-1}$$

For $n \in \mathbb{N}$ and $\boldsymbol{\gamma} = (\gamma_u)_{u \in V} \in \prod_{u \in V} \left(\mathcal{P}(X_u) \times \prod_{\alpha \in E_u^{(*)} \setminus \{\varnothing\}} \mathcal{P}(X_{\tau(\alpha_{|\alpha|})}) \right)$ define

$$\mu_n^{\boldsymbol{\gamma}} : \Omega \to \prod_{u \in V} \mathcal{P}(X_u)$$

by

$$\mu_n^{\boldsymbol{\gamma}}(\omega) = \left(\mu_{u,n}^{\gamma_u}(\omega_u)\right)_{u \in V}$$

for $\omega = (\omega_u)_{u \in V} \in \prod_{u \in V} \Omega_u = \Omega$.

Define

$$\hat{\mu}_u : \Omega_u \to \mathcal{P}(E_u^{\mathbb{N}})$$

as follows: Fix $\hat{\mu}_{u,0} \in \mathcal{P}(E_u^{\mathbb{N}})$ and let $\omega \in \Omega_u$ If $\omega \in (\Omega_u)_0$ then $\hat{\mu}_u(\omega)$ is determined by the requirement

$$(\hat{\mu}_u(\omega))([\alpha]) = \prod_{k=1}^{|\alpha|} p_{\alpha|k}(\omega) \quad \text{for } \alpha \in E_u^{(*)} \setminus \{\varnothing\}.$$

If $\omega \notin (\Omega_u)_0$ then we put

$$\hat{\mu}_u(\omega) = \hat{\mu}_{u,0}$$

Now define

$$\bar{\mu}_u(\omega) : \Omega_u \to \mathcal{P}(X_u)$$

as follows: Fix $\bar{\mu}_{u,0} \in \mathcal{P}(X_u)$ and put

$$\bar{\mu}_u(\omega) = \begin{cases} \hat{\mu}_u(\omega) \circ \pi_u(\omega)^{-1} & \text{for } \omega \in (\Omega_u)_0 \\ \bar{\mu}_{u,0} & \text{for } \omega \notin (\Omega_u)_0 \end{cases}$$

Finally define

$$\hat{\mu} : \Omega \to \prod_{u \in V} \mathcal{P}(E_u^{\mathbb{N}})$$
$$\bar{\mu} : \Omega \to \prod_{u \in V} \mathcal{P}(X_u)$$

by

$$\hat{\mu}(\omega) = (\hat{\mu}_u(\omega_u))_{u \in V}$$
$$\bar{\mu}(\omega) = (\bar{\mu}_u(\omega_u))_{u \in V}$$

for $\omega = (\omega_u)_{u \in V} \in \prod_{u \in V} \Omega_u = \Omega$.

The next theorem defines random graph directed self similar measures.

Theorem 2.3.1. *Let X_v be compact for all $v \in V$. Let $\omega = (\omega_u)_{u \in V} \in \Omega_0$. Then the following hold*

i) *Let $\gamma = (\gamma_u)_{u \in V} \in \prod_{u \in V} \left(\mathcal{P}(X_u) \times \prod_{\alpha \in E_u^{(*)} \setminus \{\varnothing\}} \mathcal{P}(X_{\tau(\alpha_{|\alpha|})})\right)$. Then the sequence*

$$\left(\mu_{u,n}^{\gamma_u}(\omega_u)\right)_n$$

is convergent in $\left(\mathcal{P}(X_u), \mathrm{w}(\mathcal{P}(X_u))\right)$; in particular the sequence

$$\left(\mu_n^{\gamma}(\omega)\right)_n$$

is convergent in $\left(\prod_{u \in V} \mathcal{P}(X_u), \prod_{u \in V} \mathrm{w}(\mathcal{P}(X_u))\right)$.

ii) *Let $\gamma = (\gamma_u)_{u \in V}, \nu = (\nu_u)_{u \in V} \in \prod_{u \in V} \left(\mathcal{P}(X_u) \times \prod_{\alpha \in E_u^{(*)} \setminus \{\varnothing\}} \mathcal{P}(X_{\tau(\alpha_{|\alpha|})})\right)$. Then*

$$\lim_n \mu_{u,n}^{\gamma_u}(\omega_u) = \lim_n \mu_{u,n}^{\nu_u}(\omega_u)$$

in particular

$$\lim_n \mu_n^{\gamma}(\omega) = \lim_n \mu_n^{\nu}(\omega)$$

Define maps

$$\mu_u : \Omega \to \mathcal{P}(X_u)$$
$$\mu : \Omega \to \prod_{u \in V} \mathcal{P}(X_u)$$

by

$$\mu_u(\omega_u) = \begin{cases} \lim_n \mu_{u,n}^{\gamma_u}(\omega_u) & \text{for } \omega_u \in (\Omega_u)_0 \\ \mu_{u,0} & \text{for } \omega_u \notin (\Omega_u)_0 \end{cases}$$
$$\mu(\omega) = \big(\mu_u(\omega_u)\big)_u$$

for $\omega = (\omega_u)_u \in \Omega$, $\gamma = (\gamma_u)_u \in \prod_{u \in V} \Big(\mathcal{P}(X_u) \times \prod_{\alpha \in E_u^{(*)} \setminus \{\varnothing\}} \mathcal{P}(X_{\tau(\alpha_{|\alpha|})})\Big)$ and $(\mu_{u,0})_u \in \prod_{u \in V} \mathcal{P}(X_u)$. It follows from Theorem 2.3.1.ii) that μ_u is independent of γ. The maps $\mu_u = \lim_n \mu_{u,n}^{\gamma_u}$ are called random graph directed self-similar measures.

The next result shows that μ_u is a random measure. If X, Y are metric spaces then $\mathcal{O}(X, Y)$ denotes the compact-open topology on $\mathrm{Con}(X, Y)$. For each $u \in V$ and $e \in E_u$ equip $\mathrm{Con}(X_{\tau(e)}, X_u)$ with the compact-open topology $\mathcal{O}(X_{\tau(e)}, X_u)$, and equip Δ_u with the topology $\mathcal{O}_u$ that Δ_u inherits from $\mathbb{R}^{\operatorname{card} E_u}$. Next, we equip $\Xi_u = \Gamma_u \times \Delta_u$ with the product topology

$$O_u := \prod_{e \in E_u} \mathcal{O}(X_{\tau(e)}, X_u) \times \mathcal{O}_u\,.$$

Finally, equip Ω_u with the product topology

$$\mathcal{T}_u := O_u \times \prod_{\alpha \in E_u^{*} \setminus \{\varnothing\}} O_{\tau(\alpha_{|\alpha|})}$$

and the Borel algebra $\Sigma_u := \mathcal{B}(\mathcal{T}_u)$ induced by $\mathcal{T}_u$.

Theorem 2.3.2. *Let X_v be compact for all $v \in V$. Then*

$$\mu_u : \Omega_u \to \mathcal{P}(X_u) \subseteq \mathcal{M}(X_u)$$

is a random measure when Ω_u is equipped with the Borel algebra $\Sigma_u := \mathcal{B}(\mathcal{T}_u)$.

The next proposition gives an explicit expression for the random measure μ_u.

Theorem 2.3.3. *Let X_v be compact for all $v \in V$. If $\omega \in (\Omega_u)_0$ satisfies*

$$S_{\alpha e}(\omega)(X_{\tau(e)}) \bigcap S_{\alpha \varepsilon}(\omega)(X_{\tau(\varepsilon)}) = \varnothing$$

for all $\alpha \in E_u^{(*)}$ and $e, \varepsilon \in E_{\tau(\alpha_{|\alpha|})}$ with $e \neq \varepsilon$, then

$$\mu_u(\omega) = \hat{\mu}_u(\omega) \circ \pi_u(\omega)^{-1}$$

2.4 The $(\lambda_u)_u$ Self-Similar Measure P_u

One of our main goals is to determine the almost sure multifractal spectrum of the random measures μ_u w.r.t. a probability measure on $\mathcal{P}(X_u)$ which in a natural way respects the self-similar structure of μ_u. This section contains the construction of a probability measure P_u on $\mathcal{P}(X_u)$ satisfying this condition. Moreover, we also prove that the measure P_u is unique.

Let $\lambda = (\lambda_u)_{u\in V} \in \prod_{u\in V} \mathcal{P}(\Xi_u)$ and put

$$\Lambda_u = \Lambda_{\lambda,u} := \lambda_u \times \prod_{\alpha\in E_u^{(*)}\setminus\{\varnothing\}} \lambda_{\tau(\alpha_{|\alpha|})} \in \mathcal{P}(\Omega_u)$$

$$\Lambda = \Lambda_\lambda := \prod_{u\in V} \Lambda_{\lambda,u} \in \mathcal{P}(\Omega)\,.$$

Next define $P_{\lambda,u} = P_u \in \mathcal{P}(\mathcal{P}(X_u))$ and $P_\lambda \in \mathcal{P}(\prod_{u\in V}\mathcal{P}(X_u))$ by

$$P_u = P_{\lambda,u} = \Lambda_{\lambda,u} \circ \mu_u^{-1}$$

$$P = P_\lambda = \Lambda_\lambda \circ \mu_u^{-1} = \left(\prod_{u\in V} \Lambda_{\lambda,u}\right)^{-1} \circ \left(\prod_{u\in V} \mu_u\right)^{-1} = \prod_{u\in V} P_{\lambda,u}\,.$$

Observe that P_u is well-defined since Theorem 2.3.2 asserts that $\mu_u : \Omega_u \to \mathcal{P}(X_u)$ is Borel measurable. The next theorem (i.e. Theorem 2.4.1) states that the measure P_u in a natural way respects the self-similar structure of the random measure μ_u. However, in order to state Theorem 2.4.1 we need some definitions. For $u \in V$ define $\varphi_u : \Xi_u \times \prod_{e\in E_u} \Omega_{\tau(e)} \to \Omega_u$ by

$$\varphi_u\Big(\big((S_e)_{e\in E_u}, (p_e)_{e\in E_u}\big), (\omega_{\tau(e)})_{e\in E_u}\Big) = \omega$$

where

$$\pi_{u,\varnothing}(\omega) = \big((S_e)_{e\in E_u}, (p_e)_{e\in E_u}\big)$$

$$\pi_{u,e\gamma}(\omega) = \pi_{\tau(e),\gamma}(\omega_{\tau(e)}) \quad \text{for} \quad e \in E_u\,,\ \gamma \in E_{\tau(e)}^{(*)}\,.$$

Define

$$\Phi_u : \Xi_u \times \prod_{e\in E_u} \mathcal{P}(X_{\tau(e)}) \to \mathcal{P}(X_u)$$

by

$$\Phi_u\Big(\big((S_e)_{e\in E_u}, (p_e)_{e\in E_u}\big), (\nu_{\tau(e)})_{e\in E_u}\Big) = \sum_{e\in E_u} p_e \nu_{\tau(e)} \circ S_e^{-1}\,.$$

For $\gamma = (\gamma_u)_{u\in V} \in \prod_{u\in V} \mathcal{P}(\Xi_u)$ define

$$T_\gamma : \prod_{u\in V} \mathcal{P}(\mathcal{P}(X_u)) \to \prod_{u\in V} \mathcal{P}(\mathcal{P}(X_u))$$

by

$$(T_\gamma((Q_u)_u))_v = \left(\gamma_v \times \prod_{e\in E_v} Q_{\tau(e)}\right) \circ \Phi_v^{-1}$$

for $(Q_u)_{u\in V} \in \prod_{u\in V} \mathcal{P}(\mathcal{P}(X_u))$. A list $Q = (Q_u)_u \in \prod_{u\in V} \mathcal{P}(\mathcal{P}(X_u))$ is called γ-self-similar if Q is a fixed point for T_γ, i.e.

$$T_\gamma(Q) = Q\,.$$

Theorem 2.4.1 asserts that $(P_u)_u$ is the unique λ-self-similar list.

Theorem 2.4.1. *The following statements hold.*

i) *The diagram below commutes*

$$\begin{array}{ccc}
\Xi_u \times \prod_{e\in E_u} (\Omega_{\tau(e)})_0 & \xrightarrow{1_{\Xi_u} \times \prod_{e\in E_u} \mu_{\tau(e)}} & \Xi_u \times \prod_{e\in E_u} \mathcal{P}(X_{\tau(e)}) \\
\varphi_u \downarrow & & \downarrow \Phi_u \\
(\Omega_u)_0 & \xrightarrow[\mu_u]{} & \mathcal{P}(X_u)
\end{array}$$

ii) *The list*

$$(P_{\lambda,u})_u \in \prod_{u\in V} \mathcal{P}(\mathcal{P}(X_u))$$

is the unique fixed point of T_λ.

iii) *If*

$$(Q_u)_{u\in V} \in \prod_{u\in V} \mathcal{P}(\mathcal{P}(X_u))$$

then

$$T_\lambda((Q_u)_u)) \xrightarrow{\text{w}} (P_u)_u$$

where $\xrightarrow{\text{w}}$ *denotes weak convergence.*

A natural condition to impose on a probability distribution on $\mathcal{P}(X_u)$ in order for that distribution to respect the self-similar structure of the random measure μ_u is to require that the two probabilities described below coincide for all Borel subsets M of $\mathcal{P}(X_u)$:

1) The probability $\mathbf{Prob}_{u,1}(M)$ that

$$\mu_u(\omega) \in M$$

where the sample point $\omega \in \Omega_u$ is chosen according to Λ_u, i.e.

$$\begin{aligned}\mathbf{Prob}_{u,1}(M) &= \Lambda_u(\mu_u \in M) \\ &= P_u(M) \qquad (2.4.1)\end{aligned}$$

2) The probability $\mathbf{Prob}_{u,2}(M)$ that

$$\sum_{e\in E_u} p_e \mu_{\tau(e)}(\omega_{\tau(e)}) \circ S_e^{-1} \in M$$

where the sample points $\omega_{\tau(e)} \in \Omega_{\tau(e)}$ are chosen independently and according to $\Lambda_{\tau(e)}$ for $e \in E_u$, and $\big((S_e)_{e\in E_u}, (p_e)_{e\in E_u}\big) \in \Xi_u$ is chosen independently of $(\omega_{\tau(e)})_{e\in E_u}$ and according to λ_u, i.e.

$$\begin{aligned}\mathbf{Prob}_{u,2}(M) &= \left(\lambda_u \times \prod_{e\in E_u} \Lambda_{\tau(e)}\right) \\ &\quad \left\{\left(((S_e)_{e\in E_u}, (p_e)_{e\in E_u}), (\omega_{\tau(e)})_{e\in E_u}\right) \in \Xi_u \times \prod_{e\in E_u} \Omega_{\tau(e)} \,\middle|\, \right. \\ &\quad \left. \sum_{e\in E_u} p_e \mu_{\tau(e)}(\omega_{\tau(e)}) \circ S_e^{-1} \in M\right\} \\ &= \left(\lambda_u \times \prod_{e\in E_u} \Lambda_{\tau(e)}\right)\left(\Phi_u \circ \left(1_{\Xi_u} \times \prod_{e\in E_u} \mu_{\tau(e)}\right)\right)^{-1}(M) \\ &= \left(\lambda_u \times \prod_{e\in E_u} \Lambda_{\tau(e)}\right)\left(\left(1_{\Xi_u} \times \prod_{e\in E_u} \mu_{\tau(e)}\right)^{-1}(\Phi_u^{-1}(M))\right) \\ &= \left(\lambda_u \times \prod_{e\in E_u} \Lambda_{\tau(e)} \circ \mu_{\tau(e)}^{-1}\right)(\Phi_u^{-1}(M)) \\ &= \left(\lambda_u \times \prod_{e\in E_u} P_{\tau(e)}\right)(\Phi_u^{-1}(M)) \\ &= T_\lambda\big((P_v)_v\big)_u(M) \qquad (2.4.2)\end{aligned}$$

Equation (2.4.1) and (2.4.2) and Theorem 2.4.2 implies that

$$\mathbf{Prob}_{u,1}(M) = P_u(M) = \mathbf{Prob}_{u,2}(M)$$

for every Borel subset M of $\mathcal{P}(X_u)$. Theorem 2.4.1.ii) can thus be interpreted as asserting that $(P_u)_u$ is the unique list in $\prod_{u\in V} \mathcal{P}(\mathcal{P}(X_u))$ which in a natural way respects the self-similar structure of the random measure μ_u.

We will now describe the connection between the random graph directed self-similar measures $\mu_u : \Omega_u \to \mathcal{P}(X_u)$ and the notion of deterministic graph directed self-similar measures introduced by e.g. Hutchinson [Hu] and Barnsley et al. [Bar]. Recall that if $\mathcal{S} = \left((S_e)_{e\in E_u}, (p_e)_{e\in E_u}\right)_{u\in V} \in \prod_{u\in V} \Xi_u$ is a fixed list, then measures $(\nu_u)_{u\in V} \in \prod_{u\in V} \mathcal{P}(X_u)$ satisfying the condition

$$\nu_u = \sum_{e\in E_u} p_e \nu_{\tau(e)} \circ S_e^{-1}$$

are called graph directed self-similar measures associated with $\mathcal{S}$. It follows from [Hu,Str] that for each list $\mathcal{S} \in \prod_{u\in V} \Xi_u$, there exists a unique list of graph directed self-similar measures $(\nu_u)_{u\in V}$ associated with $\mathcal{S}$. We will now show how to obtain this result, and thereby indicate the connection between the random graph directed self-similar measures $\mu_u : \Omega_u \to \mathcal{P}(X_u)$ and deterministic self-similar measures. Let $\left((S_e)_{e\in E_u}, (p_e)_{e\in E_u}\right)_{u\in V} \in \prod_{u\in V} \Xi_u$ and put

$$\lambda_u = \delta_{\left((S_e)_{e\in E_u}, (p_e)_{e\in E_u}\right)}.$$

Let $\lambda := (\lambda_u)_u$. Then

$$\begin{aligned}
P_{\lambda,u} &= \Lambda_{\lambda,u} \circ \mu_u^{-1} = \left(\lambda_u \times \prod_{\alpha\in E_u^{(*)}\setminus\{\varnothing\}} \lambda_{\tau(\alpha_{|\alpha|})}\right) \circ \mu_u^{-1} \\
&= \delta_{\left(\left((S_e)_{e\in E_u}, (p_e)_{e\in E_u}\right), \left((S_e)_{e\in E_{\tau(\alpha_{|\alpha|})}}, (p_e)_{e\in E_{\tau(\alpha_{|\alpha|})}}\right)_{\alpha\in E_u^{(*)}\setminus\{\varnothing\}}\right)} \circ \mu_u^{-1} \\
&= \delta_{\nu_u}
\end{aligned}$$

where

$$\nu_u := \mu_u\left(\left((S_e)_{e\in E_u}, (p_e)_{e\in E_u}\right), \left((S_e)_{e\in E_{\tau(\alpha_{|\alpha|})}}, (p_e)_{e\in E_{\tau(\alpha_{|\alpha|})}}\right)_{\alpha\in E_u^{(*)}\setminus\{\varnothing\}}\right).$$

Theorem 2.4.1.i) now yields

$$\begin{aligned}
1 =& \delta_{\nu_u}(\{\nu_u\}) = T_\lambda\left((\delta_{\nu_v})_v\right)_u(\{\nu_u\}) \\
=& \left(\lambda_u \times \prod_{e\in E_u} \delta_{\nu_{\tau(e)}}\right) \\
& \left\{\left(\left((T_e)_{e\in E_u}, (q_e)_{e\in E_u}\right), (\gamma_{\tau(e)})_{e\in E_u}\right) \in \Xi_u \times \prod_{e\in E_u} \mathcal{P}(X_{\tau(e)}) \,\middle|\, \nu_u = \sum_{e\in E_u} q_e \gamma_{\tau(e)} \circ T_e^{-1}\right\}
\end{aligned}$$

whence

$$\nu_u = \sum_{e \in E_u} p_e \nu_{\tau(e)} \circ S_e^{-1} . \tag{2.4.3}$$

Equation (2.4.3) is the deterministic definition of graph directed self-similar measures, cf. [Bar] or [Str]. Thus, Theorem 2.4.1 contains the deterministic case as a special case.

The final theorem is this section gives a probabilistic version of (2.4.3)

Theorem 2.4.2. *Define projections $\pi_u : \Omega \to \Omega_u$ by $\pi\big((\omega_v)_v\big) = \omega_u$. Then*

$$\mu_u(\pi_u) \stackrel{\mathrm{d}}{=} \sum_{e \in E_u} p_e(\pi_u) \mu_{\tau(e)}(\pi_{\tau(e)}) \circ S_e(\pi_u)^{-1}$$

where $\stackrel{\mathrm{d}}{=}$ denotes equality in distribution.

2.5 The Auxiliary Function β.

In this section we define an auxiliary function $\beta : \mathbb{R} \to \mathbb{R}$. The main theorems in section 2.6 through section 2.8 states that if certain conditions are satisfied then the P_u almost sure multifractal spectrum of $\mu \in \mathcal{P}(X_u)$ is equal to the Legendre transform of β. We also give conditions guaranteeing that the measures

$$\mathcal{H}_{\mu}^{q,\beta(q)}(\operatorname{supp}\mu), \ \mathcal{P}_{\mu}^{q,\beta(q)}(\operatorname{supp}\mu)$$

are positive and finite for P_u almost all $\mu \in \mathcal{P}(X_u)$.

However, we will first introduce six technical conditions:

(I) There exists $\Delta > 0$ such that

$$\Delta \leq p_e \qquad \text{and} \qquad \Delta \leq \operatorname{Lip}(S_e)$$

for all $u \in V$ and λ_u-a.a. $\big((S_e)_{e\in E_u}, (p_e)_{e\in E_u}\big) \in \Xi_u$.

(II) For all $u \in V$ and λ_u-a.a. $\big((S_e)_{e\in E_u}, (p_e)_{e\in E_u}\big) \in \Xi_u$, then

$$p_e < 1\,.$$

(III) There exists a number $0 < T < 1$ such that

$$p_e \leq T \qquad \text{and} \qquad \operatorname{Lip}(S_e) \leq T$$

for all $u \in V$ and λ_u-a.a. $\big((S_e)_{e\in E_u}, (p_e)_{e\in E_u}\big) \in \Xi_u$.

(IV) There exists a number $0 <. c < 1$ such that for all $u \in V$ and Λ_u-a.a. $\omega \in (\Omega_u)_0$ then

$$\operatorname{dist}\big(S_{\alpha e}(\omega)(X_{\tau(e)}), S_{\alpha\varepsilon}(\omega)(X_{\tau(\varepsilon)})\big) > c$$

for all $\alpha \in E_u^{(*)}$ and $e, \varepsilon \in E_{\tau(\alpha_{|\alpha|})}$ (and $e, \varepsilon \in E_u$ for $\alpha = \varnothing$) with $e \neq \varepsilon$.

(IV′) For all $u \in V$ and Λ_u-a.a. $\omega \in (\Omega_u)_0$ then

$$S_{\alpha e}(\omega)(\mathring{X}_{\tau(e)}) \bigcap S_{\alpha\varepsilon}(\omega)(\mathring{X}_{\tau(\varepsilon)}) = \varnothing$$

for all $\alpha \in E_u^{(*)}$ and $e, \varepsilon \in E_{\tau(\alpha_{|\alpha|})}$ (and $e, \varepsilon \in E_u$ for $\alpha = \varnothing$) with $e \neq \varepsilon$.

(V) For all $u \in V$ and for λ_u-a.a. $\big((S_e)_{e\in E_u}, (p_e)_{e\in E_u}\big) \in \Xi_u$ the contractions $(S_e)_{e\in E_u}$ are similarities.

Condition (I) will be assumed satisfied throughout the rest of of the exposition. Let $q, t \in \mathbb{R}$ and $u, v \in V$ and observe that the random variable

$$\sum_{e\in E_{uv}} p_e^q \operatorname{Lip}(S_e)^t : \Xi_u \to \mathbb{R}$$

is integrable w.r.t. λ_u (by condition (I).) Now write

$$A_{u,v}(q,t) = \mathbf{E}_{\lambda_u}\left[\sum_{e\in E_{uv}} p_e^q \operatorname{Lip}(S_e)^t\right] = \sum_{e\in E_{uv}} \mathbf{E}_{\lambda_u}[p_e^q \operatorname{Lip}(S_e)^t].$$

Observe that $A_{u,v}(q,t) > 0$ for $u,v \in V$ with $E_{uv} \neq \varnothing$. Next define a $\operatorname{card}(V) \times \operatorname{card}(V)$ matrix $A(q,t) \in \operatorname{Mat}_{\operatorname{card}(V)}(\mathbb{R})$ indexed by V, by

$$A(q,t) = \left(A_{u,v}(q,t)\right)_{u,v\in V}. \tag{2.5.1}$$

Finally, let $\Phi(q,t)$ denote the spectral radius of $A(q,t)$, i.e.

$$\Phi(q,t) = r\big(A(q,t)\big) \tag{2.5.2}$$

where $r(\cdot)$ denotes spectral radius. The next proposition summarizes most of the properties of Φ.

Proposition 2.5.1. *Assume that (E,V) is strongly connected and that conditions (I) and (II) are satisfied. Then the following statements hold.*

i) *$A(q,t)$ is irreducible.*
ii) *$\Phi : \mathbb{R}^2 \to \mathbb{R}$ is real analytic.*
iii) *If $t \in \mathbb{R}$ and $p < q$ then $\Phi(p,t) > \Phi(q,t)$.*
iv) *If $q \in \mathbb{R}$ and $s < t$ then $\Phi(q,s) > \Phi(q,t)$.*
v) *If $q \in \mathbb{R}$ then*

$$\lim_{t\to-\infty} \Phi(q,t) = \infty\,, \quad \lim_{t\to\infty} \Phi(q,t) = 0\,.$$

vi) *If $t \in \mathbb{R}$ then*

$$\lim_{q\to-\infty} \Phi(q,t) = \infty\,, \quad \lim_{q\to\infty} \Phi(q,t) = 0\,.$$

vii) *The function Φ is logarithmic convex, i.e. for $q_1, q_2, t_1, t_2 \in \mathbb{R}$ and $\alpha_1, \alpha_2 \geq 0$ with $\alpha_1 + \alpha_2 = 1$, then*

$$\Phi(\alpha_1 q_1 + \alpha_2 q_2, \alpha_1 t_1 + \alpha_2 t_2) \leq \Phi(q_1,t_1)^{\alpha_1}\Phi(q_2,t_2)^{\alpha_2}.$$

Now, for a fixed $q \in \mathbb{R}$, the function $\Phi(q,t)$ is a continuous strictly decreasing function of t, and $\lim_{t\to-\infty}\Phi(q,t) = \infty$ and $\lim_{t\to\infty}\Phi(q,t) = 0$. Therefore there exists a unique real number $\beta(q)$ such that

$$\Phi(q,\beta(q)) = 1. \tag{2.5.3}$$

This defines β implicitly as a function of q. The next proposition lists some of the elementary properties of β.

Proposition 2.5.2. *Assume (E, V) is strongly connected a that (I) and (II) are satisfied. Then the following statements hold.*

i) *The function $\beta : \mathbb{R} \to \mathbb{R}$ is real analytic.*
ii) *The function $\beta : \mathbb{R} \to \mathbb{R}$ is strictly decreasing.*
iii) $\lim_{q\to-\infty} \beta(q) = \infty\,, \quad \lim_{q\to\infty} \beta(q) = -\infty\,.$
iv) *Then function β is convex.*
v) $\beta(1) = 0$

We will now compute A, Φ and β in the deterministic case. Recall (cf. the discussion in section 2.4) that the deterministic case is obtained as follows: Let $\mathcal{S} = \big((S_e)_{e\in E_u}, (p_e)_{e\in E_u}\big)_{u\in V} \in \prod_{u\in V} \Xi_u$ and put

$$\lambda_u = \delta_{\big((S_e)_{e\in E_u}, (p_e)_{e\in E_u}\big)}\,.$$

Let $\lambda := (\lambda_u)_u$. Then

$$P_{\lambda,u} = \delta_{\nu_u}$$

where $\nu_u := \mu_u\Big(\big((S_e)_{e\in E_u}, (p_e)_{e\in E_u}\big), \big((S_e)_{e\in E_{\tau(\alpha_{|\alpha|})}}, (p_e)_{e\in E_{\tau(\alpha_{|\alpha|})}}\big)_{\alpha\in E_u^{(*)}\setminus\{\varnothing\}}\Big)$, and the measures $(\nu_u)_u$ are the unique self-similar measures associated with the list $\mathcal{S}$, i.e. the unique list of measures satisfying

$$\nu_u = \sum_{e\in E_u} p_e \nu_{\tau(e)} \circ S_e^{-1}\,.$$

The matrix $A(q,t)$ and the auxiliary function Φ corresponding to the deterministic graph directed self-similar measures $(\nu_u)_u$ associated with the list $\mathcal{S}$ are thus given by

$$A(q,t) = \big(A_{u,v}(q,t)\big)_{u,v\in V} = \left(\sum_{e\in E_{uv}} p_e^q \operatorname{Lip}(S_e)^t\right)_{u,v\in V} \tag{2.5.4}$$

$$\Phi(q,t) = r\big(A(q,t)\big) = r\left(\left(\sum_{e\in E_{uv}} p_e^q \operatorname{Lip}(S_e)^t\right)_{u,v\in V}\right)\,. \tag{2.5.5}$$

Likewise, the auxiliary function β corresponding to the deterministic self-similar measures $(\nu_u)_u$ associated with the list $\mathcal{S}$ is given by

$$1 = \Phi(q,\beta(q)) = r\left(\left(\sum_{e\in E_{uv}} p_e^q \operatorname{Lip}(S_e)^{\beta(q)}\right)_{u,v\in V}\right)\,. \tag{2.5.6}$$

The matrix $A(q,t)$ in (2.5.4) and the auxiliary functions Φ and β in (2.5.5) and (2.5.6) were first introduced by Cawley & Mauldin [Ca] and Edgar & Mauldin [Ed] in their multifractal analysis of deterministic graph directed self-similar measures in $\mathbb{R}^d$ with totally disconnected support. The matrix $A(q,t)$ in (2.5.1) and the auxiliary functions Φ and β in (2.5.2) and (2.5.3) are thus natural random analogues of the matrix $A(q,t)$ and the functions Φ and β in [Ed], and Proposition 2.5.1 and Proposition 2.5.2 may be viewed as random analogues of [Ed, Proposition 3.1 and Proposition 3.2].

We will throughout the rest of the exposition assume that (E,V) is strongly connected. Now write $A_{u,v}(q,\beta(q)) = A_{u,v}(q) = A_{u,v}$ for $u,v \in V$ and $A(q,\beta(q)) = A(q)$. The matrix $A(q)$ is irreducible and has spectral radius $\Phi(q,\beta(q)) = 1$. Perron-Frobenius theorem therefore implies that there exists unique positive left and right eigenvectors $\sigma(q) = (\sigma_u(q))_{u\in V}, \rho(q) = (\rho_u(q))_{u\in V}$ of $A(q)$ satisfying

$$A(q)\rho(q) = \rho(q) \quad \text{i.e.} \quad \sum_v \sum_{e\in E_{uv}} \mathrm{E}_{\lambda_u}[p_e^q \operatorname{Lip}(S_e)^{\beta(q)}]\rho_v(q) = \rho_u(q) \;\; \text{for} \;\; u \in V \tag{2.5.7}$$

$$\sigma(q)A(q) = \sigma(q) \quad \text{i.e.} \quad \sum_u \sum_{e\in E_{uv}} \sigma_u(q)\mathrm{E}_{\lambda_u}[p_e^q \operatorname{Lip}(S_e)^{\beta(q)}] = \sigma_v(q) \;\; \text{for} \;\; v \in V \tag{2.5.8}$$

$$\sigma_u(q), \rho_u(q) > 0 \quad \text{for} \quad u \in V$$
$$1 = \sum_u \rho_u(q) \tag{2.5.9}$$
$$1 = \sum_u \sigma_u(q)\rho_u(q) \tag{2.5.10}$$

It follows from Cramer's formula that $\sigma_u(q)$ and $\rho_u(q)$ are analytic functions of q. Write $\overline{\rho} = \overline{\rho}(q) = \max_v \rho_v(q)$ and $\underline{\rho} = \underline{\rho}(q) = \min_v \rho_v(q)$. Finally write

$$\alpha := -\beta' .$$

The next proposition (i.e. Proposition 2.5.3) investigates the asymptotic behavior of $\beta(q)$ as $q \to \pm\infty$. Assume that conditions (I) and (III) are satisfied. A finite string $e_1 \ldots e_n$ of edges is called a simple cycle if the terminal vertex $\tau(e_n)$ of the last edge e_n is equal to the initial vertex $\iota(e_1)$ of the first edge e_1, and $\tau(e_i) \neq \tau(e_j)$ for $i \neq j$ (or equivalently $\iota(e_i) \neq \iota(e_j)$ for $i \neq j$.) Write

$$Z = \{\gamma \in E^{(*)} \mid \gamma \quad \text{is a simple cycle}\} .$$

For $\gamma \in E^{(*)} \setminus \{\varnothing\}$ define $\kappa_\gamma, \chi_\gamma : \Omega_{\iota(\gamma_1)} \to [0,\infty[$ by

$$\kappa_\gamma = \log(p_{\gamma|1} \cdot \ldots \cdot p_{\gamma||\gamma|}) < 0$$
$$\chi_\gamma = \log(\operatorname{Lip}(S_{\gamma|1}) \cdot \ldots \cdot \operatorname{Lip}(S_{\gamma||\gamma|})) < 0 .$$

Put

$$\ell_\gamma = \frac{\kappa_\gamma}{\chi_\gamma}$$

and write

$$\underline{a} = \inf_{\gamma \in Z} \|\ell_\gamma\|_{-\infty}, \ \overline{a} = \sup_{\gamma \in Z} \|\ell_\gamma\|_\infty$$

where $\|\cdot\|_\infty$ denotes the essential supremum in $L^\infty(\Omega_{\iota(\gamma_1)}, \Lambda_{\iota(\gamma_1)})$, and $\|\cdot\|_{-\infty}$ denotes the essential infimum in $L^\infty(\Omega_{\iota(\gamma_1)}, \Lambda_{\iota(\gamma_1)})$.

Proposition 2.5.3. *Assume conditions (I) and (III). Then the following statements hold.*

i) *$q \to \overline{a}q + \beta(q)$ is increasing. In particular $\lim_{q\to-\infty}(\overline{a}q + \beta(q)) \in [-\infty, \infty[$ exists.*

ii)

$$\lim_{q\to-\infty} aq + \beta(q) = \begin{cases} \infty & \text{for } a < \overline{a} \\ -\infty & \text{for } \overline{a} < a \end{cases}$$

iii) *$q \to \underline{a}q + \beta(q)$ is decreasing. In particular $\lim_{q\to\infty}(\underline{a}q + \beta(q)) \in [-\infty, \infty[$ exists.*

iv)

$$\lim_{q\to\infty} aq + \beta(q) = \begin{cases} -\infty & \text{for } a < \underline{a} \\ \infty & \text{for } \underline{a} < a \end{cases}$$

Let $\zeta \in Z$ be such that $\|\ell_\zeta\|_\infty = \overline{a}$.

v) *If $\Lambda_{\iota(\zeta)}(\{\ell_\zeta = \overline{a}\}) > 0$ then*

$$\lim_{q\to-\infty} \overline{a}q + \beta(q) \in \mathbb{R}.$$

In particular β has an affine asymptote as $q \to -\infty$, viz.

$$q \to -\overline{a}q + (\lim_{p\to-\infty} \overline{a}p + \beta(p)).$$

Let $\eta \in Z$ be such that $\|\ell_\eta\|_{-\infty} = \underline{a}$.

vi) *If $\Lambda_{\iota(\eta)}(\{\ell_\eta = \underline{a}\}) > 0$ then*

$$\lim_{q\to\infty} \underline{a}q + \beta(q) \in \mathbb{R}.$$

In particular β has an affine asymptote as $q \to \infty$, viz.

$$q \to -\underline{a}q + (\lim_{p\to\infty} \underline{a}p + \beta(p)).$$

We remark that examples in Chapter 3 show that the numbers

$$\lim_{q\to-\infty} \overline{a}q + \beta(q), \quad \lim_{q\to\infty} \underline{a}q + \beta(q)$$

can take any values in $[-\infty, \infty[$. Observe that if $\lim_{q\to-\infty} \overline{a}q + \beta(q) = -\infty$ ($\lim_{q\to\infty} \underline{a}q + \beta(q) = \infty$) then the convexity of β and Proposition 2.5.3.ii) (Proposition 2.5.3.iv)) implies that β does not have an affine asymptote as $q \to -\infty$ ($q \to \infty$). This is in sharp contrast to the deterministic case in which

$$\lim_{q\to-\infty} \overline{a}q + \beta(q), \lim_{q\to\infty} \underline{a}q + \beta(q) \in [0, \infty[$$

by [Ed, Proposition 3.3].

The value of $\overline{a} - \underline{a}$ determines the behavior of β as the next result shows.

Proposition 2.5.4. *Assume conditions (I) and (III). Then the following statements are equivalent.*

i) $\underline{a} = \overline{a}$.
ii) $\beta(q) = -\beta(0)q + \beta(0)$ *for* $q \in \mathbb{R}$.

If i) or ii) are satisfied then

$$\underline{a} = \overline{a} = \beta(0).$$

It follows from the previous proposition that the behavior of β can be divided into two rather different cases, viz.

Case 1. $\underline{a} = \overline{a}$.
Case 2. $\underline{a} < \overline{a}$.

Proposition 2.5.5. *Assume that conditions (I) and (III) are satisfied.*
Case 1: $\underline{a} = \overline{a} := a$. *In this case the following statements hold.*

i) β *is an affine function of* q: $\beta(q) = a - aq$.
ii) $\alpha := -\beta' = a$ *is constant.*

Case 2: $\underline{a} < \overline{a}$. *In this case the following statements hold.*

i) β *is a strictly convex function of* q.
ii) $\alpha := -\beta'$ *is a strictly decreasing function of* q.
iii)

$$\lim_{q\to-\infty} \alpha(q) = \overline{a} < \infty$$

$$\lim_{q\to\infty} \alpha(q) = \underline{a} \geq 0$$

Below we sketch the typical shape of the graph of α.

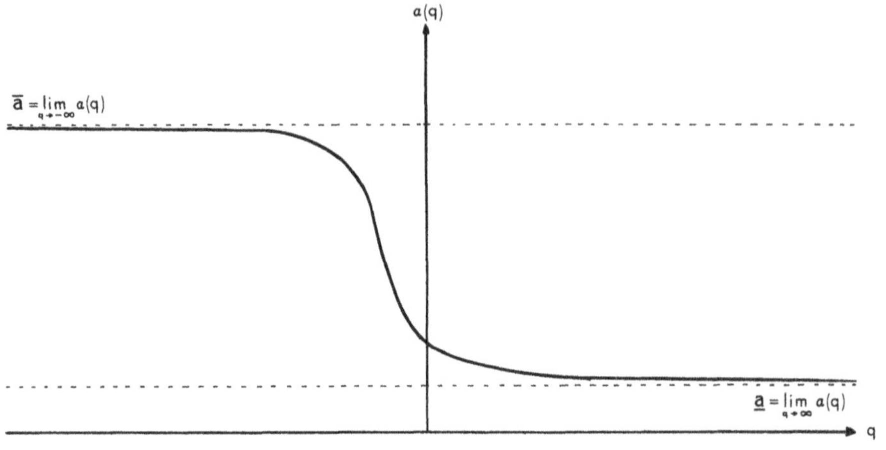

Figure 2.5.1: Figure sketching the typical shape of the graph of α.

We are now ready to state a proposition concerning the behavior of the Legendre transform of β. First define numbers $a_{\min}$, $a_{\max}$ and $\underline{e}$, $\overline{e}$ by

$$a_{\min} = \sup_{0<q}\left(-\frac{\beta(q)}{q}\right), \quad a_{\max} = \inf_{q<0}\left(-\frac{\beta(q)}{q}\right)$$
$$\underline{e} = \lim_{q\to\infty}(\underline{a}q + \beta(q)) \in [-\infty, \infty[, \quad \overline{e} = \lim_{q\to-\infty}(\overline{a}q + \beta(q)) \in [-\infty, \infty[,$$

(observe that the limits $\underline{e}$ and $\overline{e}$ exist by Proposition 2.5.3.)

Proposition 2.5.6. *Assume conditions (I) and (III). Then the following hold*
Case 1: $\underline{a} = \overline{a} := a$. *In this case*

i) $\underline{e} = \overline{e} = a$.

ii) $\beta^*(\alpha) = \begin{cases} -\infty & \text{for } \alpha \in [0,\infty[\setminus\{a\} \\ a & \text{for } \alpha = a \end{cases}$

Case 2: $\underline{a} < \overline{a}$. *In this case*

i) $\lim_{\alpha\searrow\underline{a}}\beta^*(\alpha) = \underline{e}$, $\lim_{\alpha\nearrow\overline{a}}\beta^*(\alpha) = \overline{e}$.

ii) β^* *is strictly concave and real valued on* $]\underline{a},\overline{a}[$, *and* $\beta^*(\alpha) = -\infty$ *for* $\alpha \in [0,\infty[\setminus[\underline{a},\overline{a}]$.

iii) $\underline{a} \le a_{\min} \le a_{\max} \le \overline{a}$.

iv) *If* $a_{\max} < \overline{a}$ *then there exists a unique* $-\infty < q_{\min} < 0$ *such that* $\alpha(q_{\min}) = a_{\max}$. *Moreover,* $a_{\max}$ *satisfies*

$$\beta^*(a_{\max}) = 0, \quad a_{\max} = -\frac{\beta(q_{\min})}{q_{\min}}$$

and the line $\overline{\ell} := \{(q, -a_{\max}q) \mid q \in \mathbb{R}\}$ *that passes through the origin with slope* $-a_{\max}$ *is tangent to the graph of* β *at the point* $(q_{\min}, \beta(q_{\min}))$.

v) *If* $\underline{a} < a_{\min}$ *then there exists a unique* $1 < q_{\max} < \infty$ *such that* $\alpha(q_{\max}) = a_{\min}$*. Moreover,* $a_{\min}$ *satisfies*

$$\beta^*(a_{\min}) = 0\,, \quad a_{\min} = -\tfrac{\beta(q_{\max})}{q_{\max}}$$

and the line $\underline{\ell} := \{(q, -a_{\min} q) \mid q \in \mathbb{R}\}$ *that passes through the origin with slope* $-a_{\min}$ *is tangent to the graph of* β *at the point* $(q_{\max}, \beta(q_{\max}))$*.*

vi) *If* $\alpha \in]\underline{a}, \overline{a}[$ *then*

$$\beta^*(\alpha) < 0 \Leftrightarrow \alpha \in]\underline{a}, a_{\min}[\cup]a_{\max}, \overline{a}[$$
$$\alpha \in]a_{\min}, a_{\max}[\Rightarrow \beta^*(\alpha) > 0 \Rightarrow \alpha \in [a_{\min}, a_{\max}]$$

vii) $\beta^*(\alpha(1)) = \alpha(1)$.

viii) $\beta^*(\alpha) \leq \alpha$ *for* $\alpha \geq 0$.

ix) *If* $\alpha'(1) \neq 0$ *then the line passing through the origin with slope* 1 *is tangent to the graph of* β^*.

x) *If* $\alpha'(0) \neq 0$ *then* β^* *has a unique maximum at* $\alpha(0)$ *and*

$$\beta^*(\alpha(0)) = \max_{\alpha \geq 0} \beta^*(\alpha) = \beta(0)\,.$$

If in addition the graph (V, E) *only has one vertex then the following statement holds.*

xi) $\alpha'(q) < 0$ *for all* q.

Below we sketch the typical shapes of the graphs of β and β^*.

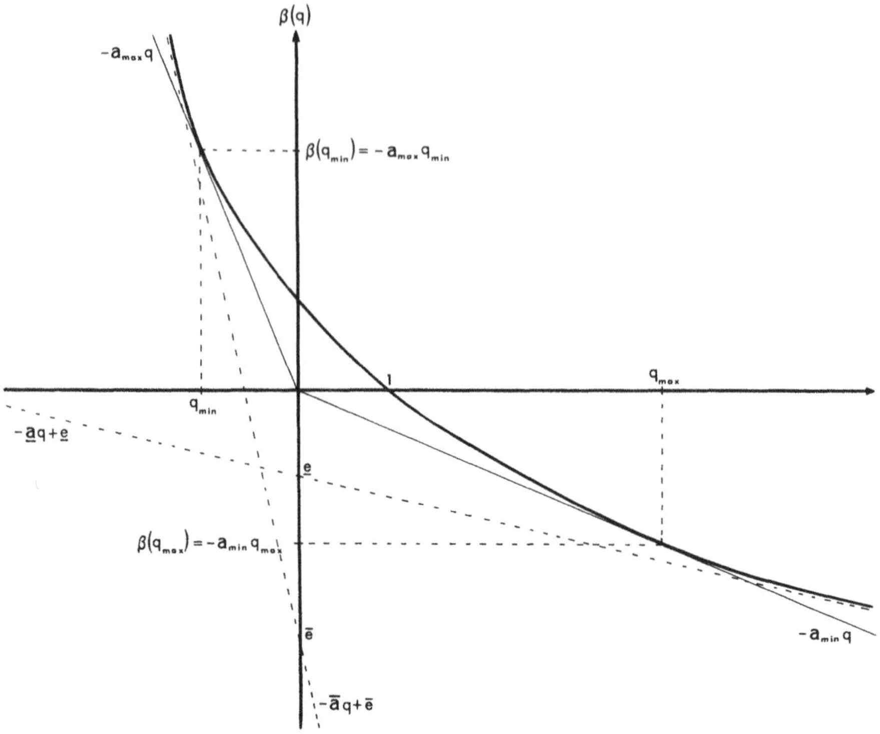

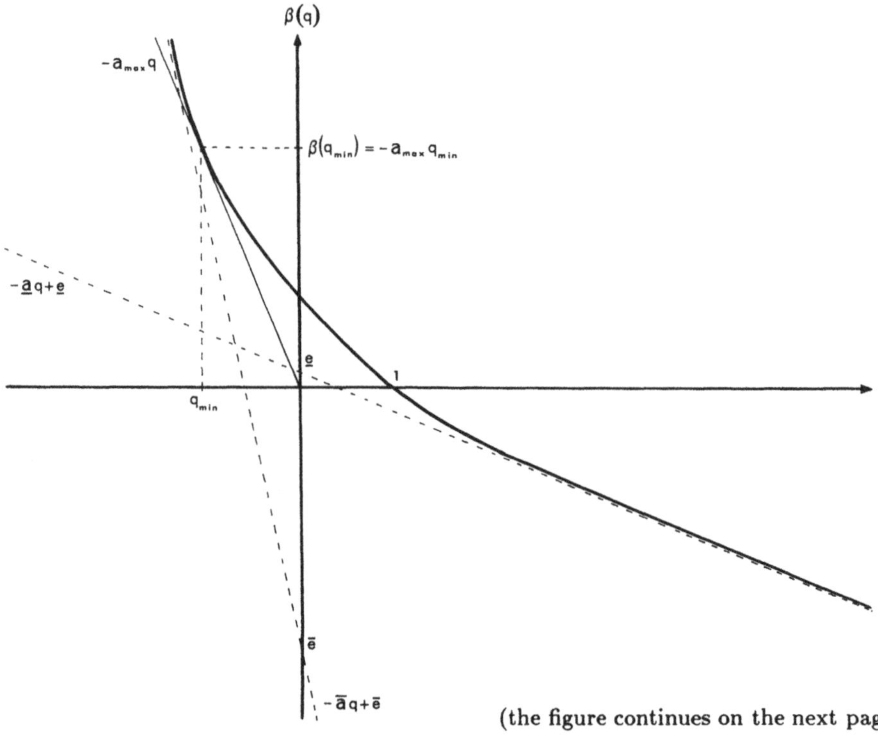

(the figure continues on the next page.)

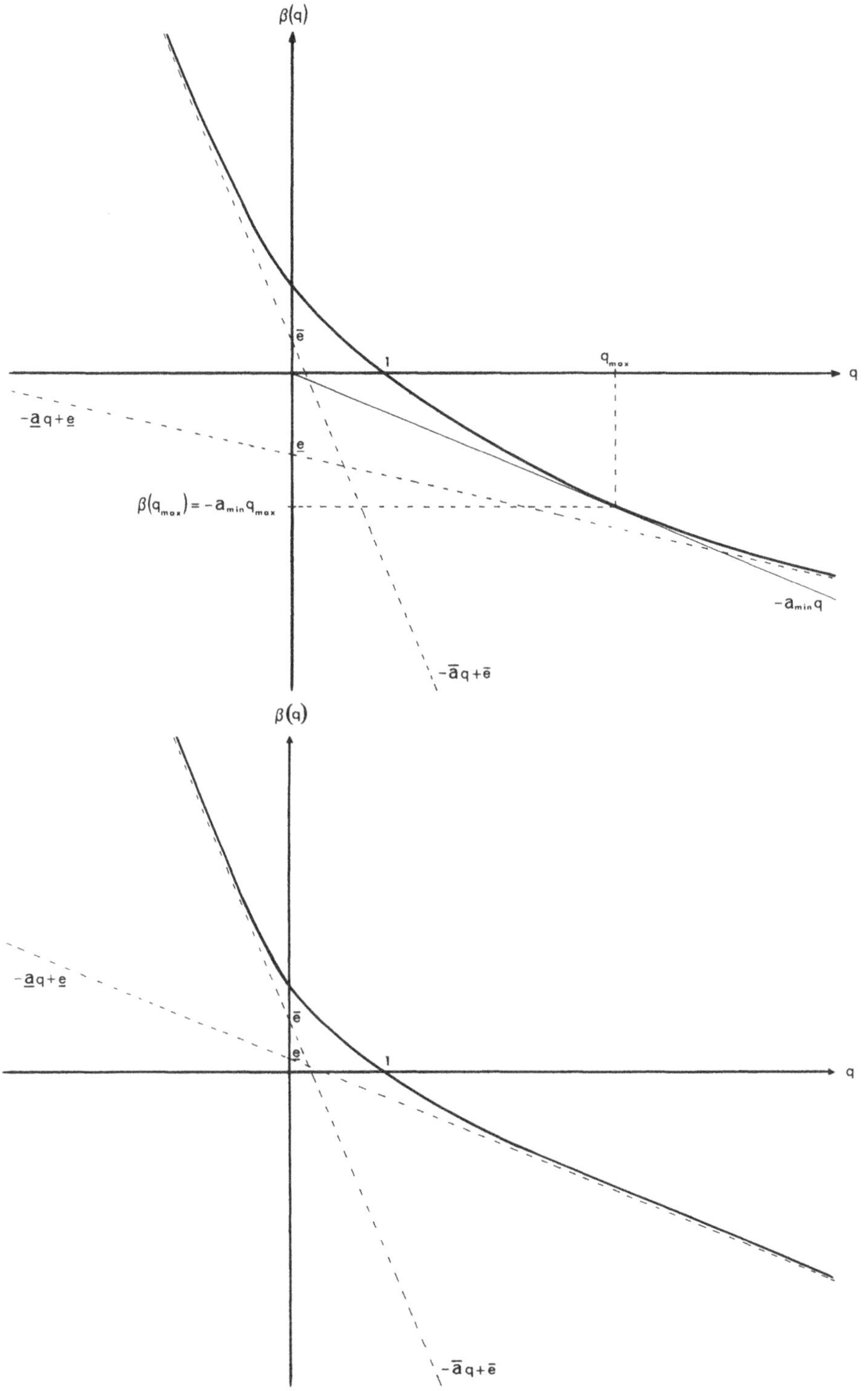

Figure 2.5.2: Figure sketching the typical shape of the graph of β.
a: The case $\underline{a} < a_{\min} < a_{\max} < \overline{a}$. b: The case $\underline{a} = a_{\min} < a_{\max} < \overline{a}$.
c: The case $\underline{a} < a_{\min} < a_{\max} = \overline{a}$. d: The case $\underline{a} = a_{\min} < a_{\max} = \overline{a}$.

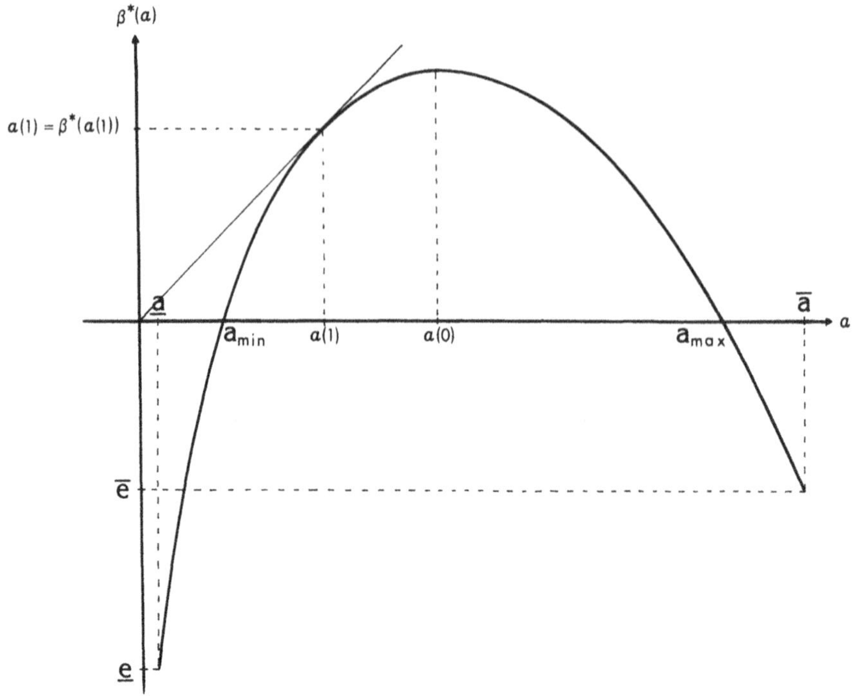

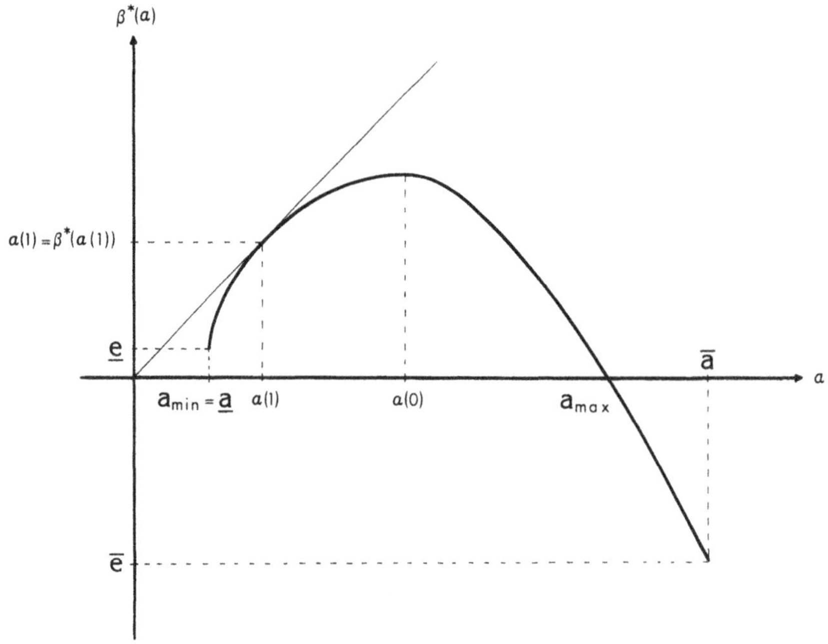

(the figure continues on the next page.)

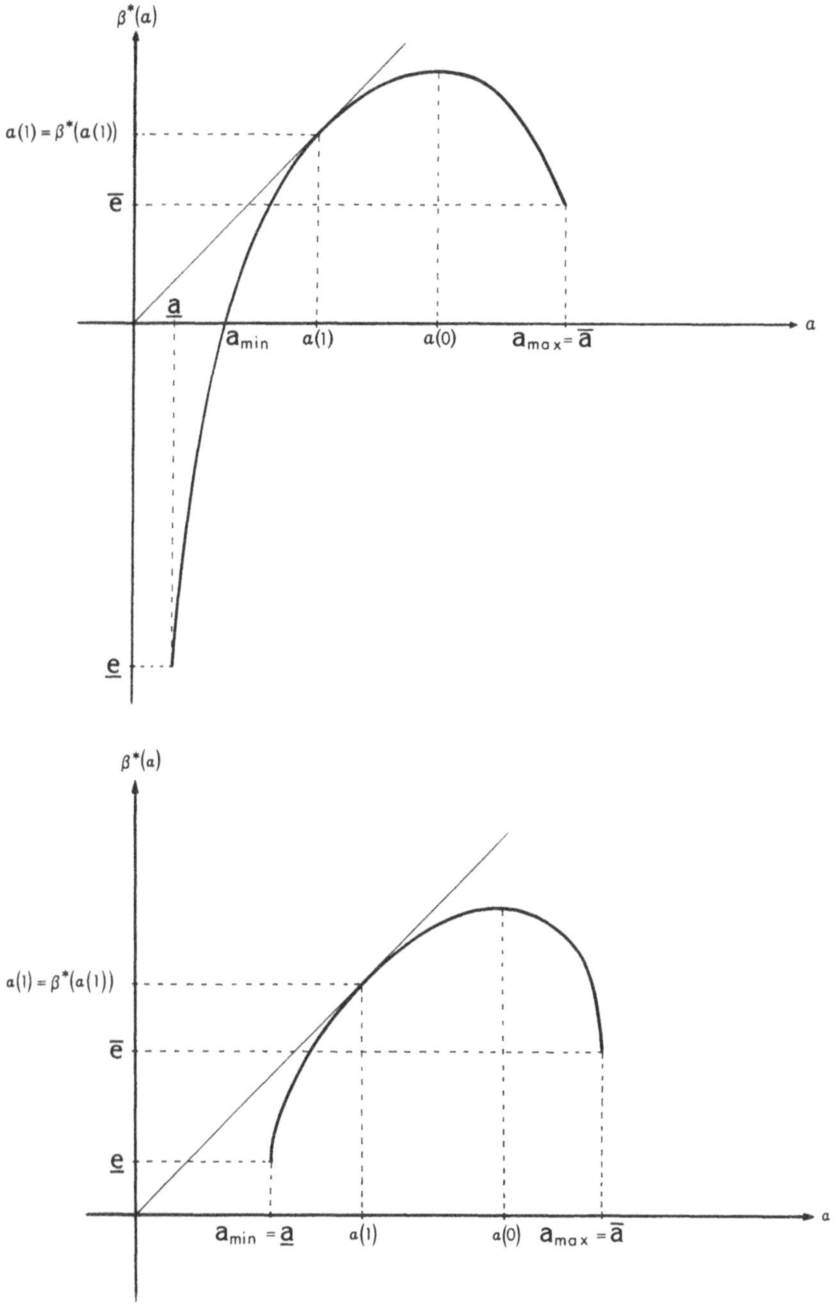

Figure 2.5.3: Figure sketching the typical shape of the graph of β^*.
a: The case $\underline{a} < a_{\min} < a_{\max} < \overline{a}$. b: The case $\underline{a} = a_{\min} < a_{\max} < \overline{a}$.
c: The case $\underline{a} < a_{\min} < a_{\max} = \overline{a}$. d: The case $\underline{a} = a_{\min} < a_{\max} = \overline{a}$.

Proposition 2.5.6 should be viewed as a probabilistic counterpart of [Ed, Proposition 3.3]. The two main differences between the random case, i.e. Proposition 2.5.6, and the deterministic case, i.e. the case in which

$$\lambda_u = \delta_{\left((S_e)_{e\in E_u},(p_e)_{e\in E_u}\right)} \quad \text{for } u \in V$$

for some fixed $\mathcal{S} = \left((S_e)_{e\in E_u},(p_e)_{e\in E_u}\right)_{u\in V} \in \prod_{u\in V} \Xi_u$ (cf. the discussion in section 2.4), are

1) the numbers $\underline{e}$ and $\overline{e}$ can, as shown by examples in Chapter 3, take any values in $[-\infty,\infty[$, in contrast to the deterministic case in which $\underline{e},\overline{e} \in [0,\infty[$ by [Ed, Proposition 3.3];
2) the numbers $a_{\min}$ and $a_{\max}$ can, as shown by examples in Chapter 3, take any values in $[\underline{a},\overline{a}]$, in contrast to the deterministic case in which $a_{\min} = \underline{a}$ and $a_{\max} = \overline{a}$ by [Ol, Theorem 5.1].

We will now make some definitions. In Case 1 we write

$$q_{\min} = -\infty\,, \qquad q_{\max} = \infty\,.$$

Assume that we are in Case 2. If $a_{\max} < \overline{a}$, then $q_{\min}$ denotes the unique number that appears in the previous proposition. If $a_{\max} = \overline{a}$, then we write

$$q_{\min} = -\infty\,.$$

If $\underline{a} < a_{\min}$, then $q_{\max}$ denotes the unique number that appears in the previous proposition. If $\underline{a} = a_{\min}$, then we write

$$q_{\max} = \infty\,.$$

2.6 The P_u Almost Sure Multifractal Spectrum of $\mu \in \mathcal{P}(X_u)$

This sections contains our main theorems concerning the P_u almost sure multifractal spectrum of $\mu \in \mathcal{P}(X_u)$. In order to state the results we need a definition. Define for $u \in V$, $q \in \mathbb{R}$ and $n \in \mathbb{N}$ the random variable $X_{u,q,n} : \Omega_u \to \mathbb{R}_+$ by

$$X_{u,q,n} = \sum_{\gamma \in E_u^{(n)}} \rho_u(q)^{-1} \left(\prod_{i=1}^{n} p_{\gamma|i}^{q} \operatorname{Lip}(S_{\gamma|i})^{\beta(q)} \right) \rho_{\tau(\gamma_n)}(q) .$$

It follows from Proposition 5.1.1 that there exists a random variable $X_{u,q} : \Omega_u \to \mathbb{R}_+$ such that

$$X_{u,q,n} \to X_{u,q} \quad \Lambda_u\text{-a.s.}.$$

If (X, Σ, μ) is a measurable space and $E \in \Sigma$, $\mu \lfloor E$ denotes the restriction of μ to E, i.e. $(\mu \lfloor E)(A) := \mu(E \cap A)$ for $A \in \Sigma$. We are now ready to state the main theorems. We will always assume that X_v is compact for all $v \in V$,

$$q, p \in]q_{\min}, q_{\max}[$$

and that conditions (I), (III), (IV) and (V) are satisfied. Furthermore, λ^d will denote the d-dimensional Lebesgue measure in $\mathbb{R}^d$.

Theorem 2.6.1. Upper bounds for $\mathcal{H}_\mu^{q,\beta(q)}$.

i) *For all v and P_v-a.a. $\mu \in \mathcal{P}(X_v)$ then*

$$b_\mu \le \beta .$$

ii) *For all v and P_v-a.a. $\mu \in \mathcal{P}(X_v)$ then*

$$\mathcal{H}_\mu^{q,\beta(q)}(\operatorname{supp} \mu) < \infty .$$

iii) *For all v and P_v-a.a. $\mu \in \mathcal{P}(X_v)$ then*

$$\Delta_\mu(\alpha(q))$$

has full $\mathcal{H}_\mu^{q,\beta(q)} \lfloor \operatorname{supp} \mu$ measure.

iv) *Assume $\alpha(q) \neq \alpha(p)$. For all v and P_v-a.a. $\mu \in \mathcal{P}(X_v)$ then*

$$\mathcal{H}_\mu^{q,\beta(q)} \lfloor \operatorname{supp} \mu \perp \mathcal{H}_\mu^{p,\beta(p)} \lfloor \operatorname{supp} \mu .$$

Theorem 2.6.2. Upper bounds for $\mathcal{P}_\mu^{q,\beta(q)}$.

i) *For all* v *and* P_v*-a.a.* $\mu \in \mathcal{P}(X_v)$ *then*

$$B_\mu \leq \Lambda_\mu \leq \beta \quad \text{on }]q_{\min}, q_{\max}[.$$

If in addition $0 < \min_v \|X_{v,q}\|_{-\infty}$ *then the following statements hold.*

ii) *For all* v *and* P_v*-a.a.* $\mu \in \mathcal{P}(X_v)$ *then*

$$\mathcal{P}_\mu^{q,\beta(q)}(\operatorname{supp}\mu) \leq \overline{\mathcal{P}}_\mu^{q,\beta(q)}(\operatorname{supp}\mu) < \infty .$$

iii) *For all* v *and* P_v*-a.a.* $\mu \in \mathcal{P}(X_v)$ *then*

$$\Delta_\mu(\alpha(q))$$

has full $\mathcal{P}_\mu^{q,\beta(q)} \lfloor \operatorname{supp}\mu$ *measure.*

iv) *Assume* $\alpha(q) \neq \alpha(p)$*. For all* v *and* P_v*-a.a.* $\mu \in \mathcal{P}(X_v)$ *then*

$$\mathcal{P}_\mu^{q,\beta(q)} \lfloor \operatorname{supp}\mu \ \perp\ \mathcal{P}_\mu^{p,\beta(p)} \lfloor \operatorname{supp}\mu .$$

Theorem 2.6.3. Lower bounds for $\mathcal{H}_\mu^{q,\beta(q)}$. *Let* $X_v \subseteq \mathbb{R}^d$ *with* $0 < \lambda^d(X_v)$ *for all* $v \in V$*.*

i) *For all* v *and* P_v*-a.a.* $\mu \in \mathcal{P}(X_v)$ *then*

$$\beta \leq b_\mu \quad \text{on} \quad]q_{\min}, q_{\max}[.$$

If in addition $\max_v \|X_{v,q}\|_\infty < \infty$ *then the following statement holds.*

ii) *For all* v *and* P_v*-a.a.* $\mu \in \mathcal{P}(X_v)$ *then*

$$0 < \mathcal{H}_\mu^{q,\beta(q)}(\Delta_\mu(\alpha(q))) .$$

Theorem 2.6.4. Lower bounds for $\mathcal{P}_\mu^{q,\beta(q)}$.

i) *For all* v *and* P_v*-a.a.* $\mu \in \mathcal{P}(X_v)$ *then*

$$\beta \leq B_\mu \quad \text{on} \quad]q_{\min}, q_{\max}[.$$

ii) *For all* v *and* P_v*-a.a.* $\mu \in \mathcal{P}(X_v)$ *then*

$$0 < \mathcal{P}_\mu^{q,\beta(q)}(\operatorname{supp}\mu) .$$

Theorem 2.6.1 through Theorem 2.6.4 are random extensions and generalizations of [Ol, Theorem 5.1].

The next theorem determines the almost sure multifractal spectrum.

Theorem 2.6.5. *Let $X_v \subseteq \mathbb{R}^d$ with $0 < \lambda^d(X_v)$ for all $v \in V$. Define for $\mu \in \mathcal{P}(X_u)$ functions $\mathcal{C}_\mu, \mathcal{I}_\mu, \mathcal{D}_\mu : \mathbb{R} \to \mathbb{R}$ by*

$$\mathcal{C}_\mu(q) := C^q_\mu(\operatorname{supp}\mu), \ \mathcal{I}_\mu(q) := (q-1)I^{q-1}_\mu, \ \mathcal{D}_\mu(q) := (1-q)D^q_\mu.$$

Then the following statements hold.
Case 1: $\underline{a} = \overline{a} := a$. In this case

i) *For P_u-a.a. $\mu \in \mathcal{P}(X_u)$ then*

$$b_\mu = B_\mu = \Lambda_\mu = \mathcal{C}_\mu = \mathcal{I}_\mu = \mathcal{D}_\mu = \beta$$

(where $\beta(q) = a - aq$ by Proposition 2.5.4.)

ii) *For P_u-a.a. $\mu \in \mathcal{P}(X_u)$ then*

$$\Delta_\mu(\alpha) = \begin{cases} \varnothing & \text{for } \alpha \neq a \\ \operatorname{supp}\mu & \text{for } \alpha = a \end{cases}$$

iii) *For P_u-a.a. $\mu \in \mathcal{P}(X_u)$ then*

$$f_\mu(a) = F_\mu(a) = a = \beta^*(a) = b^*_\mu(a) = B^*_\mu(a) = \Lambda^*_\mu(a) = \mathcal{C}^*_\mu(a) = \mathcal{I}^*_\mu(a) = \mathcal{D}^*_\mu(a).$$

Case 2: $\underline{a} < \overline{a}$. In this case

i) *For P_u-a.a. $\mu \in \mathcal{P}(X_u)$ then*

$$b_\mu = B_\mu = \hat{\beta}$$

where

$$\hat{\beta}(q) = \begin{cases} \frac{\beta(q_{\min})}{q_{\min}} q & -\infty < q \leq q_{\min} \\ \beta(q) & q_{\min} < q < q_{\max} \\ \frac{\beta(q_{\max})}{q_{\max}} q & q_{\max} \leq q < \infty \end{cases}$$

ii) *For P_u-a.a. $\mu \in \mathcal{P}(X_u)$ then*

$$\Lambda_\mu = \mathcal{C}_\mu = \mathcal{I}_\mu = \mathcal{D}_\mu = \beta \quad \text{on }]q_{\min}, q_{\max}[.$$

iii) *For P_u-a.a. $\mu \in \mathcal{P}(X_u)$ then*

$$\Delta_\mu(\alpha) = \varnothing \quad \text{for } \alpha \in [0,\infty[\setminus[a_{\min}, a_{\max}].$$

iv) *Let $\alpha \in]a_{\min}, a_{\max}[$. For P_u-a.a. $\mu \in \mathcal{P}(X_u)$ then*

$$f_\mu(\alpha) = F_\mu(\alpha) = \beta^*(\alpha) = b^*_\mu(\alpha) = B^*_\mu(\alpha) = \Lambda^*_\mu(\alpha) = \mathcal{C}^*_\mu(\alpha) = \mathcal{I}^*_\mu(\alpha) = \mathcal{D}^*_\mu(\alpha).$$

v) *Assume* $\underline{a} < a_{\min}$. *For* P_u*-a.a.* $\mu \in \mathcal{P}(X_u)$ *then*

$$f_\mu(a_{\min}) = F_\mu(a_{\min}) = \beta^*(a_{\min}) = 0\,.$$

vi) *Assume* $a_{\max} < \overline{a}$. *For* P_u*-a.a.* $\mu \in \mathcal{P}(X_u)$ *then*

$$f_\mu(a_{\max}) = F_\mu(a_{\max}) = \beta^*(a_{\max}) = 0\,.$$

vii) *For* P_u*-a.a.* $\mu \in \mathcal{P}(X_u)$ *then*

$$f_\mu \le F_\mu \le \beta^* \quad \text{on }]a_{\min}, a_{\max}[\,.$$

viii) *For* Λ_u*-a.a.* $\omega \in \Omega_u$ *then*

$$\begin{aligned}\beta(0) &= \dim \operatorname{supp} \mu_u(\omega) = \operatorname{Dim} \operatorname{supp} \mu_u(\omega)\\ &= \dim C_u(\omega) = \operatorname{Dim} C_u(\omega)\end{aligned}$$

Theorem 2.6.5 is a natural random version of [Ed, Theorem 1.6] and [Ol, Theorem 5.1].

We note that Falconer [Fa4] independently have obtained similar results for the self-similar case in $\mathbb{R}^d$, i.e. the case in which the graph (V, E) only has one vertex. However, our results are more general since we treat the general case of graph directed self-similar measures rather than confine ourselves to the sub-class of self-similar measures. We also obtain sharper results. Consider the self-similar case in which the graph (V, E) only has one vertex, $*$ say, and write $P_* := P$ and $X_* := X \subseteq \mathbb{R}^d$. Falconer [Fa4] then proves that for each fixed $\alpha_0 \in [0, \underline{a}[\cup]\underline{a}, a_{\min}[\cup]a_{\max}, \overline{a}[\cup]\overline{a}, \infty[$, $\Delta_\mu(\alpha_0) = \varnothing$ for P-a.a. $\mu \in \mathcal{P}(X)$, i.e. there exists a subset M_{α_0} of $\mathcal{P}(X)$ depending of α_0 such that

$$\begin{aligned}&P(M_{\alpha_0}) = 1 \quad \text{and}\\ &\Delta_\mu(\alpha_0) = \varnothing \quad \text{for } \mu \in M_{\alpha_0}\,,\end{aligned}$$

whereas we prove the stronger uniform statement (Theorem 2.6.5 Case 2, iii)) asserting that there exists a subset M of $\mathcal{P}(X)$ such that

$$\begin{aligned}&P(M) = 1 \quad \text{and}\\ &\Delta_\mu(\alpha) = \varnothing \quad \text{for } \mu \in M \text{ and all } \alpha \in [0, \underline{a}[\cup]\underline{a}, a_{\min}[\cup]a_{\max}, \overline{a}[\cup]\overline{a}, \infty[\,.\end{aligned}$$

We also obtain a uniform upper bound for F_μ (Theorem 2.6.5 Case 2, vii)) in the sense that there exists a subset M of $\mathcal{P}(X)$ such that

$$\begin{aligned}&P(M) = 1 \quad \text{and}\\ &F_\mu(\alpha) \le \beta^*(\alpha) \quad \text{for } \mu \in M \text{ and all } \alpha \in]a_{\min}, a_{\max}[\,.\end{aligned}$$

Observe that the statement in Theorem 2.6.5 Case 2, iv) does not imply the statement in Theorem 2.6.5 Case 2, vii). The assertion in Theorem 2.6.5 Case 2, iv) states that for each fix $\alpha_0 \in]a_{\min}, a_{\max}[$ then there exists a subset M_{α_0} of $\mathcal{P}(X_u)$ depending on α_0 such that

$$P_u(M_{\alpha_0}) = 1 \quad \text{and}$$
$$F_\mu(\alpha_0) = \beta^*(\alpha_0) \quad \text{for all} \quad \mu \in M_{\alpha_0}$$

whereas the assertion in Theorem 2.6.5 Case 2, vii) states that there exists a subset M of $\mathcal{P}(X_u)$ such that

$$P_u(M) = 1 \quad \text{and}$$
$$F_\mu(\alpha) \leq \beta^*(\alpha) \quad \text{for all} \quad \mu \in M \quad \text{and all} \quad \alpha \in]a_{\min}, a_{\max}[\,.$$

S. J. Taylor [Tay1,Tay2] has defined a fractal to be any subset E of a metric space X which satisfies

$$\dim E = \operatorname{Dim} E\,. \tag{2.6.1}$$

Olsen [Ol] suggested a natural measure-theoretical extension of this definition. A Borel probability measure μ on a metric space X is called a Taylor multifractal measure if

$$b_\mu = B_\mu\,. \tag{2.6.2}$$

It follows from Theorem 2.6.5 that P_u-a.a. $\mu \in \mathcal{P}(X_u)$ are Taylor multifractal measures. Furthermore, Theorem 2.6.5 Case 2, viii) shows that the random graph directed self-similar sets $C_u(\omega)(= \operatorname{supp} \mu_u(\omega))$ are Taylor fractals (in the sense of (2.6.1)) for Λ_u-a.a. $\omega \in \Omega_u$. This result implies, in particular, that the random self-similar sets in Graf [Gra1] are Taylor fractals almost surely provided that the strong separation condition (IV) is satisfied rather than the weaker almost sure open set condition (IV′) which is assumed in [Gra1]. However, we believe that the equation $\dim C_u(\omega) = \operatorname{Dim} C_u(\omega)$ also holds for Λ_u-a.a. $\omega \in \Omega_u$ in the case where the almost sure open set condition (IV′) is satisfied. The main reason for not obtaining this result is that our approach is aimed toward an investigation of the multifractal structure of the measures μ_u: it seems necessary to assume condition (IV), rather than condition (IV′), in order to obtain good results when working on multifractals (cf. the discussion below), and this is the main reason for assuming condition (IV) throughout the exposition – although some results, as e.g. the Λ_u-a.s. equation $\dim C_u = \operatorname{Dim} C_u$, probably still holds when condition (IV) is replaced by condition (IV′).

We have only been able to prove that

$$f_\mu(\alpha) = F_\mu(\alpha) = \beta^*(\alpha) = b^*_\mu(\alpha) = B^*_\mu(\alpha)$$

for P_u-a.a. $\mu \in \mathcal{P}(X_u)$ in the case where the strong separation condition (IV) is satisfied. It is easily seen that condition (IV) implies that the support of $\mu_u(\omega)$,

$$\operatorname{supp} \mu_u(\omega) = C_u(\omega)$$

is totally disconnected for Λ_u-a.a. $\omega \in \Omega_u$, and our results do therefore not cover random graph directed self-similar measures supported on e.g. random Sierpinski triangles, random Menger spronges and random von Koch curves. This is clearly very unsatisfactory and it would be desirable if one could replace condition (IV) with the almost sure open set condition (IV'). The almost sure open set condition (IV') is clearly satisfied in the case of random von Koch curves, random Menger spronges and random Sierpinski gaskets. We note that it suffices to assume that condition (IV') is satisfied in the study of the Hausdorff dimension $\dim C_u(\omega)$ of the support $C_u(\omega)$ of $\mu_u(\omega)$; Graf [Gra1], Graf et al. [Gra2], Falconer [Fa3] and Mauldin & Williams [Mau1] compute, in the self-similar case in which the graph (V, E) only has one vertex, $*$ say, the almost sure Hausdorff dimension $\dim C_*(\omega)$ of the support $C_*(\omega)$ of $\mu_*(\omega)$ under the assumption that the almost sure open set condition (IV') is satisfied. However, there seems to be some problems in extending these results to the multifractal case (basically due to the fact that to much mass might accumulate on the overlaps $S_e(\omega)(X_{\tau(e)}) \cap S_\varepsilon(\omega)(X_{\tau(\varepsilon)})$ where $u \in V$ and $e, \varepsilon \in E_u$ with $e \neq \varepsilon$.) These problems are present even in the case of deterministic multifractals, cf. [Ca,Ed,Fen,Ki1,Ki2,Bo,Ra,Co1,Br,Pey]. We note in particular that the multifractal spectrum of a self-similar measure in $\mathbb{R}^d$ supported on a non totally disconnected self-similar set satisfying the open set condition has not been determined; Cawley & Mauldin [Ca] and Edgar & Mauldin [Ed] only determine the multifractal spectrum for graph directed self-similar measures supported by totally disconnected graph directed self-similar sets.

The final result in this section deals with the Hausdorff dimension and the packing dimension of the measures $\mathcal{H}_\mu^{q,\beta(q)} \lfloor \operatorname{supp} \mu$ and the almost sure Hausdorff dimension and packing dimension of the measures $\mu \in \mathcal{P}(X_u)$. Recall that if μ is a probability measure on a metric space X then the Hausdorff dimension $\dim \mu$ of μ and the packing dimension $\operatorname{Dim} \mu$ of μ is defined by

$$\begin{aligned} \dim \mu &= \inf\{\dim E \mid E \subseteq X\,,\ \mu(E) > 0\} \\ \operatorname{Dim} \mu &= \inf\{\operatorname{Dim} E \mid E \subseteq X\,,\ \mu(E) > 0\} \end{aligned}$$

cf. [Haa1,Haa2,Haa3,Yo].

Theorem 2.6.6. *Let $q \in]q_{\min}, q_{\max}[$. If $\max_v \|X_{v,q}\|_\infty < \infty$ then the following statements hold.*

i) *For P_v-a.a. $\mu \in \mathcal{P}(X_v)$ then*

$$\dim \mathcal{H}_\mu^{q,\beta(q)} \lfloor \operatorname{supp} \mu = f_\mu(\alpha(q)) \,.$$

ii) *For* P_v-a.a. $\mu \in \mathcal{P}(X_v)$ *then*

$$\operatorname{Dim} \mathcal{H}_\mu^{q,\beta(q)} \lfloor \operatorname{supp} \mu = F_\mu(\alpha(q)).$$

Theorem 2.6.7. *Let* $X_v \subset \mathbb{R}^d$. *Then the following statements hold.*

i) *For* P_u-a.a. $\mu \in \mathcal{P}(X_u)$ *then*

$$\alpha_\mu = \alpha(1) \quad \mu\text{-a.e.}$$

ii) *For* P_u-a.a. $\mu \in \mathcal{P}(X_u)$ *then*

$$\dim \mu = \alpha(1) = \operatorname{Dim} \mu.$$

We will now consider the deterministic case. Recall (cf. the discussion in section 2.4) that the deterministic case is obtained as follows: Let the list $\mathcal{S} = \left((S_e)_{e \in E_u}, (p_e)_{e \in E_u}\right)_{u \in V} \in \prod_{u \in V} \Xi_u$ be given and put

$$\lambda_u = \delta_{\left((S_e)_{e \in E_u}, (p_e)_{e \in E_u}\right)}.$$

Let $\lambda := (\lambda_u)_u$. Then

$$P_{\lambda,u} = \delta_{\nu_u}$$

where $\nu_u := \mu_u\Big(\big((S_e)_{e \in E_u}, (p_e)_{e \in E_u}\big), \big((S_e)_{e \in E_{\tau(\alpha_{|\alpha|})}}, (p_e)_{e \in E_{\tau(\alpha_{|\alpha|})}}\big)_{\alpha \in E_u^{(*)} \setminus \{\varnothing\}}\Big)$, and the measures $(\nu_u)_u$ are the unique self-similar measures associated with the list $\mathcal{S}$, i.e. the unique list of measures satisfying

$$\nu_u = \sum_{e \in E_u} p_e \nu_{\tau(e)} \circ S_e^{-1}.$$

Equation (4.4.8) in Chapter 4 states that

$$\beta'(q) = -\frac{\sum_u \sum_v \sum_{e \in E_{uv}} \sigma_u(q) \int \log(p_e)\, p_e^q \operatorname{Lip}(S_e)^{\beta(q)}\, d\lambda_u\ \rho_v(q)}{\sum_u \sum_v \sum_{e \in E_{uv}} \sigma_u(q) \int \log(\operatorname{Lip}(S_e))\, p_e^q \operatorname{Lip}(S_e)^{\beta(q)}\, d\lambda_u\ \rho_v(q)}$$

whence (since $\beta(1) = 0$)

$$\alpha(1) = -\beta'(1) = \frac{\sum_u \sum_v \sum_{e \in E_{uv}} \sigma_u(1) \log(p_e)\, p_e \rho_v(1)}{\sum_u \sum_v \sum_{e \in E_{uv}} \sigma_u(1) \log(\operatorname{Lip}(S_e))\, p_e \rho_v(1)} \tag{2.6.3}$$

However, since $\sum_v \sum_{e \in E_{uv}} \mathrm{E}_{\lambda_u}[p_e^1 \operatorname{Lip}(S_e)^{\beta(1)}]\, 1 = \mathrm{E}_{\lambda_u}[\sum_{e \in E_u} p_e] = 1$ then $\rho_u(1) = \frac{1}{\operatorname{card} V}$ for all $u \in V$ (cf. (2.5.7) and (2.5.9)), and (2.6.3) therefore implies that

$$\alpha(1) = \frac{\sum_u \sum_{e \in E_u} \sigma_u(1) p_e \log(p_e)}{\sum_u \sum_{e \in E_u} \sigma_u(1) p_e \log(\operatorname{Lip}(S_e))} \tag{2.6.4}$$

Assume furthermore that the list S satisfies condition (I), (III), (IV) and (V), i.e. $0 < \min_e p_e \leq \max_e p_e < 1$, S_e is a similarity for all $e \in E$ and

$$S_e(X_{\tau(e)}) \bigcap S_\varepsilon(X_{\tau(\varepsilon)}) = \varnothing$$

for all $u \in V$ and $e, \varepsilon \in E_u$ with $e \neq \varepsilon$. Since $P_{\lambda,u} = \delta_{\nu_u}$, Theorem 2.6.7 and (2.6.4) imply that

$$\alpha_{\nu_u} = \alpha(1) = \frac{\sum_u \sum_{e \in E_u} \sigma_u(1) p_e \log(p_e)}{\sum_u \sum_{e \in E_u} \sigma_u(1) p_e \log(\mathrm{Lip}(S_e))} \quad \nu_u\text{-a.e.} \tag{2.6.5}$$

for all u. The result in (2.6.5) was first proved in the deterministic self-similar case by Geronimo & Hardin [Ge] and later in the deterministic graph directed self-similar case by Deliu et al. [Del].

2.7 The P_u Almost Sure Positivity and Finiteness of $\mathcal{H}_\mu^{q,\beta(q)}$.

This section gives (Theorem 2.7.1) a sufficient condition guaranteeing that

$$\mathcal{H}_\mu^{q,\beta(q)} \lfloor \operatorname{supp}\mu = 0 \quad \text{for} \quad P_u\text{-a.a.} \quad \mu \in \mathcal{P}(X_u)$$

and (Theorem 2.7.2) a necessary and sufficient condition guaranteeing that

$$P_u\left(\mathcal{H}_\mu^{q,\beta(q)}(\Delta_\mu(\alpha(q))) > 0\right) > 0.$$

Both conditions can be viewed as natural multifractal extensions of Graf [Gra1] and Tsujii [Tsu1,Tsu2].

Theorem 2.7.1. *Let $X_v \subseteq \mathbb{R}^d$ be compact for all v and assume that conditions (I), (III), (IV) and (V) are satisfied. Let $q \in \mathbb{R}$. Then the following holds:*
If there exists a $u \in V$ such that

$$\lambda_u\left(\rho_u^{-1}\left(\sum_{e\in E_u} p_e^q \operatorname{Lip}(S_e)^{\beta(q)}\right)\rho_{\tau(e)} \neq 1\right) > 0$$

then

$$\mathcal{H}_\mu^{q,\beta(q)}(\operatorname{supp}\mu) = 0$$

for all $v \in V$ and P_v-a.a. $\mu \in \mathcal{P}(X_v)$.

Theorem 2.7.2. *Let $X_v \subseteq \mathbb{R}^d$ be compact for all v and assume that conditions (I), (III), (IV) and (V) are satisfied. Let $q \in]q_{\min}, q_{\max}[$. Then the following statements are equivalent.*

i) *For all $u \in V$ then*

$$\rho_u^{-1}\left(\sum_{e\in E_u} p_e^q \operatorname{Lip}(S_e)^{\beta(q)}\right)\rho_{\tau(e)} = 1 \quad \Lambda_u\text{-a.s.}$$

ii) *For all $v \in V$ then*

$$P_v\left(\mathcal{H}_\mu^{q,\beta(q)}(\Delta_\mu(\alpha(q))) > 0\right) = 1.$$

iii) *For all $v \in V$ then*

$$P_v\left(\mathcal{H}_\mu^{q,\beta(q)}(\Delta_\mu(\alpha(q))) > 0\right) > 0.$$

iv) *There exists a $u \in V$ such that*

$$P_u\left(\mathcal{H}_\mu^{q,\beta(q)}(\Delta_\mu(\alpha(q))) > 0\right) = 1.$$

v) *There exists a $u \in V$ such that*

$$P_u\left(\mathcal{H}_\mu^{q,\beta(q)}(\Delta_\mu(\alpha(q))) > 0\right) > 0.$$

Theorem 2.7.1 and Theorem 2.7.2 have been proved for $q = 0$ in the self-similar case by Graf [Gra1] and in the graph directed self-similar case by Tsujii [Tsu2]; both Graf [Gra1] and Tsujii [Tsu2] considered the slightly more general setting in which condition (IV) was replaced by condition (IV′).

2.8 The P_u Almost Sure Constancy of $\mathcal{H}_\mu^{q,\beta(q)}$.

The main theorem in this section (i.e. Theorem 2.8.1) states that the measure $\mathcal{H}_\mu^{q,\beta(q)}$ treats the multifractal components $\Delta_\mu(\alpha(q))$ of the measures μ in a very uniform way, in the sense that

$$\mathcal{H}_\mu^{q,\beta(q)}(\Delta_\mu(\alpha(q)))\left(= \mathcal{H}_\mu^{q,\beta(q)}(\operatorname{supp}\mu) \quad \text{by Theorem 2.6.1}\right)$$

is constant for P_u-a.a. $\mu \in \mathcal{P}(X_u)$. This result may be viewed as a natural multifractal extension of Tsujii [Tsu1,Tsu2].

Theorem 2.8.1. *Let $X_v \subseteq \mathbb{R}^d$ be compact for all v and assume that conditions (I), (III), (IV) and (V) are satisfied. Let $q \in]q_{\min}, q_{\max}[$. For each $u \in V$ there exists a constant $c_{u,q} \geq 0$ such that*

$$\mathcal{H}_\mu^{q,\beta(q)}(\Delta_\mu(\alpha(q))) = c_{u,q}$$

for P_u-a.a. $\mu \in \mathcal{P}(X_u)$.

Theorem 2.8.1 has been proved for $q = 0$ by Tsujii [Tsu2] in the slightly more general setting in which condition (IV) is replaced by condition (IV′).

2.9 Thermodynamics.

Our results have a natural thermodynamical interpretation. Let $\mu \in \mathcal{P}(\mathbb{R}^d)$, $q \in \mathbb{R}$ and $\delta > 0$. The number

$$S^q_{\mu,\delta}(\operatorname{supp}\mu) = \sup\{\sum_i \mu(B(x_i, r_i))^q \mid \big(B(x_i,r_i)\big)_{i\in\mathbb{N}} \text{ is a centered packing of } \operatorname{supp}\mu\}$$

may clearly be interpreted as a partition function corresponding to the "inverse temperature" q; and

$$\underline{C}^q_\mu(\operatorname{supp}\mu) := \liminf_{\delta\searrow 0} \frac{\log S^q_{\mu,\delta}(\operatorname{supp}\mu)}{-\log\delta}$$

can thus be regarded as the "free energy" after passing to the thermodynamic limit at the "inverse temperature" q. Theorem 2.1.11 therefore asserts that $b_\mu(q)$ is less than the "free energy" $\underline{C}^q_\mu(\operatorname{supp}\mu)$ for $\mu \in \mathcal{P}_1(\mathbb{R}^d)$. However, equality holds in Theorem 2.1.11 in many interesting cases, e.g. for random and non-random graph directed self-similar measures in $\mathbb{R}^d$ with totally disconnected support (by Theorem 2.6.5 and [Ol,Theorem 5.1]), and for "cookie-cutter" measures in $\mathbb{R}$ (by [Ol,Theorem 6.1]). It is therefore natural to interpret $b_\mu(q)$ $\big(B_\mu(q)\big)$ as a "Hausdorff free energy" ("packing free energy") of the measure μ at the "inverse temperature" $q \in \mathbb{R}$. Observe that the "inverse temperature" q can be negative, and the parameter q can therefore not be regarded as an inverse temperature in a strict physical sense. Following the analogy with statistical physics and thermodynamics we define the Hausdorff energy $e_\mu(q)$ of μ at the "inverse temperature" q by

$$e_\mu(q) := b'_\mu(q)$$

for $q \in \{p \in \mathbb{R} \mid b_\mu \text{ is differentiable at } p\}$, and the packing energy $E_\mu(q)$ of μ at the "inverse temperature" q by

$$E_\mu(q) := B'_\mu(q)$$

for $q \in \{p \in \mathbb{R} \mid B_\mu \text{ is differentiable at } p\}$.

A physical system has, according to thermodynamics, a phase transition at the inverse temperature q if the free energy of the system displays a singularity at q. Fix $q \in \mathbb{R}$ and write

$$N_{b_\mu} := \{n \in \mathbb{N}\cup\{0\} \mid b_\mu \text{ is } n \text{ times differentiable at } q\}$$
$$N_{B_\mu} := \{n \in \mathbb{N}\cup\{0\} \mid B_\mu \text{ is } n \text{ times differentiable at } q\}$$

We now say that the measure μ has an nth-order Hausdorff phase transition at (the "inverse temperature") q if $N_{b_\mu} \neq \mathbb{N}\cup\{0\}$ and $n = \max N_{b_\mu} + 1$ (observe that the assumption $N_{b_\mu} \neq \mathbb{N} \cup \{0\}$ implies that N_{b_μ} is bounded, and $n = \max N_{b_\mu} + 1$ is therefore a well-defined integer), and we say that μ has an nth-order packing phase transition at (the "inverse temperature") q if $N_{B_\mu} \neq \mathbb{N} \cup \{0\}$ and $n = \max N_{B_\mu} + 1$ (observe that the assumption $N_{B_\mu} \neq \mathbb{N} \cup \{0\}$ implies that N_{B_μ} is bounded, and $n = \max N_{B_\mu} + 1$ is therefore a well-defined integer.) If $N_{b_\mu} = \mathbb{N} \cup \{0\}$ then we say that μ does not have a Hausdorff phase transition at (the "inverse temperature") q, and if $N_{B_\mu} = \mathbb{N} \cup \{0\}$ then we say that μ does not have a packing phase transition at (the "inverse temperature") q.

The results in section 2.6 can thus be formulated in the following way.

Theorem 2.9.1. *Let $X_v \subseteq \mathbb{R}^d$ be compact for all $v \in V$ with $0 < \lambda^d(X_v)$. Assume that conditions (I), (III), (IV) and (V) are satisfied and let $u \in V$. Then the following hold.*

Case 1: $\underline{a} = \overline{a}$. In this case

i) *The following statement holds for P_u-a.a. $\mu \in \mathcal{P}(X_u)$:*
μ does not have a Hausdorff phase transition at any $q \in \mathbb{R}$, and μ does not have a packing phase transition at any $q \in \mathbb{R}$.

Case 2: $\underline{a} < \overline{a}$. In this case

i) *The following statement holds for P_u-a.a. $\mu \in \mathcal{P}(X_u)$:*
μ does not have a Hausdorff phase transition at any $q \in \mathbb{R} \setminus \{q_{\min}, q_{\max}\}$, and μ does not have a packing phase transition at any $q \in \mathbb{R} \setminus \{q_{\min}, q_{\max}\}$.

ii) *If $-\infty < q_{\min}$ then there exists an integer $n \in \mathbb{N} \setminus \{1\}$ such that P_u-a.a. $\mu \in \mathcal{P}(X_u)$ has an n-order Hausdorff phase transition at $q_{\min}$ and an n-order packing phase transition at $q_{\min}$.*

iii) *If $-\infty < q_{\min}$ and $\alpha'(q_{\min}) < 0$ (the latter condition is according to Proposition 2.5.6 satisfied if the graph (V, E) only has one vertex) then P_u-a.a. $\mu \in \mathcal{P}(X_u)$ has a 2-order Hausdorff phase transition at $q_{\min}$ and a 2-order packing phase transition at $q_{\min}$.*

iv) *If $q_{\max} < \infty$ then there exists an integer $n \in \mathbb{N} \setminus \{1\}$ such that P_u-a.a. $\mu \in \mathcal{P}(X_u)$ has an n-order Hausdorff phase transition at $q_{\max}$ and an n-order packing phase transition at $q_{\max}$.*

v) *If $q_{\min} < \infty$ and $\alpha'(q_{\max}) < 0$ (the latter condition is according to Proposition 2.5.6 satisfied if the graph (V, E) only has one vertex) then P_u-a.a. $\mu \in \mathcal{P}(X_u)$ has a 2-order Hausdorff phase transition at $q_{\max}$ and a 2-order packing phase transition at $q_{\max}$.*

Somewhat similar observations concerning the relation between phase transitions and multifractal box dimensions (rather than the generalized multifractal dimension functions b_μ and B_μ) have been noticed by the physicists Halsey et al. [Ha] and Lee et al. [Le1,Le2], and by the mathematicians Lopes [Lo2,Lo4,Lo5] and Mendes–France & Tenenbaum [Me] for certain types of non-random measures.

Chapter 3
Examples

This chapter contains some examples which illustrate the theorems in Chapter 2. Throughout this chapter, λ^d will denote the d-dimensional Lebesgue measure in $\mathbb{R}^d$.

3.1 Example 1

Let $I = [0,1]$ and $0 \leq s_1 < t_1 < s_2 < t_2 < s_3 < t_3 \leq 1$. Put

$$J_j = [s_j, t_j] .$$

Let $P_1, P_2, P_3 \in]0,1[$ with $\sum_{i=1}^{3} P_i = 1$. Now construct a random probability measure μ on $[0,1]$ in the following inductive way. The measure μ will be the weak limit of a sequence $(\mu_n)_{n\in\mathbb{N}}$ of random probability measures satisfying,

1) μ_n is supported by 2^n disjoint closed subintervals $(I_{i_1\ldots i_n})_{i_1,\ldots,i_n=1,2}$ of I;
2) for all $i_1,\ldots,i_n \in \{1,2\}^n$, the measure μ_n has a constant density (w.r.t. Lebesgue measure) on the interval $I_{i_1\ldots i_n}$.

We will construct the sequence $(\mu_n)_n$ by induction on n. Let

$$\Delta_0 := \{(q_1,q_2) \in \mathbb{R}^2 | \tfrac{1}{4} \leq q_i \leq \tfrac{3}{4} \, , \, \sum_{i=1}^{2} q_i = 1\} .$$

The start of the induction. Consider the subset $J_1 \cup J_2 \cup J_3$ of I. First we remove exactly one of the intervals J_1, J_2 or J_3 from the union $J_1 \cup J_2 \cup J_3$. We remove J_1 with probability P_1, we remove J_2 with probability P_2 and we remove J_3 with probability P_3. The remaining set, C_1, is thus

$$C_1 = I_1 \bigcup I_2$$

where $I_1 = J_{j_1}$ and $I_2 = J_{j_2}$ for some $j_1, j_2 \in \{1,2,3\}$ with $j_1 \neq j_2$. Next choose a point $(q_1,q_2) \in \Delta_0$ with respect to the uniform distribution on Δ_0. Finally let μ_1 be the measure supported on C_1 which has density $\frac{q_i}{\operatorname{diam} I_i}$ on I_i for $i = 1,2$; i.e. we put

$$\mu_1 = \sum_{i=1,2} \frac{q_i}{\operatorname{diam} I_i} (\lambda^1 \lfloor I_i) .$$

The inductive step. Suppose that the measure μ_n has been constructed and that

$$\mu_n = \sum_{i_1,\ldots,i_n=1,2} c_{i_1\ldots i_n}(\lambda^1 \lfloor I_{i_1\ldots i_n}).$$

Now fix $i_1,\ldots,i_n \in \{1,2\}$ and write $I_{i_1\ldots i_n} := [u_{i_1\ldots i_n}, v_{i_1\ldots i_n}]$. Put

$$J_{i_1\ldots i_n j} := [u_{i_1\ldots i_n} + (v_{i_1\ldots i_n} - u_{i_1\ldots i_n})s_j, u_{i_1\ldots i_n} + (v_{i_1\ldots i_n} - u_{i_1\ldots i_n})t_j].$$

Consider the subset $J_{i_1\ldots i_n 1} \cup J_{i_1\ldots i_n 2} \cup J_{i_1\ldots i_n 3}$ of $I_{i_1\ldots i_n}$. First we remove exactly one of the intervals $J_{i_1\ldots i_n 1}$, $J_{i_1\ldots i_n 2}$ or $J_{i_1\ldots i_n 3}$ from the union $J_{i_1\ldots i_n 1} \cup J_{i_1\ldots i_n 2} \cup J_{i_1\ldots i_n 3}$. We remove $J_{i_1\ldots i_n 1}$ with probability P_1, we remove $J_{i_1\ldots i_n 2}$ with probability P_2 and we remove $J_{i_1\ldots i_n 3}$ with probability P_3. The remaining set, $C_{i_1\ldots i_n}$, is thus

$$C_{i_1\ldots i_n} = I_{i_1\ldots i_n 1} \bigcup I_{i_1\ldots i_n 2}$$

where $I_{i_1\ldots i_n 1} = J_{i_1\ldots i_n j_1}$ and $I_{i_1\ldots i_n 2} = J_{i_1\ldots i_n j_2}$ for some $j_1, j_2 \in \{1,2,3\}$ with $j_1 \neq j_2$. Next choose a point $(q_{i_1\ldots i_n 1}, q_{i_1\ldots i_n 2}) \in \Delta_0$ with respect to the uniform distribution on Δ_0. Finally we let μ_{n+1} be the measure supported on the set $C_{n+1} := \cup_{i_1,\ldots,i_n} C_{i_1\ldots i_n} = \cup_{i_1,\ldots,i_n i_{n+1}} I_{i_1\ldots i_n i_{n+1}}$ which has density $\frac{c_{i_1\ldots i_n} q_{i_1\ldots i_n i_{n+1}}}{\operatorname{diam} I_{i_1\ldots i_n i_{n+1}}}$ on $I_{i_1\ldots i_n i_{n+1}}$ for $i_1,\ldots,i_n,i_{n+1} = 1,2$; i.e. we put

$$\mu_{n+1} = \sum_{i_1,\ldots,i_n,i_{n+1}=1,2} \frac{c_{i_1\ldots i_n} q_{i_1\ldots i_n i_{n+1}}}{\operatorname{diam} I_{i_1\ldots i_n i_{n+1}}}(\lambda^1 \lfloor I_{i_1\ldots i_n i_{n+1}}).$$

This completes the construction of the sequence $(\mu_n)_n$.

Let μ be the weak limit of the sequence $(\mu_n)_n$. We will now compute the almost sure multifractal spectrum of μ. The measure μ can be modeled as a random graph directed self-similar measure corresponding to a graph $(V,E) = (\{*\},\{1,2\})$ with one vertex $*$ and two edges 1 and 2 as follows. Let $X_* = I$ and define $T_1, T_2, T_3 : I \to I$ by

$$T_i(x) = (t_i - s_i)x + s_i.$$

Next define $\lambda := \lambda_* \in \mathcal{P}(\Xi_*)$ by

$$\lambda = \gamma \times L$$

where

$$\gamma = P_1 \delta_{(T_2,T_3)} + P_2 \delta_{(T_1,T_3)} + P_3 \delta_{(T_1,T_2)}$$

and L is the normalized line measure (i.e. the normalized 1-dimensional Hausdorff measure) on Δ_0. The auxiliary function β is in this case given by

$$1 = \int \left(p_1^q \operatorname{Lip}(S_1)^{\beta(q)} + p_2^q \operatorname{Lip}(S_2)^{\beta(q)}\right) d\lambda((S_1,S_2),(p_1,p_2))$$

$$= \int p_1^q \, dL(p_1,p_2) \int \mathrm{Lip}(S_1)^{\beta(q)} \, d\gamma(S_1,S_2) +$$

$$\int p_2^q \, dL(p_1,p_2) \int \mathrm{Lip}(S_2)^{\beta(q)} \, d\gamma(S_1,S_2)$$

$$= \begin{cases} \frac{2}{q+1}\left(\left(\frac{3}{4}\right)^{q+1} - \left(\frac{1}{4}\right)^{q+1}\right)\Big(P_1\Big((t_2-s_2)^{\beta(q)} + (t_3-s_3)^{\beta(q)}\Big) + \\ \qquad P_2\Big((t_1-s_1)^{\beta(q)} + (t_3-s_3)^{\beta(q)}\Big) + \\ \qquad P_3\Big((t_1-s_1)^{\beta(q)} + (t_2-s_2)^{\beta(q)}\Big)\Big) & \text{for } q \neq -1\,; \\ 2\log 3\Big(P_1\Big((t_2-s_2)^{\beta(q)} + (t_3-s_3)^{\beta(q)}\Big) + \\ \qquad P_2\Big((t_1-s_1)^{\beta(q)} + (t_3-s_3)^{\beta(q)}\Big) + \\ \qquad P_3\Big((t_1-s_1)^{\beta(q)} + (t_2-s_2)^{\beta(q)}\Big)\Big) & \text{for } q = -1\,. \end{cases} \tag{3.1.1}$$

Assume further that

$$t_1 - s_1 = t_2 - s_2 = t_3 - s_3 := a\ , \quad P_1 = P_2 = P_3 = \tfrac{1}{3}\,.$$

It follows from (3.1.1) that

$$\beta(q) = \begin{cases} \log\left(4^q \frac{q+1}{3^{q+1}-1}\right) \Big/ \log a & \text{for } q \neq -1\,; \\ -\log(4\log 3)/\log a & \text{for } q = -1\,. \end{cases}$$

Also

$$\underline{a} = \inf_{i=1,2} \left\| \frac{\log p_i}{\log(\mathrm{Lip}\, S_i)} \right\|_{-\infty} = \frac{\log \frac{3}{4}}{\log a}\,,$$

$$\overline{a} = \sup_{i=1,2} \left\| \frac{\log p_i}{\log(\mathrm{Lip}\, S_i)} \right\|_{\infty} = \frac{\log \frac{1}{4}}{\log a}\,.$$

Clearly $\overline{a}q + \beta(q) = \log\left(\frac{q+1}{3^{q+1}-1}\right) \Big/ \log a$ for $q \neq -1$, whence

$$\overline{e} = \lim_{q \to -\infty} (\overline{a}q + \beta(q)) = -\infty\,. \tag{3.1.2}$$

It follows from (3.1.2) and Proposition 2.5.3 that β does not have an asymptote as $q \to -\infty$; also by Proposition 2.5.6

$$\lim_{\alpha \nearrow \overline{a}} \beta^*(\alpha) = \overline{e} = -\infty\,.$$

It is easily seen that $\underline{a}q + \beta(q) = \log\left(3^q \frac{q+1}{3^{q+1}-1}\right) \Big/ \log a$ for $q \neq -1$, whence

$$\underline{e} = \lim_{q \to \infty} (\underline{a}q + \beta(q)) = -\infty\,. \tag{3.1.3}$$

It follows from (3.1.3) and Proposition 2.5.3 that β does not have an asymptote as $q \to \infty$; also by Proposition 2.5.6

$$\lim_{\alpha \searrow \underline{a}} \beta^*(\alpha) = \underline{e} = -\infty\,.$$

Below we sketch the graphs of β and β^*.

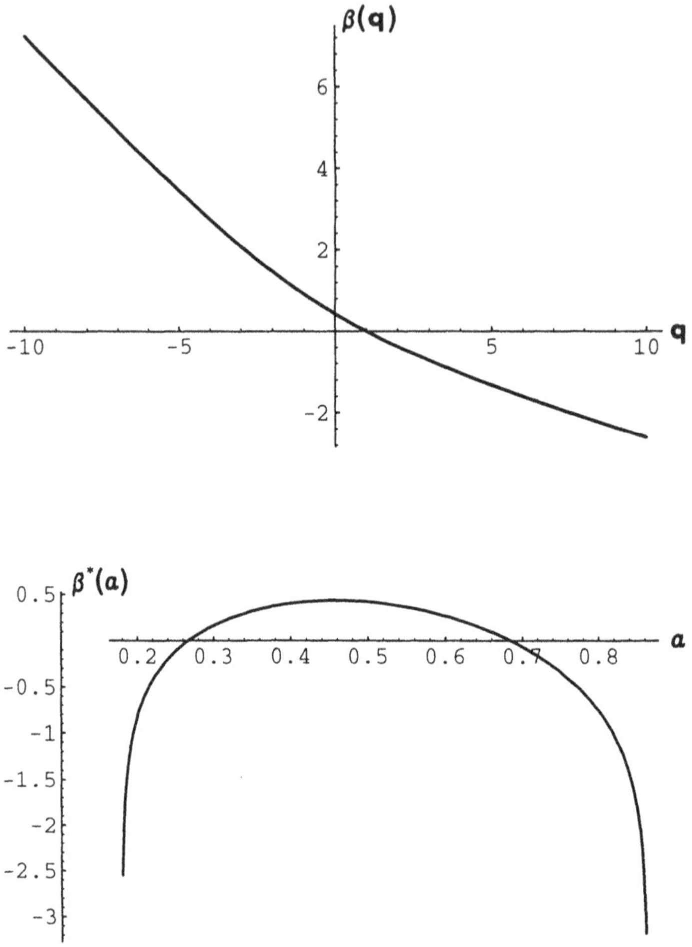

Figure 3.1.1: The graphs of β and β^* for $a = \dfrac{1}{5}$. We have:

$q_{\min} \approx -4.11\,, \quad q_{\max} \approx 6.05388$

$\underline{a} = \dfrac{\log 3/4}{\log 1/5} \approx 0.178747\,, \quad a_{\min} \approx 0.26654$

$\overline{a} = \dfrac{\log 1/4}{\log 1/5} \approx 0.861353\,, \quad a_{\max} \approx 0.68478$

$\underline{e} = \overline{e} = -\infty\,.$

Write $P := P_* \in \mathcal{P}(\mathcal{P}(X_*)) = \mathcal{P}(\mathcal{P}([0,1]))$. It follows from Theorem 2.6.5 that for each fixed $\alpha \in]a_{\min}, a_{\max}[$ then

$$f_\mu(\alpha) = F_\mu(\alpha) = \beta^*(\alpha) \quad \text{for } P\text{-a.a. } \mu \in \mathcal{P}([0,1])\,.$$

Furthermore, P-a.a. $\mu \in \mathcal{P}([0,1])$ satisfies

$$\Delta_\mu(\alpha) = \varnothing \quad \text{for } \alpha \in [0,\infty[\setminus[a_{\min}, a_{\max}]\,.$$

Clearly

$$\sum_{i=1,2} p_i^q \operatorname{Lip}(S_i)^{\beta(q)} = (p_1^q + p_2^q)a^{\beta(q)} = (p_1^q + p_2^q)4^{q\frac{q+1}{3^{q+1}-1}} \quad \lambda\text{-a.s.}\,,$$

whence

$$\left.\begin{aligned} &\sum_{i=1,2} p_i^q \operatorname{Lip}(S_i)^{\beta(q)} = 1 \quad \lambda\text{-a.s.} && \text{for } q = 0,1 \\ &\lambda\left(\sum_{i=1,2} p_i^q \operatorname{Lip}(S_i)^{\beta(q)} \neq 1\right) > 0 && \text{for } q \neq 0,1 \end{aligned}\right\} \tag{3.1.4}$$

Equation (3.1.4), Theorem 2.7.1, Theorem 2.7.2 and Theorem 2.8.1 imply that for each $q \in]q_{\min}, q_{\max}[$ there exists a number c_q with $c_q > 0$ for $q = 0,1$ and $c_q = 0$ for $q \neq 0,1$, such that

$$\mathcal{H}_\mu^{q,\beta(q)}(\Delta_\mu(\alpha(q))) = \begin{cases} c_q > 0 & \text{for } P\text{-a.a. } \mu \in \mathcal{P}([0,1]) \text{ for } q = 0,1\,; \\ c_q = 0 & \text{for } P\text{-a.a. } \mu \in \mathcal{P}([0,1]) \text{ for } q \neq 0,1\,. \end{cases}$$

3.2 Example 2

Let $I := [0,1]$, $p \in]0,1[$ and $(r_1,r_2),(s_1,s_2) \in]0,1[^2$ with $\sum_i r_i = 1 = \sum_i s_i$ and $r_1 \neq r_2$ and $s_1 \neq s_2$. Define $T_1, T_2 : I \to I$ by

$$T_1(x) = \frac{x}{3}\,, \quad T_2(x) = \frac{x}{3} + \frac{2}{3}\,.$$

For $i_1, \dots, i_n \in \{1,2\}^n$ write

$$I_{i_1 \dots i_n} = T_{i_n} \circ \cdots \circ T_{i_n}(I)\,.$$

For each integer n write

$$C_n := \bigcup_{i_1,\dots,i_n} I_{i_1 \dots i_n}\,.$$

The set

$$C := \bigcap_n \bigcup_{i_1,\dots,i_n} I_{i_1 \dots i_n}$$

is just the classical ternary Cantor set. Now construct a random probability measure measure μ on C in the following inductive way. The measure μ will be the weak limit of a sequence of random probability measures $(\mu_n)_{n\in\mathbb{N}}$ satisfying,

1) μ_n is supported by the set C_n;
2) for all $i_1, \dots, i_n \in \{1,2\}^n$ the measure μ_n has a constant density (w.r.t. Lebesgue measure) on the interval $I_{i_1 \dots i_n}$.

We will construct the sequence $(\mu_n)_n$ by induction on n.

The start of the induction. Assign with probability p, the density $\frac{r_1}{3^{-1}}$ to I_1 and the density $\frac{r_2}{3^{-1}}$ to I_2, and assign with probability $1-p$, the density $\frac{s_1}{3^{-1}}$ to I_1 and the density $\frac{s_2}{3^{-1}}$ to I_2; i.e. we let

$$\mu_1 = \sum_{i=1,2} \frac{q_i}{3^{-1}} (\lambda^1 \lfloor I_i)$$

where $(q_1, q_2) = (r_1, r_2)$ with probability p and $(q_1, q_2) = (s_1, s_2)$ with probability $1-p$.

The inductive step. Suppose that the measure μ_n has been constructed and that

$$\mu_n = \sum_{i_1,\dots,i_n=1,2} c_{i_1 \dots i_n} (\lambda^1 \lfloor I_{i_1 \dots i_n})\,.$$

Now fix $i_1, \dots, i_n \in \{1,2\}$. Assign with probability p, the density $\frac{c_{i_1 \dots i_n} r_1}{3^{-(n+1)}}$ to $I_{i_1 \dots i_n 1}$ and the density $\frac{c_{i_1 \dots i_n} r_2}{3^{-(n+1)}}$ to $I_{i_1 \dots i_n 2}$, and assign with probability $1-p$, the density $\frac{c_{i_1 \dots i_n} s_1}{3^{-(n+1)}}$ to $I_{i_1 \dots i_n 1}$ and the density $\frac{c_{i_1 \dots i_n} s_2}{3^{-(n+1)}}$ to $I_{i_1 \dots i_n 2}$; i.e. we let

$$\mu_{n+1} = \sum_{i_1,\dots,i_n,i_{n+1}=1,2} \frac{c_{i_1 \dots i_n} q_{i_1 \dots i_n i_{n+1}}}{3^{-(n+1)}} (\lambda^1 \lfloor I_{i_1 \dots i_n i_{n+1}})$$

where $(q_{i_1\dots i_n 1}, q_{i_1\dots i_n 2}) = (r_1, r_2)$ with probability p and $(q_{i_1\dots i_n 1}, q_{i_n\dots i_n 2}) = (s_1, s_2)$ with probability $1-p$. This completes the construction of the sequence $(\mu_n)_n$.

Let μ be the weak limit of the sequence $(\mu_n)_n$. We will now compute the almost sure multifractal spectrum of μ. The measure μ can be modeled as a random graph directed self-similar measure corresponding to a graph $(V, E) = (\{*\}, \{1,2\})$ with one vertex $*$ and two edges 1 and 2 as follows. Let $X_* = I$ and define $\lambda = \lambda_* \in \mathcal{P}(\Xi_*)$ by

$$\lambda = \delta_{(T_1,T_2)} \times \gamma$$

where

$$\gamma = p\delta_{(r_1,r_2)} + (1-p)\delta_{(s_1,s_2)}\,.$$

The auxiliary function β is in this case given by

$$\begin{aligned}
1 &= \int \left(p_1^q \operatorname{Lip}(S_1)^{\beta(q)} + p_2^q \operatorname{Lip}(S_2)^{\beta(q)}\right) d\lambda((S_1,S_2),(p_1,p_2)) \\
&= \int p_1^q \, d\gamma(p_1,p_2) \int \operatorname{Lip}(S_1)^{\beta(q)} \, d\delta_{(T_1,T_2)}(S_1,S_2) + \\
&\qquad \int p_2^q \, d\gamma(p_1,p_2) \int \operatorname{Lip}(S_2)^{\beta(q)} \, d\delta_{(T_1,T_2)}(S_1,S_2) \\
&= (p(r_1^q + r_2^q) + (1-p)(s_1^q + s_2^q)) \left(\tfrac{1}{3}\right)^{\beta(q)} ,
\end{aligned}$$

whence

$$\beta(q) = \frac{\log\left(p(r_1^q + r_2^q) + (1-p)(s_1^q + s_2^q)\right)}{\log 3}\,.$$

Also

$$\begin{aligned}
\underline{a} &= \inf_{i=1,2} \left\| \frac{\log p_i}{\log(\operatorname{Lip} S_i)} \right\|_{-\infty} = \frac{\log \max\{r_1, r_2, s_1, s_2\}}{\log \frac{1}{3}}\,, \\
\overline{a} &= \sup_{i=1,2} \left\| \frac{\log p_i}{\log(\operatorname{Lip} S_i)} \right\|_{\infty} = \frac{\log \min\{r_1, r_2, s_1, s_2\}}{\log \frac{1}{3}}\,.
\end{aligned}$$

Write $m := \min\{r_1, r_2, s_1, s_2\}$ and $M := \max\{r_1, r_2, s_1, s_2\}$. Clearly $\overline{a}q + \beta(q) = \log\left(p\left(\left(\frac{r_1}{m}\right)^q + \left(\frac{r_2}{m}\right)^q\right) + (1-p)\left(\left(\frac{s_1}{m}\right)^q + \left(\frac{s_2}{m}\right)^q\right)\right) \big/ \log 3$, whence

$$\begin{aligned}
\overline{e} := \overline{e}(p) &= \lim_{q\to-\infty} (\overline{a}q + \beta(q)) \\
&= \begin{cases} \log p/\log 3 & \text{if } m = r_i \text{ for some } i \text{ and } m < s_1, s_2\,; \\ \log(1-p)/\log 3 & \text{if } m = s_i \text{ for some } i \text{ and } m < r_1, r_2\,; \\ \log 1/\log 3 = 0 & \text{if } r_i = m = s_j \text{ for some } i \text{ and } j\,. \end{cases}
\end{aligned}$$

The function

$$q \to -\overline{a}q + \overline{e}$$

is thus an asymptote for β as $q \to -\infty$; also by Proposition 2.5.6

$$\lim_{\alpha \nearrow \overline{a}} \beta^*(\alpha) = \overline{e}\,.$$

Observe that $\overline{e}(p) \leq 0$ for all $p \in]0,1[$.

Clearly $\underline{a}q + \beta(q) = \log\left(p\left(\left(\frac{r_1}{M}\right)^q + \left(\frac{r_2}{M}\right)^q\right) + (1-p)\left(\left(\frac{s_1}{M}\right)^q + \left(\frac{s_2}{M}\right)^q\right)\right) \Big/ \log 3$, whence

$$\begin{aligned}\underline{e} := \underline{e}(p) &= \lim_{q\to\infty} (\underline{a}q + \beta(q)) \\ &= \begin{cases} \log p/\log 3 & \text{if } M = r_i \text{ for some } i \text{ and } s_1, s_2 < M\,; \\ \log(1-p)/\log 3 & \text{if } M = s_i \text{ for some } i \text{ and } r_1, r_2 < M\,; \\ \log 1/\log 3 = 0 & \text{if } r_i = M = s_j \text{ for some } i \text{ and } j\,. \end{cases}\end{aligned}$$

The function

$$q \to -\underline{a}q + \underline{e}$$

is thus an asymptote for β as $q \to \infty$; also by Proposition 2.5.6

$$\lim_{\alpha \searrow \underline{a}} \beta^*(\alpha) = \underline{e}\,.$$

Observe that $\underline{e}(p) \leq 0$ for all $p \in]0,1[$.

Below we sketch the graphs of β and β^*.

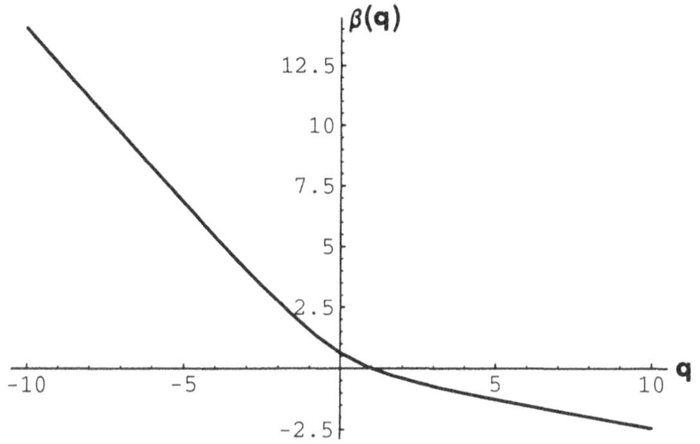

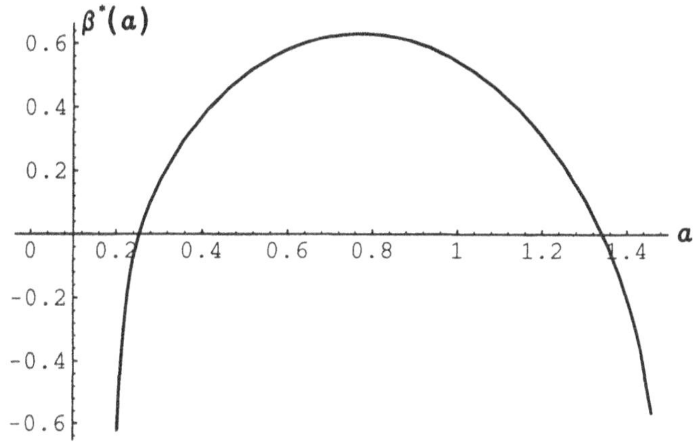

Figure 3.2.1: The graphs of β and β^* for $p = \frac{1}{2}$, $r_1 = \frac{1}{5}$, $r_2 = 1 - r_1$, $s_1 = \frac{3}{10}$ and $s_2 = 1 - s_1$.

We have:

$$q_{\min} \approx -2.90158\,, \quad q_{\max} \approx 4.59091$$

$$\underline{a} = \frac{\log 4/5}{\log 1/3} \approx 0.203114\,, \quad a_{\min} \approx 0.253076$$

$$\overline{a} = \frac{\log 1/5}{\log 1/3} \approx 1.46497\,, \quad a_{\max} \approx 1.34229$$

$$\underline{e} = \overline{e} = \frac{\log 1/2}{\log 3} \approx -0.63093\,.$$

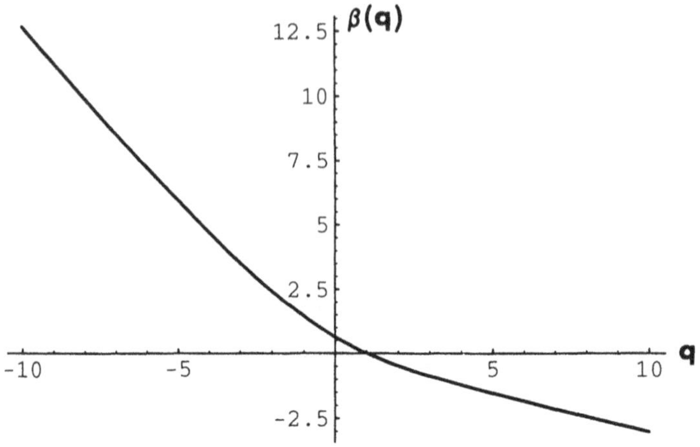

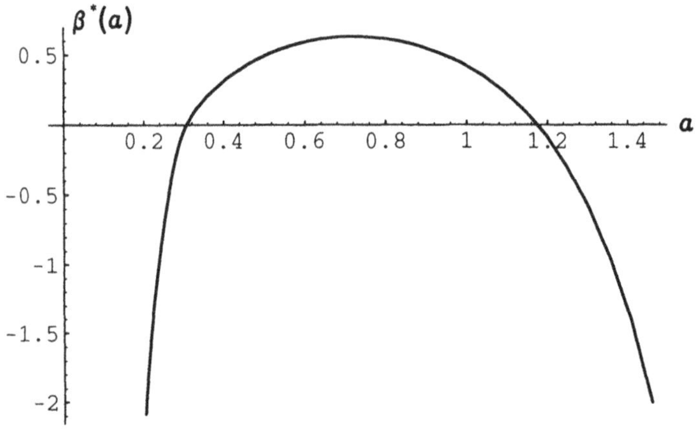

Figure 3.2.2: The graphs of β and β^* for $p = \frac{1}{10}$, $r_1 = \frac{1}{5}$, $r_2 = 1 - r_1$, $s_1 = \frac{3}{10}$ and $s_2 = 1 - s_1$.

We have:

$$q_{\min} \approx -3.45\,, \quad q_{\max} \approx 5.69463$$

$$\underline{a} = \frac{\log 4/5}{\log 1/3} \approx 0.203114\,, \quad a_{\min} \approx 0.306371$$

$$\overline{a} = \frac{\log 1/5}{\log 1/3} \approx 1.46497\,, \quad a_{\max} \approx 1.17641$$

$$\underline{e} = \overline{e} = \frac{\log 1/10}{\log 3} \approx -2.0959\,.$$

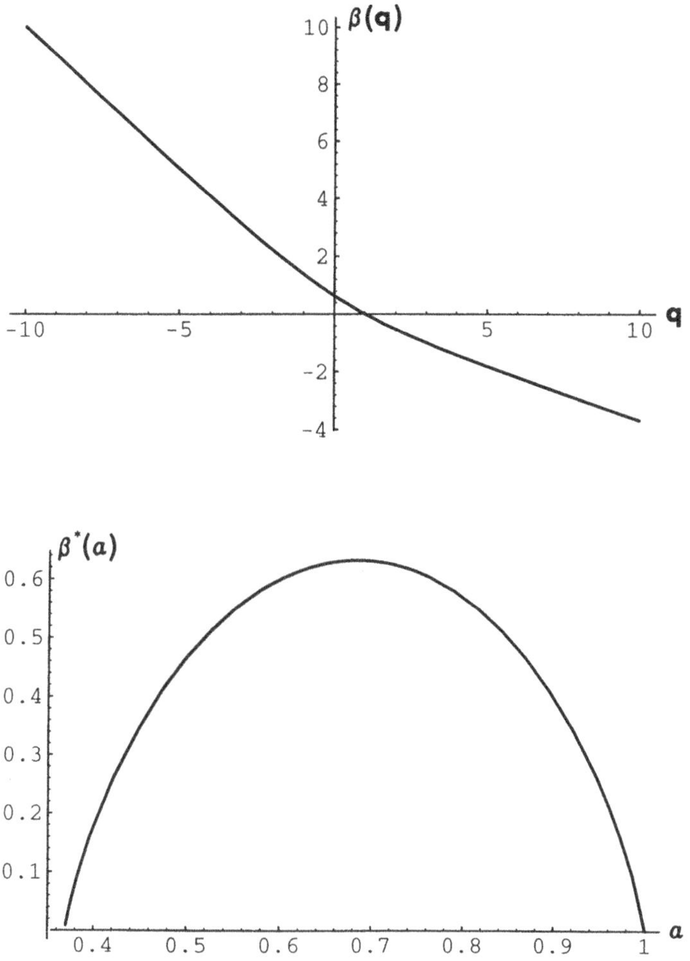

Figure 3.2.3: The graphs of β and β^* for $p = \frac{1}{2}$, $r_1 = \frac{1}{3}$, $r_2 = 1 - r_1$, $s_1 = \frac{2}{3}$ and $s_2 = 1 - s_1$.

We have:

$$\underline{a} = a_{\min} = \frac{\log 2/3}{\log 1/3} \approx 0.36907\,, \quad \overline{a} = a_{\max} = 1$$

$$\underline{e} = \overline{e} = 0\,.$$

Write $P := P_* \in \mathcal{P}(\mathcal{P}(X_*)) = \mathcal{P}(\mathcal{P}([0,1]))$. It follows from Theorem 2.6.5 that for each fixed $\alpha \in]a_{\min}, a_{\max}[$ then

$$f_\mu(\alpha) = F_\mu(\alpha) = \beta^*(\alpha) \quad \text{for } P\text{-a.a. } \mu \in \mathcal{P}([0,1]).$$

Moreover, P-a.a. $\mu \in \mathcal{P}([0,1])$ satisfies

$$\Delta_\mu(\alpha) = \varnothing \quad \text{for } \alpha \in [0,\infty[\setminus]a_{\min}, a_{\max}[.$$

Clearly

$$\begin{aligned}\sum_{i=1,2} p_i^q \operatorname{Lip}(S_i)^{\beta(q)} &= (p_1^q + p_2^q)\left(\tfrac{1}{3}\right)^{\beta(q)} \\ &= (p_1^q + p_2^q)\Big/\left(p(r_1^q + r_2^q) + (1-p)(s_1^q + s_2^q)\right) \quad \lambda\text{-a.s.},\end{aligned}$$

whence

$$\left.\begin{aligned}&\sum_{i=1,2} p_i^q \operatorname{Lip}(S_i)^{\beta(q)} = 1 \quad \lambda\text{-a.s.} && \text{for } r_1^q + r_2^q = s_1^q + s_2^q \\ &\lambda\left(\sum_{i=1,2} p_i^q \operatorname{Lip}(S_i)^{\beta(q)} \neq 1\right) > 0 && \text{for } r_1^q + r_2^q \neq s_1^q + s_2^q\end{aligned}\right\} \tag{3.2.1}$$

Equation (3.2.1), Theorem 2.7.1, Theorem 2.7.2 and Theorem 2.8.1 imply that for each $q \in]q_{\min}, q_{\max}[$ and $p \in]0,1[$ there exists a number $c_q := c_q(r_1, r_2, s_1, s_2)$ with $c_q > 0$ for $r_1^q + r_2^q = s_1^q + s_2^q$ and $c_q = 0$ for $r_1^q + r_2^q \neq s_1^q + s_2^q$ such that

$$\mathcal{H}_\mu^{q,\beta(q)}(\Delta_\mu(\alpha(q))) = \begin{cases} c_q(r_1, r_2, s_1, s_2) > 0 & \text{for } P\text{-a.a. } \mu \in \mathcal{P}([0,1]) \\ & \qquad \text{for } r_1^q + r_2^q = s_1^q + s_2^q; \\ c_q(r_1, r_2, s_1, s_2) = 0 & \text{for } P\text{-a.a. } \mu \in \mathcal{P}([0,1]) \\ & \qquad \text{for } r_1^q + r_2^q \neq s_1^q + s_2^q. \end{cases}$$

3.3 Example 3

Let $a > 0$ and $I = [0,2]$. Now construct a random probability measure μ on $[0,2]$ in the following inductive way. The measure μ will be the weak limit of a sequence $(\mu_n)_{n\in\mathbb{N}}$ of random probability measures satisfying,

1) μ_n is supported by 2^n disjoint closed subintervals $(I_{i_1\dots i_n})_{i_1,\dots,i_n=1,2}$ of I;
2) for all $i_1,\dots,i_n \in \{1,2\}^n$ the measure μ_n has a constant density (w.r.t. Lebesgue measure) on the interval $I_{i_1\dots i_n}$.

We will construct the sequence $(\mu_n)_n$ by induction on n. Let

$$\Delta_0 := [\tfrac{1}{4}, \tfrac{3}{4}].$$

The start of the induction. Pick a point x in Δ_0 with respect to the uniform distribution on Δ_0. Write

$$I_1 := [0, x^a] \, , \quad I_2 := [2 - (1-x)^a, 2] \, .$$

and let $\mu_1 \in \mathcal{P}(I)$ be the measure supported on $C_1 := I_1 \cup I_2$ which has density $\frac{x}{\operatorname{diam} I_1}$ on I_1 and density $\frac{1-x}{\operatorname{diam} I_2}$ on I_2; i.e. we put

$$\mu_1 = \frac{x}{\operatorname{diam} I_1}(\lambda^1 \lfloor I_1) + \frac{1-x}{\operatorname{diam} I_2}(\lambda^1 \lfloor I_2) \, .$$

The inductive step. Suppose that the measure μ_n has been constructed and that

$$\mu_n = \sum_{i_1,\dots,i_n=1,2} c_{i_1\dots i_n}(\lambda^1 \lfloor I_{i_1\dots i_n}) \, .$$

For each $i_1,\dots,i_n \in \{1,2\}$ write $I_{i_1\dots i_n} = [u_{i_1\dots i_n}, v_{i_1\dots i_n}]$. Now fix $i_1,\dots,i_n \in \{1,2\}$ and choose a point $x_{i_1\dots i_n}$ in Δ_0 with respect to the uniform distribution on Δ_0, and define intervals $I_{i_1\dots i_n 1}$ and $I_{i_1\dots i_n 2}$ by

$$I_{i_1\dots i_n 1} = [u_{i_1\dots i_n}, u_{i_1\dots i_n} + (v_{i_1\dots i_n} - u_{i_1\dots i_n})x^a_{i_1\dots i_n}]$$
$$I_{i_1\dots i_n 2} = [u_{i_1\dots i_n} - (v_{i_1\dots i_n} - u_{i_1\dots i_n})(1 - x_{i_1\dots i_n})^a, v_{i_1\dots i_n}]$$

Next let μ_{n+1} be the measure supported on $C_{n+1} := \bigcup_{i_1,\dots,i_n i_{n+1}} I_{i_1\dots i_n i_{n+1}}$ which has density $\frac{x_{i_1\dots i_n} c_{i_1\dots i_n}}{\operatorname{diam} I_{i_n\dots i_n 1}}$ on $I_{i_1\dots i_n 1}$ and density $\frac{(1-x_{i_1\dots i_n})c_{i_1\dots i_n}}{\operatorname{diam} I_{i_n\dots i_n 2}}$ on $I_{i_1\dots i_n 2}$; i.e. we put

$$\mu_{n+1} = \sum_{i_1,\dots,i_n=1,2} \left(\frac{x_{i_1\dots i_n} c_{i_1\dots i_n}}{\operatorname{diam} I_{i_1\dots i_n 1}}(\lambda^1 \lfloor I_{i_1\dots i_n 1}) + \frac{(1 - x_{i_1\dots i_n})c_{i_1\dots i_n}}{\operatorname{diam} I_{i_1\dots i_n 2}}(\lambda^1 \lfloor I_{i_1\dots i_n 2}) \right) \, .$$

This completes the construction of the sequence $(\mu_n)_n$.

Let μ be the weak limit of the sequence $(\mu_n)_n$. We will now compute the almost sure multifractal spectrum of μ. The measure μ can be modeled as a random graph directed self-similar measure corresponding to a graph $(V,E) = (\{*\},\{1,2\})$ with one vertex $*$ and two edges 1 and 2 as follows. For $x \in [0,1]$ define $S_{1,x}, S_{2,x} : I \to I$ by

$$S_{1,x}(t) = \tfrac{x^a}{2}t\ ,\quad S_{2,x}(t) = \tfrac{(1-x)^a}{2}t + 2 - (1-x)^a\,.$$

Let $X_* = I$ and define $\lambda = \lambda_* \in \mathcal{P}(\Xi_*)$ by

$$\lambda = 2(\lambda^1 \lfloor [\tfrac{1}{4},\tfrac{3}{4}]) \circ F^{-1}$$

where $F : [\frac{1}{4},\frac{3}{4}] \to \Xi_*$ is defined by

$$F(x) = \big((S_{1,x}, S_{2,x}), (x, 1-x)\big)\,.$$

The auxiliary function β is in this case given by

$$\begin{aligned} 1 &= \int \Big(p_1^q \operatorname{Lip}(S_1)^{\beta(q)} + p_2^q \operatorname{Lip}(S_2)^{\beta(q)}\Big)\, d\lambda((S_1,S_2),(p_1,p_2)) \\ &= 2\int_{\frac{1}{4}}^{\frac{3}{4}} \left(x^q \left(\tfrac{x^a}{2}\right)^{\beta(q)} + (1-x)^q \left(\tfrac{(1-x)^a}{2}\right)^{\beta(q)}\right) dx \\ &= \begin{cases} 4\dfrac{\left(\frac{3}{4}\right)^{q+a\beta(q)+1} - \left(\frac{1}{4}\right)^{q+a\beta(q)+1}}{(q+a\beta(q)+1)2^{\beta(q)}} & \text{for } q + a\beta(q) + 1 \neq 0\,; \\ 4\dfrac{\log 3}{2^{\beta(q)}} & \text{for } q + a\beta(q) + 1 = 0\,. \end{cases} \end{aligned}$$

Also

$$\begin{aligned} \underline{a} &= \inf_{i=1,2} \left\| \frac{\log p_i}{\log(\operatorname{Lip} S_i)} \right\|_{-\infty} = \frac{\log \frac{4}{3}}{\log 4^a} = \frac{1}{a}\left(1 - \frac{\log 3}{\log 4}\right), \\ \overline{a} &= \sup_{i=1,2} \left\| \frac{\log p_i}{\log(\operatorname{Lip} S_i)} \right\|_{\infty} = \frac{\log 4}{\log\left(\frac{4}{3}\right)^a} = \frac{1}{a}\left(1 - \frac{\log 3}{\log 4}\right)^{-1}. \end{aligned}$$

Write $P := P_* \in \mathcal{P}(\mathcal{P}(X_*)) = \mathcal{P}(\mathcal{P}([0,2]))$. It follows from Theorem 2.6.5 that for each fixed $\alpha \in]a_{\min}, a_{\max}[$ then

$$f_\mu(\alpha) = F_\mu(\alpha) = \beta^*(\alpha) \quad \text{for } P\text{-a.a. } \mu \in \mathcal{P}([0,2])\,.$$

Furthermore, P-a.a. $\mu \in \mathcal{P}([0,2])$ satisfies

$$\Delta_\mu(\alpha) = \varnothing \quad \text{for } \alpha \in [0,\infty[\setminus]a_{\min}, a_{\max}[\,.$$

Clearly

$$\left.\begin{array}{ll}\displaystyle\sum_{i=1,2} p_i^q \operatorname{Lip}(S_i)^{\beta(q)} = 1 \quad \lambda\text{-a.s.} & \text{for } q = 1 \\ \lambda\left(\displaystyle\sum_{i=1,2} p_i^q \operatorname{Lip}(S_i)^{\beta(q)} \neq 1\right) > 0 & \text{for } q \neq 1\end{array}\right\} \tag{3.3.1}$$

Equation (3.3.1), Theorem 2.7.1, Theorem 2.7.2 and Theorem 2.8.1 imply that for each $q \in]q_{\min}, q_{\max}[$ there exists a number c_q with $c_q > 0$ for $q = 1$ and $c_q = 0$ for $q \neq 1$, such that

$$\mathcal{H}_\mu^{q,\beta(q)}(\Delta_\mu(\alpha(q))) = \begin{cases} c_q > 0 & \text{for } P\text{-a.a. } \mu \in \mathcal{P}([0,2]) \text{ for } q = 1\,; \\ c_q = 0 & \text{for } P\text{-a.a. } \mu \in \mathcal{P}([0,2]) \text{ for } q \neq 1\,. \end{cases}$$

3.4 Example 4

Let $I = [0,1]$ and $\delta, c > 0$ with $\delta + c < 1 - \delta$. Write

$$T = \{(x,y) \in \mathbb{R}^2 \mid \delta \le x \ , \ y \le 1-\delta \ , \ c \le y - x \ \}.$$

Now construct a random probability measure μ in $[0,1]$ in the following inductive way. The measure μ will be the weak limit of a sequence $(\mu_n)_{n\in\mathbb{N}}$ of random probability measures satisfying,

1) μ_n is supported by 2^n disjoint closed subintervals $(I_{i_1\dots i_n})_{i_1,\dots,i_n=1,2}$ of I;
2) for all $i_1,\dots,i_n \in \{1,2\}^n$ the measure μ_n has a constant density (w.r.t. Lebesgue measure) on the interval $I_{i_1\dots i_n}$.

We will construct the sequence $(\mu_n)_n$ by induction on n.

The start of the induction. Pick a point (x,y) in T with respect to the uniform distribution on T. Let

$$I_1 := [0,x] \ , \ \ I_2 := [y,1] .$$

and assign the density $\frac{1}{2x}$ to I_1 and the density $\frac{1}{2(1-y)}$ to I_2; i.e. we put

$$\mu_1 = \frac{1}{2\,\mathrm{diam}\, I_1}(\lambda^1 \lfloor I_1) + \frac{1}{2\,\mathrm{diam}\, I_2}(\lambda^1 \lfloor I_2) .$$

The inductive step. Suppose that the measure μ_n has been constructed and that

$$\mu_n = \sum_{i_1,\dots,i_n=1,2} c_{i_1\dots i_n}(\lambda^1 \lfloor I_{i_1\dots i_n}) .$$

For each $i_1,\dots,i_n \in \{1,2\}$ write $I_{i_1\dots i_n} = [u_{i_1\dots i_n}, v_{i_1\dots i_n}]$. Now fix $i_1,\dots,i_n \in \{1,2\}$ and choose a point $(x_{i_1\dots i_n}, y_{i_1\dots i_n})$ in T with respect to the uniform distribution on T, and define intervals $I_{i_1\dots i_n 1}$ and $I_{i_1\dots i_n 2}$ by

$$I_{i_1\dots i_n 1} = [u_{i_1\dots i_n}, u_{i_1\dots i_n} + (v_{i_1\dots i_n} - u_{i_1\dots i_n})x_{i_1\dots i_n}]$$
$$I_{i_1\dots i_n 2} = [u_{i_1\dots i_n} - (v_{i_1\dots i_n} - u_{i_1\dots i_n})(1 - y_{i_1\dots i_n}), v_{i_1\dots i_n}]$$

Next let μ_{n+1} be the measure supported on $\cup_{i_1,\dots,i_n i_{n+1}} I_{i_1\dots i_n i_{n+1}}$ which has density $\frac{c_{i_1\dots i_n}}{2\,\mathrm{diam}\, I_{i_n\dots i_n 1}}$ on $I_{i_1\dots i_n 1}$ and density $\frac{c_{i_1\dots i_n}}{2\,\mathrm{diam}\, I_{i_n\dots i_n 2}}$ on $I_{i_1\dots i_n 2}$; i.e. we put

$$\mu_{n+1} = \sum_{i_1,\dots,i_n=1,2} \left(\frac{c_{i_1\dots i_n}}{2\,\mathrm{diam}\, I_{i_1\dots i_n 1}}(\lambda^1 \lfloor I_{i_1\dots i_n 1}) + \frac{c_{i_1\dots i_n}}{2\,\mathrm{diam}\, I_{i_1\dots i_n 2}}(\lambda^1 \lfloor I_{i_1\dots i_n 2}) \right) .$$

This completes the construction of the sequence $(\mu_n)_n$.

Let μ be the weak limit of the sequence $(\mu_n)_n$. We will now compute the almost sure multifractal spectrum of μ. The measure μ can be modeled as a random graph directed self-similar measure corresponding to a graph $(V, E) = (\{*\}, \{1, 2\})$ with one vertex $*$ and two edges 1 and 2 as follows. For $(x, y) \in T$ define $S_{1,(x,y)}, S_{2,(x,y)} : I \to I$ by

$$S_{1,(x,y)}(t) = xt \;, \quad S_{2,(x,y)}(t) = (1-y)t + y$$

Let $X_* = I$ and define $\lambda = \lambda_* \in \mathcal{P}(\Xi_*)$ by

$$\lambda = \frac{2}{(1-2\delta-c)^2}(\lambda^2 \lfloor T) \circ F^{-1} \times \delta_{(\frac{1}{2},\frac{1}{2})}$$

where $F : T \to \mathrm{Con}(I, I)^2$ is defined by

$$F(x, y) = \left(S_{1,(x,y)}, S_{2,(x,y)}\right).$$

The auxiliary function β is in this case given by

$$\begin{aligned} 1 &= \int \left(p_1^q \,\mathrm{Lip}(S_1)^{\beta(q)} + p_2^q \,\mathrm{Lip}(S_2)^{\beta(q)}\right) d\lambda((S_1, S_2), (p_1, p_2)) \\ &= \frac{2}{(1-2\delta-c)^2}\left(\int p_1^q \, d\delta_{(\frac{1}{2},\frac{1}{2})}(p_1, p_2) \int_{\delta}^{1-\delta-c} \left(\int_{x+c}^{1-\delta} x^\beta \, dy\right) dx \,+ \right. \\ &\qquad \left. \int p_2^q \, d\delta_{(\frac{1}{2},\frac{1}{2})}(p_1, p_2) \int_{\delta+c}^{1-\delta} \left(\int_{\delta}^{y-c} (1-y)^\beta \, dx\right) dy\right), \end{aligned}$$

whence

$$q = \left(\log\left(\frac{4}{(1-2\delta-c)^2}\right) + \log\left(f(\beta(q))\right)\right) \Big/ \log 2$$

where

$$f(t) = \begin{cases} (1-\delta-c)\frac{(1-\delta-c)^{t+1}-\delta^{t+1}}{t+1} - \frac{(1-\delta-c)^{t+2}-\delta^{t+2}}{t+2} & \text{for } t \neq -1, -2; \\ (1-\delta-c)\log\left(\frac{1-\delta-c}{\delta}\right) - (1-c) & \text{for } t = -1; \\ (1-\delta-c)\frac{(1-\delta-c)^3-\delta^3}{3}\log\left(\frac{1-\delta-c}{\delta}\right) & \text{for } t = -2. \end{cases}$$

Also

$$\underline{a} = \inf_{i=1,2}\left\|\frac{\log p_i}{\log(\mathrm{Lip}\, S_i)}\right\|_{-\infty} = \frac{\log\frac{1}{2}}{\log\delta},$$

$$\overline{a} = \sup_{i=1,2}\left\|\frac{\log p_i}{\log(\mathrm{Lip}\, S_i)}\right\|_{\infty} = \frac{\log\frac{1}{2}}{\log(1-\delta-c)}.$$

Below we sketch the graphs of β and β^*.

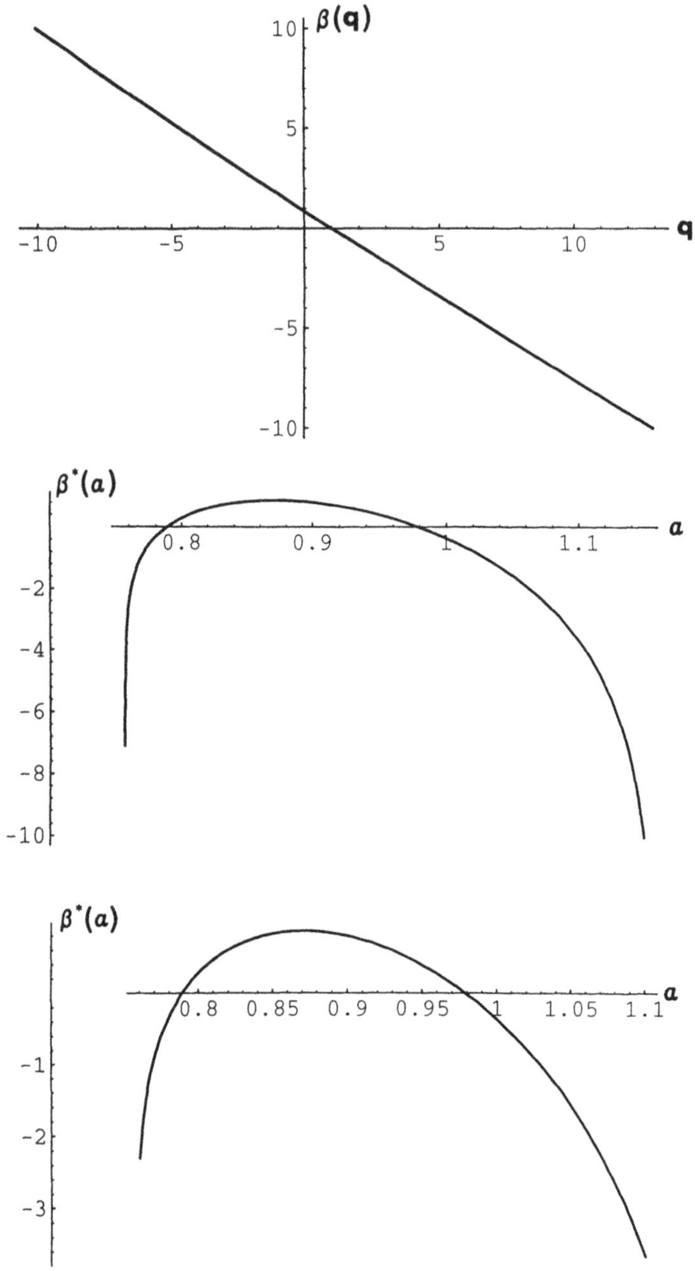

Figure 3.4.1: The graphs of β and β^* for $\delta = \dfrac{2}{5}$ and $c = \dfrac{1}{20}$. We have:

$$\underline{a} = \frac{\log 1/2}{\log 2/5} \approx 0.756471\,, \qquad a_{\min} \approx 0.788656$$

$$\overline{a} = \frac{\log 1/2}{\log 11/20} \approx 1.15943\,, \qquad a_{\max} \approx 0.978811\,.$$

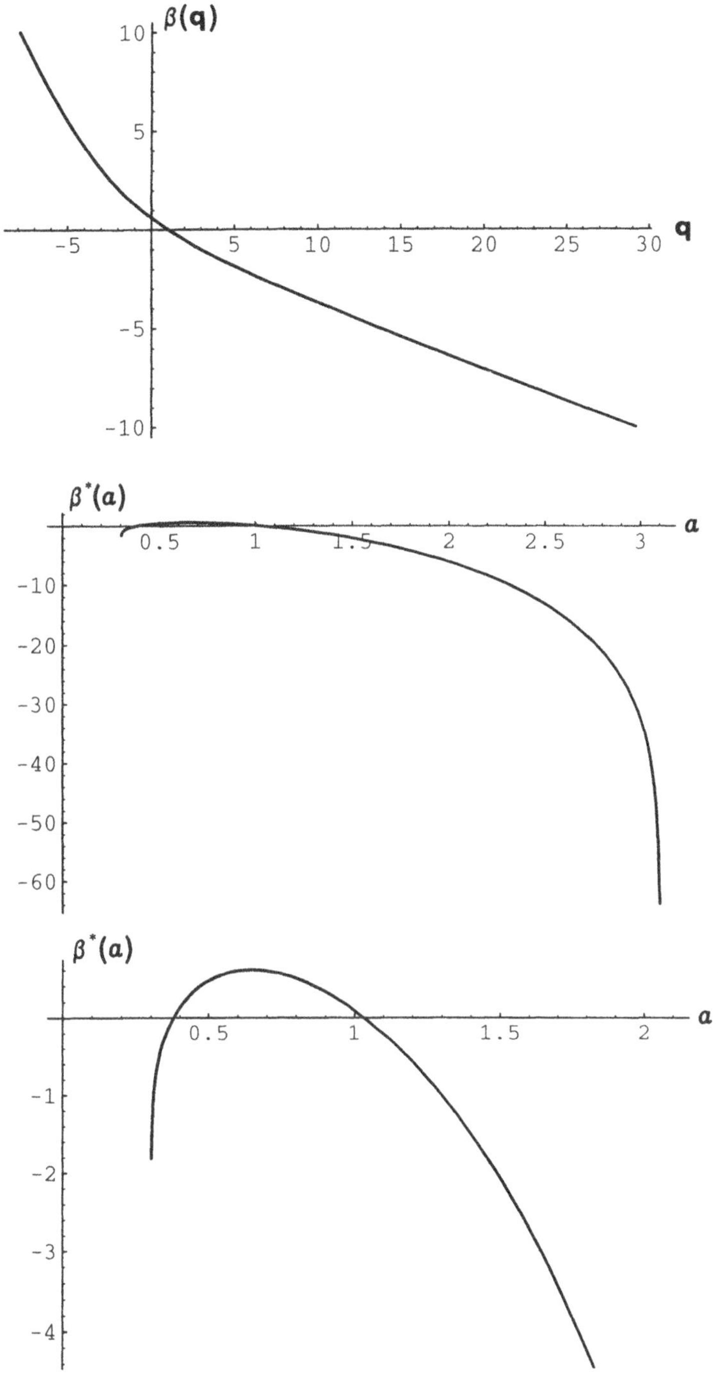

Figure 3.4.2: The graphs of β and β^* for $\delta = c = \dfrac{1}{10}$. We have:

$$\underline{a} = \frac{\log 1/2}{\log 1/10} \approx 0.30103\,, \quad a_{\min} \approx 0.377766$$

$$\overline{a} = \frac{\log 1/2}{\log 4/5} \approx 3.10628\,, \quad a_{\max} \approx 1.03305\,.$$

Write $P := P_* \in \mathcal{P}(\mathcal{P}(X_*)) = \mathcal{P}(\mathcal{P}([0,1]))$. It follows from Theorem 2.6.5 that for each fixed $\alpha \in]a_{\min}, a_{\max}[$ then

$$f_\mu(\alpha) = F_\mu(\alpha) = \beta^*(\alpha) \quad \text{for } P\text{-a.a. } \mu \in \mathcal{P}([0,1])\,.$$

Furthermore, P-a.a. $\mu \in \mathcal{P}([0,1])$ satisfies

$$\Delta_\mu(\alpha) = \varnothing \quad \text{for } \alpha \in [0,\infty[\setminus]a_{\min}, a_{\max}[\,.$$

Clearly

$$\sum_{i=1,2} p_i^q \operatorname{Lip}(S_i)^{\beta(q)} = \left(\tfrac{1}{2}\right)^q \operatorname{Lip}(S_1)^{\beta(q)} + \left(\tfrac{1}{2}\right)^q \operatorname{Lip}(S_2)^{\beta(q)} \quad \lambda\text{-a.s.}\,,$$

whence

$$\left.\begin{aligned} \sum_{i=1,2} p_i^q \operatorname{Lip}(S_i)^{\beta(q)} = 1 \quad \lambda\text{-a.s.} \qquad & \text{for } q = 1 \\ \lambda\left(\sum_{i=1,2} p_i^q \operatorname{Lip}(S_i)^{\beta(q)} \neq 1 \right) > 0 \qquad & \text{for } q \neq 1 \end{aligned}\right\} \tag{3.4.1}$$

Equation (3.4.1), Theorem 2.7.1, Theorem 2.7.2 and Theorem 2.8.1 imply that for each $q \in]q_{\min}, q_{\max}[$ there exists a number c_q with $c_q > 0$ for $q = 1$ and $c_q = 0$ for $q \neq 1$, such that

$$\mathcal{H}_\mu^{q,\beta(q)}(\Delta_\mu(\alpha(q))) = \begin{cases} c_q > 0 & \text{for } P\text{-a.a. } \mu \in \mathcal{P}([0,1]) \text{ for } q = 1\,; \\ c_q = 0 & \text{for } P\text{-a.a. } \mu \in \mathcal{P}([0,1]) \text{ for } q \neq 1\,. \end{cases}$$

3.5 Random Conservative Cascades

Let $N \in \mathbb{N}$ and put $\Sigma = \cup_{n=0}^{\infty}\{1,\ldots,N\}$. Let $X \subset \mathbb{R}^d$ be compact with $\lambda^d(X) > 0$, and $r \in]0,1[$. Write $m = \frac{\lambda^d \lfloor X}{\lambda^d(X)}$, i.e. m is the normalized restriction of the Lebesgue measure λ^d to X. A strongly bounded, separated and conservative (N,r)-cascade (or simply an (N,r)-cascade) is a list

$$\mathcal{W} = \big((S,\mathcal{S},Q),(W_\sigma)_{\sigma\in\Sigma},W,(T_i)_{i=1,\ldots,N}\big)$$

where

1) $(S,\mathcal{S},Q)$ is a probability space;
2) $(W_\sigma)_{\sigma\in\Sigma}$ is an identically distributed family of positive random variables on $(S,\mathcal{S},Q)$, W is a random variable on $(S,\mathcal{S},Q)$ and $W_\sigma \overset{\mathrm{d}}{=} W$ for all $\sigma \in \Sigma$ (where $\overset{\mathrm{d}}{=}$ denotes equality in distribution);
3) For each $i \in \{1,\ldots,N\}$, $T_i : X \to X$ is a similarity with Lipschitz constant equal to r;
4) The strongly bounded axiom. There exists numbers $0 < \Delta \leq T < r^{-d}$ such that
$$\Delta \leq W_\sigma \leq T \quad P\text{-a.s.}$$
for all $\sigma \in \Sigma$;
5) The separation axiom. There exists a positive number $c > 0$ such that
$$\operatorname{dist}(T_i(X),T_j(X)) \geq c$$
for all i,j with $i \neq j$;
6) The conservation (mass preserving) axiom. For each $\sigma \in \Sigma$,
$$\sum_{i=1}^{N} W_{\sigma i} = r^{-d}.$$

Define a sequence $(\mu_n)_n$ of random measures by

$$\begin{aligned}
\mu_1 &= \sum_{i_1=1,\ldots,N} W_{i_1}\; m\lfloor T_{i_1}(X)\\
\mu_2 &= \sum_{i_1,i_2=1,\ldots,N} W_{i_1}W_{i_1i_2}\; m\lfloor T_{i_1}\circ T_{i_2}(X)\\
&\vdots\\
\mu_n &= \sum_{i_1,\ldots,i_n=1,\ldots,N} W_{i_1}W_{i_1i_2}\cdot\ldots\cdot W_{i_1\ldots i_n}\; m\lfloor T_{i_1}\circ T_{i_2}\circ\cdots\circ T_{i_n}(X)\\
&\vdots
\end{aligned} \tag{3.5.1}$$

The measure μ_n is thus supported by $\cup_{i_1,\ldots,i_n=1,\ldots,N} T_{i_1} \circ \cdots \circ T_{i_n}(X)$; and for each list $i_1, \ldots, i_n \in \{1, \ldots, N\}$, μ_n has the constant density $W_{i_1} W_{i_1 i_2} \cdot \ldots \cdot W_{i_1 \ldots i_n}$ w.r.t. m on the set $T_{i_1} \circ \cdots \circ T_{i_n}(X)$.

We have the following result.

Theorem 3.5.1. *Let $\mathcal{W} = ((S, \mathcal{S}, Q), (W_\sigma)_{\sigma \in \Sigma}, W, (T_i)_{i=1,\ldots,N})$ be an (N, r)-cascade. Then the sequence $(\mu_n)_n$ in (3.5.1) converges weakly Q-a.s. to a random measure*

$$\mu : S \to \mathcal{P}(X).$$

Define the function $\tau : \mathbb{R} \to \mathbb{R}$ by

$$\tau(q) = -\frac{\log N}{\log r} - dq - \frac{\log \mathbf{E}_Q[W^q]}{\log r}.$$

Then the following statements hold.

i) *Assume $\alpha \geq 0$ with $\tau^*(\alpha) > 0$. For Q-a.a. $s \in S$ then*

$$f_{\mu(s)}(\alpha) = F_{\mu(s)}(\alpha) = \tau^*(\alpha).$$

ii) *For Q-a.a. $s \in S$ then*

$$\Delta_{\mu(s)}(\alpha) = \varnothing \quad \text{for } \alpha \in \{a \geq 0 \mid \tau^*(a) < 0\}.$$

We will make a few comments before proving Theorem 3.5.1.

We first remark that related results on random cascades in a somewhat different setting have been obtained by Holley & Waymire [Ho]. The three most important distinctions between our setting and the setting in Holley & Waymire are given below.

1) Holley & Waymire [Ho] assume that $(W_\sigma)_{\sigma \in \Sigma}$ is an i.i.d. family. The conservation condition 6) implies that $(W_\sigma)_{\sigma \in \Sigma}$ never constitute an independent family for the cascades which we are considering.
2) Holley & Waymire [Ho] are using a different conservation condition, viz.

$$\mathbf{E}_Q[W] = 1$$

instead of 6).
3) Holley & Waymire [Ho] are working in the unit interval $[0, 1]$ and assume the open set condition (rather than the stronger separation condition stated in 5).) Specifically Holley & Waymire define $T_i : [0, 1] \to [0, 1]$ by

$$T_i(x) := \tfrac{1}{N}x + \tfrac{i-1}{N}$$

for $i = 1, \ldots, N$. In this case $T_i(]0, 1[) \cap T_j(]0, 1[) = \varnothing$ for all i, j with $i \neq j$, and the open set condition (1.1.1) is therefore satisfied.

Because of the open set condition Holley & Waymire [Ho] have to change the definition of the local dimension in order to obtain satisfactory results, cf. also the discussion preceding Theorem 2.6.6. Holley & Waymire thus define the upper and lower local dimensions $\overline{\theta}_\mu(x)$ resp. $\underline{\theta}_\mu(x)$ of a measure $\mu \in \mathcal{P}([0,1])$ at a point $x \in [0,1]$ by

$$\overline{\theta}_\mu(x) = \limsup_n \frac{\log \mu I_n(x)}{\log N^{-n}} = \limsup_n \frac{\log \mu I_n(x)}{-n \log N}$$

and

$$\underline{\theta}_\mu(x) = \liminf_n \frac{\log \mu I_n(x)}{\log N^{-n}} = \liminf_n \frac{\log \mu I_n(x)}{-n \log N},$$

where $I_n(x)$ denotes the unique half-closed N-adic interval of order n that contains x, i.e. $I_n(x)$ is the unique interval of the form $[\frac{k}{N^n}, \frac{k+1}{N^n}[$ with $k \in \{0, \ldots, N^n - 1\}$ that contains x. However, we believe that the local dimensions $\overline{\alpha}_\mu$ and $\underline{\alpha}_\mu$ introduced in (1.0.1) and (1.0.2), and hence also the corresponding multifractal spectra functions f_μ and F_μ, are more natural than $\overline{\theta}_\mu$ and $\underline{\theta}_\mu$.

The theory of random cascades was introduced by Mandelbrot [Ma1,Ma2,Ma3] in 1974 as a model for turbulence. The random cascades introduced in [Ma1,Ma2,Ma3] are conservative in the sense of condition 6). A rigorous theory of random cascades was subsequently developed in a series of papers by Kahane and Peyrière [Kah1,Kah2,Kah3,Kah4,Kah5,Kah6,Kah7,Pey1,Pey2] (under assumptions that are slightly different from axiom 1) through axiom 6)) using the notion of so-called T-martingales. The papers [Kah1,Kah3,Kah4,Kah7,Pey1,Pey2] give necessary and sufficient conditions guaranteeing that the limit measure μ is non-degenerated (i.e. μ is not almost surely equal to the zero-measure), and that the random variable $Z := \mu(X)$ has finite moment of order h for $h > 0$. Finally [Kah1,Kah3,Kah4,Kah7, Pey1,Pey2] determine the almost sure Hausdorff dimension $\dim \mu$ of μ. We further remark that the papers [Kah1,Kah2,Kah3,Kah4,Kah5,Kah6,Kah7,Pey1,Pey2] do not investigate the multifractal structure of μ. However, the multifractal spectrum of the some strongly bounded random cascades has (as mentioned above) recently been investigated by Holley & Waymire [Ho], whereas Waymire & Williams [Way1,Way2] study the dimension distribution

$$F(t) := \sup\{\mu(B) \mid B \text{ is Borel, } \dim B \leq t\}, \quad 0 < t < \infty$$

(rather than the multifractal spectra functions) of some somewhat more general random cascade measures μ.

We will now prove the assertions in Theorem 3.5.1. The (N, r)-cascade $\mathcal{W}$ can be modeled as a random graph directed self-similar measure corresponding to a graph

$(V,E) = (\{*\},\{1,\dots,N\})$ with one vertex $*$ and N edges $1,\dots,N$ as follows. Let $X_* := X$ and define $\lambda := \lambda_* \in \mathcal{P}(\Xi_*)$ by

$$\lambda = \delta_{(T_1,\dots,T_N)} \times \nu$$

where

$$\nu = Q \circ (r^d W_1, \dots, r^d W_N)^{-1}.$$

Write $\Omega := \Omega_*$ and $\Lambda := \Lambda_*$, where (as usual)

$$\Omega_* = \prod_{\Sigma} \Xi_*,$$
$$\Lambda_* = \prod_{\Sigma} \lambda_*.$$

Also, μ_* denotes the random graph directed self-similar measure corresponding to the graph $(\{*\},\{1,\dots,N\})$ and the measure λ_*. Finally, write $P := P_*$ (where, as usual, $P_* := \Lambda_* \circ \mu_*^{-1} \in \mathcal{P}(\mathcal{P}(X_*))$.) The auxiliary function β is given by

$$\begin{aligned}
1 &= \mathbf{E}_\lambda \left[\sum_{i=1}^{N} p_i^q \operatorname{Lip}(S_i)^{\beta(q)} \right] \\
&= \sum_{i=1}^{N} \int p_i^q \, d\nu \int \operatorname{Lip}(S_i)^{\beta(q)} \, d\delta_{(T_1,\dots,T_N)} \\
&= r^{\beta(q)} \sum_{i=1}^{N} \int p_i^q \, d(Q \circ (r^d W_1, \dots, r^d W_N)^{-1}) \\
&= r^{\beta(q)} \sum_{i=1}^{N} r^{dq} \int W_i^q \, dQ \\
&= r^{\beta(q)} N r^{dq} \mathbf{E}_Q[W^q],
\end{aligned}$$

whence

$$\beta = \tau. \tag{3.5.2}$$

Let

$$\gamma_{i_1 \dots i_n} = \frac{m\lfloor (T_{i_1} \circ \cdots \circ T_{i_n}(X))}{r^{dn}} \in \mathcal{P}(X)$$

for $n \in \mathbb{N}$ and $i_1, \dots, i_n \in \{1, \dots, N\}$. Write $\boldsymbol{\gamma} = (\gamma_{i_1 \dots i_n})_{i_1,\dots,i_n = 1,\dots,N; n \in \mathbb{N}}$. Using the notation in section 2.3 we have

$$\mu_{*,n}^{\gamma} = \sum_{1_1,\dots,i_n=1,\dots,N} p_{i_1} p_{i_1 i_2} \cdot \dots \cdot p_{i_1 \dots i_n} \frac{m\lfloor (T_{i_1} \circ \cdots \circ T_{i_n}(X))}{r^{dn}} \circ (S_{i_1} \circ S_{i_1 i_2} \circ \dots \circ S_{i_1 \dots i_n})^{-1}$$

for $n \in \mathbb{N}$. Next define $\Phi : S \to \Omega$ by

$$\Phi(s) = \big((T_1, \ldots, T_N), (r^d W_{\sigma 1}(s), \ldots, r^d W_{\sigma N}(s))\big)_{\sigma \in \Sigma}.$$

It is easily seen that the diagram below commutes,

$$\begin{array}{ccc} S & \xrightarrow{\mu_n} & \mathcal{P}(X) \\ {\scriptstyle \Phi}\downarrow & & \| \\ \Omega & \xrightarrow[\mu^{\gamma}_{*,n}]{} & \mathcal{P}(X) \end{array} \tag{3.5.3}$$

and that

$$Q \circ \Phi^{-1} = \Lambda. \tag{3.5.4}$$

Diagram (3.5.3), Theorem 2.3.1 and equation (3.5.4) imply that

$$\mu_n(s) = \mu^{\gamma}_{*,n}(\Phi(s)) \xrightarrow{\text{w}} \mu_*(\Phi(s))$$

(where $\xrightarrow{\text{w}}$ denotes weak convergence) for Q-a.a. $s \in S$, i.e. the weak limit $\mu = \lim_n \mu_n$ exists Q-a.s., and

$$\mu := \lim_n \mu_n = \mu_* \circ \Phi \quad Q\text{-a.s.} \tag{3.5.5}$$

The results in Theorem 3.5.1 follow immediately from (3.5.2), (3.5.4), (3.5.5) and Theorem 2.6.5.

We will now compute $\underline{a}, \overline{a}$ and $\underline{e}, \overline{e}$. We have

$$\begin{aligned} \underline{a} &= \inf_{i=1,\ldots,N} \left\| \frac{\log p_i}{\log(\operatorname{Lip} S_i)} \right\|_{-\infty} \\ &= \inf_{i=1,\ldots,N} \left\| \frac{\log(r^d W_i)}{\log r} \right\|_{-\infty} \\ &= d + \frac{\log(\|W\|_\infty)}{\log r} \end{aligned}$$

and

$$\begin{aligned} \overline{a} &= \sup_{i=1,\ldots,N} \left\| \frac{\log p_i}{\log(\operatorname{Lip} S_i)} \right\|_{\infty} \\ &= \sup_{i=1,\ldots,N} \left\| \frac{\log(r^d W_i)}{\log r} \right\|_{\infty} \end{aligned}$$

$$= d + \frac{\log(\|W\|_{-\infty})}{\log r}.$$

Hence

$$\underline{a}q + \beta(q) = -\frac{1}{\log r} \log \left(N \frac{\mathbf{E}_Q[W^q]}{\|W\|_\infty^q} \right),$$

$$\overline{a}q + \beta(q) = -\frac{1}{\log r} \log \left(N \frac{\mathbf{E}_Q[W^q]}{\|W\|_{-\infty}^q} \right).$$

A straightforward calculation shows that

$$\lim_{q \to \infty} \frac{\mathbf{E}_Q[W^q]}{\|W\|_\infty^q} = Q(W = \|W\|_\infty), \quad \lim_{q \to -\infty} \frac{\mathbf{E}_Q[W^q]}{\|W\|_{-\infty}^q} = Q(W = \|W\|_{-\infty})$$

whence

$$\underline{e} = \lim_{q \to \infty} (\underline{a}q + \beta(q)) = -\frac{\log(Q(W = \|W\|_\infty)N)}{\log r}$$

$$\overline{e} = \lim_{q \to -\infty} (\overline{a}q + \beta(q)) = -\frac{\log(Q(W = \|W\|_{-\infty})N)}{\log r}$$

where we put $\log 0 = -\infty$.

3.6 Random Dubins-Freedman related Distribution Functions

This example studies some random distribution functions similar to those introduced by Dubins & Freedman [Du1,Du2] in 1963. We obtain results analogous to the results in Kinney & Pitcher [Kin].

Let $C^0([0,1])$ denote the family of continuous real valued functions on $[0,1]$. Write

$$D := \{F \in C^0([0,1]) \mid \text{F is increasing}, \ F(0) = 0, \ F(1) = 1\}$$

i.e. D is the family of continuous distribution functions F with $F(0) = 0$ and $F(1) = 1$. Let $\delta, c > 0$ with $\delta + c < 1 - \delta$. Put

$$T := [\delta, 1 - \delta - c] \times [\delta, 1 - \delta] \subseteq [0,1]^2$$

and let $\mu \in \mathcal{P}([0,1]^2)$ with $\mu(T) = 1$. We first define a sequence $(F_n)_n$ of random distribution functions in D satisfying the following two conditions:

1) There exist points $x_{n,1}, \dots, x_{n,2^n} \in [0,1]$ with

$$0 = x_{n,1} < x_{n,2} < \cdots < x_{n,2^n-1} < x_{n,2^n} = 1$$

such that $F_n(x_{n,2k}) = F_n(x_{n,2k+1})$ for $k = 1, \dots, 2^{n-1} - 1$, and the restriction of F_n to $[x_{n,k}, x_{n,k+1}]$ is affine for all k;

2) $F_n \in D$.

We will construct the sequence $(F_n)_n$ be induction on n.

The start of the induction. Choose a point $(u_{1,1}, v_{1,1})$ in T according to the measure μ, and let

$$x_{1,1} := 0, \ x_{1,2} := u_{1,1}, \ x_{1,3} := u_{1,1} + c, \ x_{1,4} := 1.$$

Now define $F_1 : [0,1] \to [0,1]$ to be the unique piecewise affine function in D such that:

$$F_1(x_{1,1}) := 0, \ F_1(x_{1,1}) := v_{1,1}, \ F_1(x_{1,3}) := v_{1,1}, \ F_1(x_{1,4}) := 1,$$

the restriction of F_{n+1} to $[x_{n+1,k}, x_{n+1,k+1}]$ is affine for all k.

The inductive step. Assume that the function F_n has been constructed. Next choose 2^n points $(u_{n+1,1}, v_{n+1,1}), \dots, (u_{n+1,2^n}, v_{n+1,2^n})$ in T independently and according to μ. Let

$$\left.\begin{aligned} x_{n+1,2k} &:= x_{n,k} \\ x_{n+1,2k+1} &:= x_{n,k+1} \end{aligned}\right\} \qquad \text{for } k \text{ even}$$

$$\left.\begin{aligned} x_{n+1,2k} &:= (x_{n,k+1} - x_{n,k})u_{n+1,k} + x_{n,k} \\ x_{n+1,2k+1} &:= (x_{n,k+1} - x_{n,k})(u_{n+1,k} + c) + x_{n,k} \end{aligned}\right\} \qquad \text{for } k \text{ odd}$$

and define F_{n+1} to be the unique piecewise affine function in D such that:

$$\left.\begin{aligned} F_{n+1}(x_{n+1,2k}) &:= F_n(x_{n,k}) \\ F_{n+1}(x_{n+1,2k+1}) &:= F_n(x_{n,k+1}) \end{aligned}\right\} \qquad \text{for } k \text{ even}$$

$$\left.\begin{aligned} F_{n+1}(x_{n+1,2k}) &= (F_n(x_{n,k+1}) - F_n(x_{n,k}))v_{n+1,k} + F_n(x_{n,k}) \\ &= F_{n+1}(x_{n+1,2k+1}) \end{aligned}\right\} \qquad \text{for } k \text{ odd}$$

the restriction of F_{n+1} to $[x_{n+1,k}, x_{n+1,k+1}]$ is affine for all k.

This completes the construction of the sequence $(F_n)_n$.

The random functions F_n can be modeled as the distribution functions of a sequence $(\mu^{\gamma}_{*,n})_n$ of random measures converging weakly to a graph directed self-similar measure μ_* corresponding to a graph $(V, E) = (\{*\}, \{1, 2\})$ with one vertex $*$ and two edges 1 and 2 as follows. Let $X_* := [0,1]$ and define a map

$$\Phi : [0,1]^2 \to \Xi_* = \mathrm{Con}([0,1])^2 \times \{(p_1, p_2) \in [0,1]^2 \mid p_1 + p_2 = 1\}$$

by

$$\Phi(x,y) = ((S_{1,(x,y)}, S_{2,(x,y)}), (y, 1-y))$$

where

$$\begin{aligned} S_{1,(x,y)}(t) &= xt\,, \\ S_{2,(x,y)}(t) &= (1-x-c)t + (x+c)\,. \end{aligned}$$

Next define $\lambda_* \in \mathcal{P}(\Xi_*)$ by

$$\lambda_* := (\lambda^2 \lfloor [0,1]^2) \circ \Phi^{-1}\,.$$

As usual we write

$$\Omega_* = \prod_{E_*^{(*)}} \Xi_*\,,$$

$$\Lambda_* = \prod_{E_*^{(*)}} \lambda_*$$

where $E_*^{(*)} = \cup_{n=0}^{\infty} \{1,2\}^n$. Let $C_* : \Omega_* \to \mathcal{C}([0,1])$ denote the graph directed self similar set associated with (V, E) and λ_*. For $\sigma \in E_*^{(*)}$ write $\gamma_\sigma := \lambda^1 \lfloor [0,1] \in \mathcal{P}([0,1])$ and put $\gamma := (\gamma_\sigma)_{\gamma \in E_*^{(*)}}$. Using notation as in section 2.3 we have

$$\mu^{\gamma}_{*,n} = \sum_{\sigma \in E_*^{(n)}} p_{\gamma|1} \cdot \ldots \cdot p_{\gamma|n} \lambda^1 \circ (S_{\sigma|1} \circ \cdots \circ S_{\sigma|n})^{-1}\,.$$

It follows from Theorem 2.3.1 that the weak limit

$$\mu_* = \lim_n \mu^{\gamma}_{*,n} \tag{3.6.1}$$

exists Λ_*-a.s.. Let F be the (random) distribution function of μ_*. It is easily seen that

$$F_n(t) = \mu^{\gamma}_{*,n}([0,t]) \quad \text{for } t \in [0,1]$$

i.e.:

$$F_n \text{ is the distribution function of } \mu^{\gamma}_{*,n}\,.$$

Equation (3.6.1) therefore implies that Λ_*-a.s.,

$$F_n(t) \to F(t)$$

for every continuity point t of F. However, Lemma 4.3.4 implies that Λ_*-a.s., $\mu_*(\{x\}) = 0$ for all $x \in [0,1]$. The distribution function F is therefore continuous Λ_*-a.s., whence

$$F_n \to F \quad \text{pointwise } \Lambda_*\text{-a.s.} \tag{3.6.2}$$

For each $x \in [0,1[$ and $n \in \mathbb{N}$ there exists a unique $k \in \{1,\ldots,2^n-1\}$ such that $x \in [x_{n,k}, x_{n,k+1}[$; now write

$$I(n,x) := [x_{n,k}, x_{n,k+1}[\,.$$

For each $x \in [0,1[$ and $n \in \mathbb{N}$ there exists a unique $l \in \{1,\ldots,2^{n-1}-1\}$ such that $x \in [v_{n,l}, v_{n,l+1}[$; now write

$$J(n,x) := [v_{n,l}, v_{n,l+1}[\,.$$

Define for $t \in [0,1[$ the function $h_t : [0,1]^2 \to \mathbb{R}$ by $h_t(x,y) := y\log x + (1-y)\log(1-x-t)$ (we put $0\log 0 = 0$), and let $\Upsilon : [0,1]^2 \to [0,1]^2$ be defined by $\Upsilon(x,y) = (y,y)$. Then the following results hold.

Theorem 3.6.1.

i) *Λ_*-a.s. then*

$$\lim_n \frac{\log \operatorname{diam} I(n,x)}{n} = \mathbf{E}_\mu[h_c] \quad \textit{for } \mu_*\textit{-a.a. } x \in [0,1[.$$

ii) *Λ_*-a.s. then*

$$\lim_n \frac{\log \operatorname{diam} J(n,x)}{n} = \mathbf{E}_\mu[h_0(\Upsilon)] \quad \textit{for } \mu_*\textit{-a.a. } x \in [0,1[.$$

iii) *Λ_*-a.s. then*

$$\dim \mu_* = \operatorname{Dim} \mu_* = \frac{\mathbf{E}_\mu[h_c]}{\mathbf{E}_\mu[h_0(\Upsilon)]}\,.$$

We note that the random distribution functions F_n for the case $\delta = c = 0$ were introduced by Dubins & Freedman [Du1,Du2] in 1963, and that results similar to Theorem 3.6.1 (except for the assertion involving $\operatorname{Dim}\mu_*$) were proved for the case $c = \delta = 0$ by Kinney & Pitcher [Kin] in 1964. A detailed discussion of the pointwise limit $\lim_n F_n$ of the random distributions F_n for the case $\delta = c = 0$ can be found in [Gra3].

We will now prove Theorem 3.6.1. First define the subset L of $[0,1]$ by $L := (\cup_n\{x_{n,1},\dots,x_{n,2^n}\}) \cup (\cup_n\{v_{n,1},\dots,v_{n,2^{n-1}}\})$. Observe that

$$\mu_*(L) = 0 \quad \Lambda_*\text{-a.s.} \tag{3.6.3}$$

by Lemma 4.3.4 since L is countable. Next observe that

$$\operatorname{diam} I(n,x) = \operatorname{Lip}(S_{\sigma|1}(\omega)\circ\cdots\circ S_{\sigma|n}(\omega)) = \prod_{k=1}^{n}\operatorname{Lip}(S_{\sigma|k}(\omega))$$

$$\operatorname{diam} J(n,x) = \prod_{k=1}^{n} p_{\sigma|k}(\omega)$$

for $\omega \in \Omega_*$, $\sigma \in E_*^{\mathbb{N}}$ and $x := \pi_*(\omega)(\sigma) \in C_*(\omega)\setminus L$ (where, as usual, $\pi_* : \Omega_* \to [0,1]$ denotes the projection defined in section 2.2.) Lemma 6.3.3 therefore implies that the following assertions hold for Λ_*-a.a. $\omega \in \Omega_*$:

$$\begin{aligned}\lim_n \frac{\log\operatorname{diam} I(n,x)}{n} &= \sum_i \int \left(p_i^1 \operatorname{Lip}(S_i)^{\beta(1)} \log(\operatorname{Lip}(S_i))\right) d\lambda_* \\ &= \int \left(v\log u + (1-v)\log(1-u-c)\right) d\mu(u,v) \\ &\quad \text{for } \mathcal{M}_{*,1}(\omega)\text{-a.a. } x \in C_*(\omega)\setminus L \end{aligned} \tag{3.6.4}$$

$$\begin{aligned}\lim_n \frac{\log\operatorname{diam} J(n,x)}{n} &= \sum_i \int \left(p_i^1 \operatorname{Lip}(S_i)^{\beta(1)} \log p_i\right) d\lambda_* \\ &= \int \left(v\log v + (1-v)\log(1-v)\right) d\mu(u,v) \\ &\quad \text{for } \mathcal{M}_{*,1}(\omega)\text{-a.a. } x \in C_*(\omega)\setminus L \end{aligned} \tag{3.6.5}$$

where the random measure $\mathcal{M}_{*,1} : \Omega_* \to \mathcal{P}([0,1])$ is defined in section 6.1. However, it follows from (6.1.7) that $\mathcal{M}_{*,1}$ and μ_* are equivalent Λ_*-a.s. (in the sense that $\mathcal{M}_{*,1} \ll \mu_*$ Λ_*-a.s. and $\mu_* \ll \mathcal{M}_{*,1}$ Λ_*-a.s.), and the assertions in Theorem 3.6.1 i) and ii) therefore follow from (3.6.3), (3.6.4) and (3.6.5).

Finally, it follows from Theorem 2.6.7 that

$$\dim\mu_* = \operatorname{Dim}\mu_* = \alpha(1) = -\beta'(1) \quad \Lambda_*\text{-a.s.} \tag{3.6.6}$$

where (by (4.4.8), cf. also the discussion leading to (2.6.5))

$$
\begin{aligned}
-\beta'(1) &= \frac{\sum_i \int \left(p_i^1 \operatorname{Lip}(S_i)^{\beta(1)} \log p_i\right) \, d\lambda_*}{\sum_i \int \left(p_i^1 \operatorname{Lip}(S_i)^{\beta(1)} \log(\operatorname{Lip}(S_i))\right) \, d\lambda_*} \\
&= \frac{\int \left(v \log u + (1-v) \log(1-u-c)\right) d\mu(u,v)}{\int \left(v \log v + (1-v) \log(1-v)\right) d\mu(u,v)} \\
&= \frac{\mathbf{E}_\mu[h_c]}{\mathbf{E}_\mu[h_0(\Upsilon)]}
\end{aligned}
$$

which together with (3.6.6) give the desired result.

Chapter 4
Proofs of Auxiliary Results

4.1 Proofs of the Theorems in Section 2.2

Lemma 4.1.1. *Let $\omega \in (\Omega_u)_0$ and $\varepsilon > 0$. Then there exists an integer $N \in \mathbb{N}$ such that*

$$\sup_{\alpha \in E_u^{(n)}} \prod_{k=1}^{n} \mathrm{Lip}(S_{\alpha|k}(\omega)) \le \varepsilon \quad \text{for } n \ge N$$

Proof. The set

$$G_n := \left\{ \sigma \in E_u^{\mathbb{N}} \,\middle|\, \prod_{k=1}^{n} \mathrm{Lip}(S_{\alpha|k}(\omega)) \le \varepsilon \right\}, \quad n \in \mathbb{N}$$

is clearly open in $E_u^{\mathbb{N}}$ (when $E_u^{\mathbb{N}}$ is equipped with the trace topology inherited from the product topology on $E^{\mathbb{N}}$ induced by the discrete topology on E.) Since $\omega \in (\Omega_u)_0$, $(G_n)_n$ is an open cover of $E_u^{\mathbb{N}}$ which clearly satisfies $G_1 \subseteq G_2 \subseteq \ldots$. Now, $E_u^{\mathbb{N}}$ is easily seen to be closed and thus compact. We can therefore choose N such that $E_u^{\mathbb{N}} = G_N$. This proves the lemma. □.

Proof of Theorem 2.2.1

In order to simplify notation write $S_\gamma(\omega_u) = S_\gamma^{(u)}$ for $\gamma \in E_u^{(*)}$. We also write $\mathbf{K}_u = \left(K_\varnothing^{(u)}, (K_\alpha^{(u)})_{\alpha \in E_u^* \setminus \{\varnothing\}}\right)$.

i) We first prove that the sequence

$$(C_{u,n}^{\mathbf{K}_u}(\omega))_n$$

is Cauchy in $(\mathcal{C}(X_u), D_{X_u})$. Let $\varepsilon > 0$. It follows from Lemma 4.1.1 that there exists an integer N such that

$$\sup_{\alpha \in E_u^{(n)}} \prod_{k=1}^{n} \mathrm{Lip}(S_{\alpha|k}^{(u)}) \le \varepsilon / (\max_v (\mathrm{diam}\, X_v) + 1)$$

for $n \geq N$. Now let $n \geq N$ and $m \in \mathbb{N}$. For each $\alpha \in E_u^{(n)}$ choose any path $\gamma(\alpha) \in E_{\tau(\alpha_{|\alpha|})}^{(m)}$. Then

$$\begin{aligned}
&\sup_{x \in C_{u,n}^{\mathbf{K}_u}(\omega_u)} \operatorname{dist}(x, C_{u,n+m}^{\mathbf{K}_u}(\omega_u)) \\
&= \sup_{x \in \bigcup_{\alpha \in E_u^{(n)}} S_{\alpha|1}^{(u)} \circ \cdots \circ S_{\alpha|n}^{(u)}(K_\alpha^{(u)})} \operatorname{dist}\Big(x, \bigcup_{\beta \in E_u^{(n)}} \bigcup_{\gamma \in E_{\tau(\beta_{|\beta|})}^{(m)}} S_{\beta|1}^{(u)} \circ \cdots \circ S_{\beta|n}^{(u)} \circ \\
&\qquad S_{\beta(\gamma|1)}^{(u)} \circ \cdots \circ S_{\beta\gamma}^{(u)}(K_{\beta\gamma}^{(u)})\Big) \\
&= \sup_{\alpha \in E_u^{(n)}} \sup_{x \in S_{\alpha|1}^{(u)} \circ \cdots \circ S_{\alpha|n}^{(u)}(K_\alpha^{(u)})} \operatorname{dist}\Big(x, \bigcup_{\beta \in E_u^{(n)}} \bigcup_{\gamma \in E_{\tau(\beta_{|\beta|})}^{(m)}} S_{\beta|1}^{(u)} \circ \cdots \circ S_{\beta|n}^{(u)} \circ \\
&\qquad S_{\beta(\gamma|1)}^{(u)} \circ \cdots \circ S_{\beta\gamma}^{(u)}(K_{\beta\gamma}^{(u)})\Big) \\
&\leq \sup_{\alpha \in E_u^{(n)}} \sup_{x \in S_{\alpha|1}^{(u)} \circ \cdots \circ S_{\alpha|n}^{(u)}(K_\alpha^{(u)})} \operatorname{dist}\Big(x, S_{\alpha|1}^{(u)} \circ \cdots \circ S_{\alpha|n}^{(u)} \circ \\
&\qquad S_{\alpha(\gamma(\alpha)|1)}^{(u)} \circ \cdots \circ S_{\alpha\gamma(\alpha)}^{(u)}(K_{\alpha\gamma(\alpha)}^{(u)})\Big) \\
&\leq \sup_{\alpha \in E_u^{(n)}} \sup_{x \in S_{\alpha|1}^{(u)} \circ \cdots \circ S_{\alpha|n}^{(u)}(K_\alpha^{(u)})} \operatorname{diam}\big(S_{\alpha|1}^{(u)} \circ \cdots \circ S_{\alpha|n}^{(u)}(K_\alpha^{(u)})\big) \\
&= \sup_{\alpha \in E_u^{(n)}} \left(\prod_{k=1}^{n} \operatorname{Lip}(S_{\alpha|k}^{(u)})\right) \max_v(\operatorname{diam} X_v) \leq \varepsilon .
\end{aligned} \tag{4.1.1}$$

It follows by a similar argument that

$$\sup_{x \in C_{u,n+m}^{\mathbf{K}_u}(\omega_u)} \operatorname{dist}(x, C_{u,n}^{\mathbf{K}_u}(\omega_u)) \leq \varepsilon . \tag{4.1.2}$$

Now inequalities (4.1.1) and (4.1.2) imply that

$$D_{X_u}\big(C_{u,n}^{\mathbf{K}_u}(\omega_u), C_{u,n+m}^{\mathbf{K}_u}(\omega_u)\big) \leq \varepsilon$$

for all $n \geq N$ and $m \in \mathbb{N}$. This proves that the sequence $(C_{u,n}^{\mathbf{K}_u}(\omega))_n$ is Cauchy and thus convergent.

ii) Proof of $\lim_n C_{u,n}^{\mathbf{K}_u}(\omega_u) \subseteq C_u(\omega_u)$: Clearly

$$\begin{aligned}
\forall m \in \mathbb{N} : \forall n \geq m : \bigcup_{\alpha \in E_u^{(n)}} S_{\alpha|1}^{(u)} \circ \cdots \circ S_{\alpha|n}^{(u)}(K_\alpha^{(u)}) \subseteq \\
\bigcup_{\beta \in E_u^{(m)}} \overline{S_{\beta|1}^{(u)} \circ \cdots \circ S_{\beta|m}^{(u)}(X_{\tau(\beta_m)})}
\end{aligned}$$

whence

$$\forall m \in \mathbb{N} : \lim_n C^{\mathbf{K}_u}_{u,n}(\omega_u) \subseteq \bigcup_{\beta \in E_u^{(m)}} \overline{S^{(u)}_{\beta|1} \circ \cdots \circ S^{(u)}_{\beta|m}(X_{\tau(\beta_m)})}\,.$$

Thus

$$\lim_n C^{\mathbf{K}_u}_{u,n}(\omega_u) \subseteq \bigcap_m \bigcup_{\beta \in E_u^{(m)}} \overline{S^{(u)}_{\beta|1} \circ \cdots \circ S^{(u)}_{\beta|m}(X_{\tau(\beta_m)})} = C_u(\omega_u)\,.$$

Proof of $C_u(\omega_u) \subseteq \lim_n C_{K_u,u,n}(\omega_u)$: Assume, in order to get a contradiction, that

$$x \in C_u(\omega_u) \setminus \lim_n C^{\mathbf{K}_u}_{u,n}(\omega_u)\,.$$

Put $\delta = \operatorname{dist}\left(x, \lim_n C^{\mathbf{K}_u}_{u,n}(\omega_u)\right)$. Observe that $\delta > 0$ since $\lim_n C^{\mathbf{K}_u}_{u,n}(\omega_u)$ is compact. An easy compactness argument shows that there exists a $\sigma \in E_u^{\mathbb{N}}$ satisfying

$$x \in S^{(u)}_{\sigma|1} \circ \cdots \circ S^{(u)}_{\sigma|m}(X_{\tau(\sigma_m)})$$

for all $m \in \mathbb{N}$. Since $\omega_u \in (\Omega_u)_0$ there exists an integer $M \in \mathbb{N}$ such that

$$\begin{aligned}
\frac{1}{2} \operatorname{dist}\left(x, \lim_n C^{\mathbf{K}_u}_{u,n}(\omega_u)\right) &> \left(\prod_{k=1}^{m} \operatorname{Lip}(S^{(u)}_{\sigma|k})\right) \max_v (\operatorname{diam} X_v) \\
&\geq \operatorname{diam}\left(S^{(u)}_{\sigma|1} \circ \cdots \circ S^{(u)}_{\sigma|m}(X_{\tau(\sigma_m)})\right) \qquad \text{for} \quad m \geq M\,.
\end{aligned}$$

Hence

$$S^{(u)}_{\sigma|1} \circ \cdots \circ S^{(u)}_{\sigma|m}(X_{\tau(\sigma_m)}) \subseteq B\left(x, \frac{1}{2} \operatorname{dist}\left(x, \lim_n C^{\mathbf{K}_u}_{u,n}(\omega_u)\right)\right) \qquad \text{for} \quad m \geq M$$

and so

$$\begin{aligned}
& D_{X_u}\left(\lim_n C^{\mathbf{K}_u}_{u,n}(\omega_u), C^{\mathbf{K}_u}_{u,m}(\omega_u)\right) \\
=& D_{X_u}\left(\lim_n C^{\mathbf{K}_u}_{u,n}(\omega_u), \bigcup_{\alpha \in E_u^{(m)}} S_{\alpha|1} \circ \cdots \circ S_{\alpha|m}(K^{(u)}_\alpha)\right) \\
\geq& \frac{1}{2} \operatorname{dist}\left(x, \lim_n C^{\mathbf{K}_u}_{u,n}(\omega_u)\right) = \frac{1}{2}\delta \qquad \text{for} \quad m \geq M
\end{aligned}$$

which is a contradiction. □

4.2 Proofs of the Theorems in Section 2.3

For a compact metric space X define $L_X : \mathcal{P}(X) \times \mathcal{P}(X) \to \mathbb{R}_+$ by

$$L_X(\mu,\nu) = \sup_{f \in \mathrm{Lip}_1(X)} \left| \int f\, d\mu - \int f\, d\nu \right|$$

where $\mathrm{Lip}_1(X)$ denotes the class of Lipschitz functions $f : X \to \mathbb{R}$ with $\mathrm{Lip}(f) \leq 1$. It follows from Hutchinson [Hu] that L_X is a complete metric on $\mathcal{P}(X)$ and that L_X induces the weak topology on $\mathcal{P}(X)$.

Proof of Theorem 2.3.1

i) We will now prove that the sequence

$$(\mu_{u,n}^{\gamma_u}(\omega_u))_n \tag{4.2.1}$$

is Cauchy w.r.t. L_{X_u}. Let $\gamma_u = \left(\gamma_\varnothing^{(u)}, (\gamma_\alpha^{(u)})_{\alpha \in E_u^{(*)} \setminus \{\varnothing\}}\right)$. In order to simplify notation write $S_\gamma(\omega_u) = S_\gamma^{(u)}$ and $p_\gamma(\omega_u) = p_\gamma^{(u)}$ for $\gamma \in E_u^{(*)}$. Let $\varepsilon > 0$. By Lemma 4.1.1 there exists an integer N such that

$$\sup_{\alpha \in E_u^{(n)}} \prod_{k=1}^{n} \mathrm{Lip}(S_{\alpha|k}^{(u)}) \leq \varepsilon / (\max_v(\mathrm{diam}\, X_v) + 1)$$

for $n \geq N$. Now let $n \geq N$ and $m \in \mathbb{N}$. Fix $f \in \mathrm{Lip}_1(X_u)$. For each $\alpha \in E_u^{(n)}$ write

$$\underline{k}_\alpha^{(u)} := \inf f(S_{\alpha|1}^{(u)} \circ \cdots \circ S_{\alpha|n}^{(u)}(X_{\tau(\alpha_u)})) \in \mathbb{R},$$

$$\overline{k}_\alpha^{(u)} := \sup f(S_{\alpha|1}^{(u)} \circ \cdots \circ S_{\alpha|n}^{(u)}(X_{\tau(\alpha_u)})) \in \mathbb{R}.$$

Then clearly

$$\underline{k}_\alpha^{(u)} \leq \int f(S_{\alpha|1}^{(u)} \circ \cdots \circ S_{\alpha|n}^{(u)})\, d\gamma_\alpha^{(u)} \leq \overline{k}_\alpha^{(u)}$$

$$\underline{k}_\alpha^{(u)} \leq \sum_{\beta \in E_{\tau(\alpha_{|\alpha|})}^{(m)}} p_{\alpha(\beta|1)}^{(u)} \cdots p_{\alpha\beta}^{(u)} \int f(S_{\alpha|1}^{(u)} \circ \cdots \circ S_{\alpha|n}^{(u)} \circ S_{\alpha(\beta|1)}^{(u)} \circ \cdots \circ S_{\alpha\beta}^{(u)})\, d\gamma_{\alpha\beta}^{(u)} \leq \overline{k}_\alpha^{(u)}$$

(since $\sum_{\beta \in E_{\tau(\alpha_{|\alpha|})}^{(m)}} p_{\alpha(\beta|1)}^{(u)} \cdots p_{\alpha\beta}^{(u)} = 1$) whence

$$\Bigg| \int f(S_{\alpha|1}^{(u)} \circ \cdots \circ S_{\alpha|n}^{(u)})\, d\gamma_\alpha^{(u)} -$$

$$\sum_{\beta \in E_{\tau(\alpha_{|\alpha|})}^{(m)}} p_{\alpha(\beta|1)}^{(u)} \cdots p_{\alpha\beta}^{(u)} \int f(S_{\alpha|1}^{(u)} \circ \cdots \circ S_{\alpha|n}^{(u)} \circ S_{\alpha(\beta|1)}^{(u)} \circ \cdots \circ S_{\alpha\beta}^{(u)})\, d\gamma_{\alpha\beta}^{(u)} \Bigg|$$

$$\leq \overline{k}_\alpha^{(u)} - \underline{k}_\alpha^{(u)} = \mathrm{diam}\, f(S_{\alpha|1}^{(u)} \circ \cdots \circ S_{\alpha|n}^{(u)}(X_{\tau(\alpha_n)})) \tag{4.2.2}$$

Hence

$$\left|\int f\, d(\mu_{u,n}^{\gamma_u}(\omega_u)) - \int f\, d(\mu_{u,n+m}^{\gamma_u}(\omega_u))\right|$$

$$= \left| \sum_{\alpha\in E_u^{(n)}} p_{\alpha|1}^{(u)} \cdots p_{\alpha|n}^{(u)} \int f(S_{\alpha|1}^{(u)} \circ \cdots \circ S_{\alpha|n}^{(u)}) d\gamma_\alpha^{(u)} - \right.$$

$$\left. \sum_{\sigma\in E_u^{(n+m)}} p_{\sigma|1}^{(u)} \cdots p_{\sigma|n+m}^{(u)} \int f(S_{\sigma|1}^{(u)} \circ \cdots \circ S_{\sigma|n+m}^{(u)}) d\gamma_\sigma^{(u)} \right|$$

$$= \left| \sum_{\alpha\in E_u^{(n)}} p_{\alpha|1}^{(u)} \cdots p_{\alpha|n}^{(u)} \int f(S_{\alpha|1}^{(u)} \circ \cdots \circ S_{\alpha|n}^{(u)}) d\gamma_\alpha^{(u)} - \right.$$

$$\sum_{\alpha\in E_u^{(n)}} \sum_{\beta\in E_{\tau(\alpha_n)}^{(m)}} p_{\alpha|1}^{(u)} \cdots p_{\alpha|n}^{(u)} p_{\alpha(\beta|1)}^{(u)} \cdots p_{\alpha\beta}^{(u)} \cdot$$

$$\left. \int f(S_{\alpha|1}^{(u)} \circ \cdots \circ S_{\alpha|n}^{(u)} \circ S_{\alpha(\beta|1)}^{(u)} \circ \cdots \circ S_{\alpha\beta}^{(u)}) d\gamma_{\alpha\beta}^{(u)} \right|$$

$$\le \sum_{\alpha\in E_u^{(n)}} p_{\alpha|1}^{(u)} \cdots p_{\alpha|n}^{(u)} \left| \int f(S_{\alpha|1}^{(u)} \circ \cdots \circ S_{\alpha|n}^{(u)}) d\gamma_\alpha^{(u)} - \right.$$

$$\left. \sum_{\beta\in E_{\tau(\alpha_n)}^{(m)}} p_{\alpha(\beta|1)}^{(u)} \cdots p_{\alpha\beta}^{(u)} \int f(S_{\alpha|1}^{(u)} \circ \cdots \circ S_{\alpha|n}^{(u)} \circ S_{\alpha(\beta|1)}^{(u)} \circ \cdots \circ S_{\alpha\beta}^{(u)}) d\gamma_{\alpha\beta}^{(u)} \right|$$

[by (4.2.2)]

$$\le \sum_{\alpha\in E_u^{(n)}} p_{\alpha|1}^{(u)} \cdots p_{\alpha|n}^{(u)} \operatorname{diam} f(S_{\alpha|1}^{(u)} \circ \cdots \circ S_{\alpha|n}^{(u)}(X_{\tau(\alpha_n)}))$$

$$\le \sum_{\alpha\in E_u^{(n)}} p_{\alpha|1}^{(u)} \cdots p_{\alpha|n}^{(u)} \operatorname{Lip}(f) \prod_{k=1}^{n} \operatorname{Lip}(S_{\alpha|k}^{(u)}) \max_v(\operatorname{diam} X_v)$$

$$\le \left(\sum_{\alpha\in E_u^{(u)}} p_{\alpha|1}^{(u)} \cdots p_{\alpha|n}^{(u)} \right) \varepsilon = \varepsilon$$

Since $f \in \operatorname{Lip}_1(X_u)$ was arbitrary this inequality shows that

$$L_{X_u}(\mu_{u,n}^{\gamma_u}(\omega_u), \mu_{u,n+m}^{\gamma_u}(\omega_u)) \le \varepsilon$$

Hence, the sequence in (4.2.1) is Cauchy and thus convergent w.r.t. the metric L_{X_u}, and therefore convergent w.r.t. the weak topology.

ii) Let $\gamma_u = (\gamma_\varnothing^{(u)}, (\gamma_\alpha^{(u)})_{\alpha\in E_u^{(*)}\setminus\{\varnothing\}})$ and $\boldsymbol{\nu}_u = (\nu_\varnothing^{(u)}, (\nu_\alpha^{(u)})_{\alpha\in E_u^{(*)}\setminus\{\varnothing\}})$. Write $S_\gamma(\omega_u) = S_\gamma^{(u)}$ and $p_\gamma(\omega_u) = p_\gamma^{(u)}$ for $\gamma \in E_u^{(*)}$. It is clearly sufficient to prove that

$$L_{X_u}(\mu_{u,n}^{\gamma_u}(\omega_u), \mu_{u,n}^{\boldsymbol{\nu}_u}(\omega_u)) \to 0 \quad \text{as} \quad n \to \infty\,. \tag{4.2.3}$$

Let $\varepsilon > 0$. It follows from Lemma 4.1.1 that there exists an integer $N \in \mathbb{N}$ such that

$$\sup_{\alpha\in E_u^{(n)}} \prod_{k=1}^{n} \operatorname{Lip}(S_{\alpha|k}^{(u)}) \leq \varepsilon/(\max_v(\operatorname{diam} X_v) + 1)$$

for $n \geq N$. Now fix $f \in \operatorname{Lip}_1(X_u)$ and $n \geq N$. For each $\alpha \in E_u^{(n)}$ write

$$\underline{k}_\alpha^{(u)} := \inf f(S_{\alpha|1}^{(u)} \circ \cdots \circ S_{\alpha|n}^{(u)}(X_{\tau(\alpha_n)})) \in \mathbb{R}\,,$$
$$\overline{k}_\alpha^{(u)} := \sup f(S_{\alpha|1}^{(u)} \circ \cdots \circ S_{\alpha|n}^{(u)}(X_{\tau(\alpha_n)})) \in \mathbb{R}\,.$$

Then clearly

$$\underline{k}_\alpha^{(u)} \leq \int f(S_{\alpha|1}^{(u)} \circ \cdots \circ S_{\alpha|n}^{(u)})\, d\gamma_\alpha^{(u)} \leq \overline{k}_\alpha^{(u)}$$
$$\underline{k}_\alpha^{(u)} \leq \int f(S_{\alpha|1}^{(u)} \circ \cdots \circ S_{\alpha|n}^{(u)})\, d\nu_\alpha^{(u)} \leq \overline{k}_\alpha^{(u)}$$

whence

$$\left|\int f(S_{\alpha|1}^{(u)} \circ \cdots \circ S_{\alpha|n}^{(u)})\, d\gamma_\alpha^{(u)} - \int f(S_{\alpha|1}^{(u)} \circ \cdots \circ S_{\alpha|n}^{(u)})\, d\nu_\alpha^{(u)}\right|$$
$$\leq \overline{k}_\alpha^{(u)} - \underline{k}_\alpha^{(u)} = \operatorname{diam} f(S_{\alpha|1}^{(u)} \circ \cdots \circ S_{\alpha|n}^{(u)}(X_{\alpha_n}))\,. \tag{4.2.4}$$

Hence

$$\left|\int f\, d(\mu_{u,n}^{\gamma_u}(\omega_u)) - \int f\, d(\mu_{u,n}^{\boldsymbol{\nu}_u}(\omega_u))\right|$$
$$= \Bigg|\sum_{\alpha\in E_u^{(n)}} p_{\alpha|1}^{(u)} \cdots p_{\alpha|n}^{(u)} \int f(S_{\alpha|1}^{(u)} \circ \cdots \circ S_{\alpha|n}^{(u)})\, d\gamma_\alpha^{(u)} -$$
$$\sum_{\alpha\in E_u^{(n)}} p_{\alpha|1}^{(u)} \cdots p_{\alpha|n}^{(u)} \int f(S_{\alpha|1}^{(u)} \circ \cdots \circ S_{\alpha|n}^{(u)})\, d\nu_\alpha^{(u)}\Bigg|$$
$$\leq \sum_{\alpha\in E_u^{(n)}} p_{\alpha|1}^{(u)} \cdots p_{\alpha|n}^{(u)} \Bigg| \int f(S_{\alpha|1}^{(u)} \circ \cdots \circ S_{\alpha|n}^{(u)})\, d\gamma_\alpha^{(u)} -$$
$$\int f(S_{\alpha|1}^{(u)} \circ \cdots \circ S_{\alpha|n}^{(u)})\, d\nu_\alpha^{(u)}\Bigg|$$

$$
\begin{aligned}
&[\text{by } (4.2.4)]\\
&\leq \sum_{\alpha\in E_u^{(n)}} p_{\alpha|1}^{(u)} \cdots p_{\alpha|n}^{(u)} \operatorname{diam} f(S_{\alpha|1}^{(u)} \circ \cdots \circ S_{\alpha|n}^{(u)}(X_{\alpha_n}))\\
&= \sum_{\alpha\in E_u^{(n)}} p_{\alpha|1}^{(u)} \cdots p_{\alpha|n}^{(u)} \operatorname{Lip}(f) \prod_{k=1}^{n} \operatorname{Lip}(S_{\alpha|k}^{(u)}) \operatorname{diam}(X_{\alpha_n})\\
&\leq \left(\sum_{\alpha\in E_u^{(n)}} p_{\alpha|1}^{(u)} \cdots p_{\alpha|n}^{(u)} \right) \varepsilon\\
&= \varepsilon
\end{aligned}
$$

which proves (4.2.3). □

Our next goal is to prove Theorem 2.3.2. In order to prove Theorem 2.3.2 we need the next three small lemmas

Lemma 4.2.1. *Let $u \in V$ and $\alpha \in E_u^{(n)}$. The maps*

$$
\Omega_u \to \operatorname{Con}(X_{\tau(\alpha_n)}, X_{\tau(\alpha_{n-1})}) \times \cdots \times \operatorname{Con}(X_{\tau(\alpha_1)}, X_u)
$$
$$
\Omega_u \to \mathbb{R}^n
$$

defined by

$$
\omega \to (S_{\alpha|n}(\omega), \ldots, S_{\alpha|1}(\omega))
$$
$$
\omega \to (p_{\alpha|n}(\omega), \ldots, p_{\alpha|1}(\omega))
$$

are continuous.

Proof. Obvious. □

The next lemma represents a minor extension of [Gra1, Lemma 3.4].

Lemma 4.2.2. *Let $n \in \mathbb{N}$ and $(X_1, d_1), \ldots, (X_{n+1}, d_{n+1})$ be metric spaces. The map*

$$
\prod_{i=1}^{n} \operatorname{Con}(X_i, X_{i+1}) \to \operatorname{Con}(X_1, X_{n+1})
$$

defined by

$$
(f_1, \ldots, f_n) \to f_n \circ \cdots \circ f_1
$$

is $\prod_{i=1}^{n} \mathcal{O}(X_i, X_{i+1})$-$\mathcal{O}(X_1, X_{n+1})$ continuous (recall that if X, Y are metric spaces, then $\mathcal{O}(X, Y)$ denotes the compact-open topology on $\operatorname{Con}(X, Y)$, i.e. the topology of pointwise convergence.)

Proof. Let $(f_{1,m}, \ldots, f_{n,m})_m$ be a sequence in $\prod_{i=1}^{n} \operatorname{Con}(X_i, X_{i+1})$ and $(f_1, \ldots, f_n) \in \prod_{i=1}^{n} \operatorname{Con}(X_i, X_{i+1})$ with $(f_{1,m}, \ldots, f_{n,m}) \to (f_1, \ldots, f_n)$ w.r.t. $\prod_{i=1}^{n} \mathcal{O}(X_i, X_{i+1})$

as $m \to \infty$. Let $x \in X_1$. Then $f_{i,m}(f_{i-1} \circ \cdots \circ f_1(x)) \to f_i(f_{i-1} \circ \cdots \circ f_1(x))$ for $i = 1, \ldots, n$ as $m \to \infty$. Hence

$$
\begin{aligned}
& d_{n+1}\big(f_{n,m} \circ \cdots \circ f_{1,m}(x), f_n \circ \cdots \circ f_1(x)\big) \\
& \leq \sum_{i=1}^{n} d_{n+1}\big(f_{n,m} \circ \cdots \circ f_{i+1,m} \circ f_{i,m} \circ f_{i_1} \circ \cdots \circ f_1(x), \\
& \qquad\qquad f_{n,m} \circ \cdots \circ f_{i+1,m} \circ f_i \circ f_{i_1} \circ \cdots \circ f_1(x)\big) \\
& \leq \sum_{i=1}^{n} \mathrm{Lip}(f_{n,m}) \cdot \cdots \cdot \mathrm{Lip}(f_{i+1,m}) \cdot \\
& \qquad d_{i+1}\big(f_{i,m}(f_{i-1} \circ \cdots \circ f_1(x)), f_i(f_{i-1} \circ \cdots \circ f_1(x))\big) \\
& \leq \sum_{i=1}^{n} d_{i+1}\big(f_{i,m}(f_{i-1} \circ \cdots \circ f_1(x)), f_i(f_{i-1} \circ \cdots \circ f_1(x))\big) \to 0. \ \square
\end{aligned}
$$

Lemma 4.2.3. *Let X, Y be metric spaces, $S : X \to Y$ a continuous map and $a \in \mathbb{R}_+$. The following maps are continuous w.r.t. the vague topologies*

$$
\begin{aligned}
\mathcal{M}(X) \to \mathcal{M}(Y) &: \mu \to \mu \circ S^{-1} \\
\mathcal{M}(X) \to \mathcal{M}(X) &: \mu \to a\mu \\
\mathcal{M}(X) \times \mathcal{M}(X) \to \mathcal{M}(X) &: (\mu, \nu) \to \mu + \nu .
\end{aligned}
$$

Proof. Obvious. $\square$

We are now ready to prove Theorem 2.3.2

Proof of Theorem 2.3.2

It is obvious that $(\Omega_u)_0$ is a Borel set, i.e. $(\Omega_u)_0 \in \mathcal{B}(\mathcal{T}_u) = \Sigma_u$. If $\omega \in (\Omega_u)_0$ then

$$\mu_u(\omega) = \lim_n \mu_{u,n}^{\boldsymbol{\gamma}}(\omega)$$

for $\boldsymbol{\gamma} = \big(\gamma_\varnothing, (\gamma_\alpha)_{\alpha \in E_u^{(*)} \setminus \{\varnothing\}}\big) \in \mathcal{P}(X_u) \times \prod_{\alpha \in E_u^{(*)} \setminus \{\varnothing\}} \mathcal{P}(X_{\tau(\alpha_{|\alpha|})})$, where

$$\mu_{u,n}^{\boldsymbol{\gamma}}(\omega) = \sum_{\alpha \in E_u^{(n)}} p_{\alpha|1}(\omega) \cdot \cdots \cdot p_{\alpha|n}(\omega) \gamma_\alpha \circ (S_{\alpha|1}(\omega) \circ \cdots \circ S_{\alpha|n}(\omega))^{-1} .$$

It follows from Lemma 4.2.1 through Lemma 4.2.3 that $\mu_{u,n}^{\boldsymbol{\gamma}}$ is $\mathcal{T}_u$-w$(\mathcal{P}(X_u))$ continuous on $(\Omega_u)_0$, and $\mu_u = \lim_n \mu_{u,n}^{\boldsymbol{\gamma}}$ is thus Borel measurable on the Borel set $(\Omega_u)_0$. This proves the assertion since μ_u is constant on $\Omega_u \setminus (\Omega_u)_0$. $\square$.

Proof of Theorem 2.3.3

Write $S_\gamma(\omega) = S_\gamma$ and $p_\gamma(\omega) = p_\gamma$ for $\gamma \in E_u^{(*)}$, and let $\boldsymbol{\gamma} = \big(\gamma_\varnothing, (\gamma_\alpha)_{\alpha \in E_u^{(*)} \setminus \{\varnothing\}}\big) \in \mathcal{P}(X_u) \times \prod_{\alpha \in E_u^{(*)} \setminus \{\varnothing\}} \mathcal{P}(X_{\tau(\alpha_{|\alpha|})})$. We divide the proof into three steps.

Step 1. For $\alpha \in E_u^{(*)}$,

$$\lim_n \Big((\mu_{u,n}^{\gamma}(\omega))(S_{\alpha|1} \circ \cdots \circ S_{\alpha||\alpha|}(X_{\tau(\alpha_{|\alpha|})})) \Big)$$
$$=(\hat{\mu}_u(\omega) \circ \pi_u(\omega)^{-1})(S_{\alpha|1} \circ \cdots \circ S_{\alpha||\alpha|}(X_{\tau(\alpha_{|\alpha|})}))$$

Proof of Step 1. Let $\alpha \in E_u^{(*)}$ and $n \geq |\alpha|$. Since $S_{\beta e}(X_{\tau(e)}) \cap S_{\beta \varepsilon}(X_{\tau(\varepsilon)}) = \varnothing$ for all $\beta \in E_u^{(*)}$ and $e, \varepsilon \in E_{\tau(\alpha_{|\alpha|})}$ with $e \neq \varepsilon$,

$$(\mu_{u,n}^{\gamma}(\omega))(S_{\alpha|1} \circ \cdots \circ S_{\alpha||\alpha|}(X_{\tau(\alpha_{|\alpha|})}))$$
$$= \sum_{\beta \in E_u^{(|\alpha|)}} \sum_{\sigma \in E_{\tau(\beta_{|\alpha|})}^{(n-|\alpha|)}} p_{\beta|1} \cdots p_{\beta||\alpha|} p_{\beta(\sigma|1)} \cdots p_{\beta\sigma} \gamma_{\beta\sigma} \circ$$
$$(S_{\beta(\sigma|1)} \circ \cdots \circ S_{\beta\sigma})^{-1} \Big((S_{\beta|1} \circ \cdots \circ S_{\beta||\alpha|})^{-1} (S_{\alpha|1} \circ \cdots \circ S_{\alpha||\alpha|}(X_{\tau(\alpha_{|\alpha|})})) \Big)$$
$$= \sum_{\sigma \in E_{\tau(\alpha_{|\alpha|})}^{(n-|\alpha|)}} p_{\alpha|1} \cdots p_{\alpha||\alpha|} p_{\alpha(\sigma|1)} \cdots p_{\alpha\sigma} \gamma_{\alpha\sigma}((S_{\alpha(\sigma|1)} \circ \cdots \circ S_{\alpha\sigma})^{-1}(X_{\tau(\alpha_{|\alpha|})}))$$
$$= \sum_{\sigma \in E_{\tau(\alpha_{|\alpha|})}^{(n-|\alpha|)}} p_{\alpha|1} \cdots p_{\alpha||\alpha|} p_{\alpha(\sigma|1)} \cdots p_{\alpha\sigma} \gamma_{\alpha\sigma}(X_{\tau(\alpha\sigma_{|\alpha\sigma|})})$$
$$= p_{\alpha|1} \cdots p_{\alpha||\alpha|} \cdot$$

Hence

$$\lim_n \Big((\mu_{u,n}^{\gamma}(\omega))(S_{\alpha|1} \circ \cdots \circ S_{\alpha||\alpha|}(X_{\tau(\alpha_{|\alpha|})})) \Big) = p_{\alpha|1} \cdots p_{\alpha||\alpha|} \tag{4.2.5}$$

Also, since $S_{\beta e}(X_{\tau(e)}) \cap S_{\beta \varepsilon}(X_{\tau(\varepsilon)}) = \varnothing$ for all $\beta \in E_u^{(*)}$ and $e, \varepsilon \in E_{\tau(\alpha_{|\alpha|})}$ with $e \neq \varepsilon$, $\pi_u(\omega_u)^{-1}(S_{\beta|1} \circ \cdots \circ S_{\beta||\beta|}(X_{\tau(\beta_{|\beta|})})) = [\beta]$ for $\beta \in E_u^{(*)}$, whence

$$(\hat{\mu}_u(\omega) \circ \pi_u(\omega)^{-1})(S_{\alpha|1} \circ \cdots \circ S_{\alpha||\alpha|}(X_{\tau(\alpha_{|\alpha|})})) = \hat{\mu}_u(\omega)([\alpha])$$
$$= p_{\alpha|1} \cdots p_{\alpha||\alpha|} \tag{4.2.6}$$

The desired result now follows from equations 4.2.5 and 4.2.6. This proves step 1.

Step 2. If $G \subseteq C_u(\omega)$ is open (relative to $C_u(\omega)$) then

$$(\mu_u(\omega))(G) = (\hat{\mu}_u(\omega) \circ \pi_u(\omega)^{-1})(G)$$

Proof of Step 2. Write $\mu_n := \mu_{u,n}^{\gamma}(\omega)$. Let $A = \{\alpha \in E_u^{(*)} \mid S_{\alpha|1} \circ \cdots \circ S_{\alpha||\alpha|}(X_{\tau(\alpha_{|\alpha|})}) \subseteq G\}$. Since G is open and $\omega \in (\Omega_u)_0$, $G = \cup_{\alpha \in A} S_{\alpha|1} \circ \cdots \circ S_{\alpha||\alpha|}(X_{\tau(\alpha_{|\alpha|})})$. Now, we need only cover G once: if $\alpha, \beta \in A$ and $[\alpha] \cap [\beta] \neq \varnothing$ then one of then is contained in the other, so we may discard the smaller one. So there is a set $A_0 \subseteq A$ such that

$$G = \bigcup_{\alpha \in A_0} S_{\alpha|1} \circ \cdots \circ S_{\alpha||\alpha|}(X_{\tau(\alpha_{|\alpha|})})$$

and $[\alpha] \cap [\beta] = \varnothing$ for $\alpha, \beta \in A_0$ with $\alpha \neq \beta$. Since $S_{\alpha e}(X_{\tau(e)}) \cap S_{\alpha \varepsilon}(X_{\tau(\varepsilon)}) = \varnothing$ for all $\alpha \in E_u^{(*)}$ and $e, \varepsilon \in E_{\tau(\alpha_{|\alpha|})}$ with $e \neq \varepsilon$,

$$S_{\alpha|1} \circ \cdots \circ S_{\alpha||\alpha|}(X_{\tau(\alpha_{|\alpha|})}) \bigcap S_{\beta|1} \circ \cdots \circ S_{\beta||\beta|}(X_{\tau(\beta_{|\beta|})}) = \varnothing \tag{4.2.7}$$

for $\alpha, \beta \in A_0$ with $\alpha \neq \beta$. Hence

$$\begin{aligned}
(\mu_u(\omega))(G) &\leq \liminf_n \mu_n(G) \qquad \left[\text{since} \quad \mu_n := \mu^{\gamma}_{u,n}(\omega) \xrightarrow{\text{w}} \mu_u(\omega)\right] \\
&= \liminf_n \mu_n \left(\bigcup_{\alpha \in A_0} S_{\alpha|1} \circ \cdots \circ S_{\alpha||\alpha|}(X_{\tau(\alpha_{|\alpha|})}) \right) \\
&\left[\text{by } (4.2.7)\right] \\
&= \liminf_n \sum_{\alpha \in A_0} \mu_n \left(S_{\alpha|1} \circ \cdots \circ S_{\alpha||\alpha|}(X_{\tau(\alpha_{|\alpha|})}) \right) \\
&\leq \limsup_n \sum_{\alpha \in A_0} \mu_n \left(S_{\alpha|1} \circ \cdots \circ S_{\alpha||\alpha|}(X_{\tau(\alpha_{|\alpha|})}) \right) \\
&\leq \sum_{\alpha \in A_0} \limsup_n \mu_n \left(S_{\alpha|1} \circ \cdots \circ S_{\alpha||\alpha|}(X_{\tau(\alpha_{|\alpha|})}) \right) \\
&\left[\text{since} \quad \mu_n := \mu^{\gamma}_{u,n}(\omega) \xrightarrow{\text{w}} \mu_u(\omega)\right] \\
&\leq \sum_{\alpha \in A_0} \mu_u(\omega) \left(S_{\alpha|1} \circ \cdots \circ S_{\alpha||\alpha|}(X_{\tau(\alpha_{|\alpha|})}) \right) \\
&\left[\text{by } (4.2.7)\right] \\
&= \mu_u(\omega) \left(\bigcup_{\alpha \in A_0} S_{\alpha|1} \circ \cdots \circ S_{\alpha||\alpha|}(X_{\tau(\alpha_{|\alpha|})}) \right) \\
&= (\mu_u(\omega_u))(G).
\end{aligned}$$

Thus

$$\begin{aligned}
(\mu_u(\omega))(G) &= \sum_{\alpha \in A_0} \limsup_n \mu_n \left(S_{\alpha|1} \circ \cdots \circ S_{\alpha||\alpha|}(X_{\tau(\alpha_{|\alpha|})}) \right) \\
&\left[\text{by step } 1\right] \\
&= \sum_{\alpha \in A_0} (\hat{\mu}_u(\omega) \circ \pi_u(\omega)^{-1}) \left(S_{\alpha|1} \circ \cdots \circ S_{\alpha||\alpha|}(X_{\tau(\alpha_{|\alpha|})}) \right)
\end{aligned}$$

$$\begin{aligned}
&\text{[by (4.2.7)]}\\
&= (\hat{\mu}_u(\omega) \circ \pi_u(\omega)^{-1}) \left(\bigcup_{\alpha \in A_0} S_{\alpha|1} \circ \cdots \circ S_{\alpha||\alpha|}(X_{\tau(\alpha_{|\alpha|})}) \right)\\
&= (\hat{\mu}_u(\omega) \circ \pi_u(\omega)^{-1})(G)
\end{aligned}$$

which proves Step 2.

Step 3.

$$\mu_u(\omega) \doteq \hat{\mu}_u(\omega) \circ \pi_u(\omega)^{-1}$$

Proof of Step 3. Follows from step 2 by outer regularity. □

4.3 Proofs of the Theorems in Section 2.4

We begin with two small and a somewhat larger lemma.

Lemma 4.3.1. *For* $u \in V$,

$$\left(\lambda_u \times \prod_{e\in E_u} \Lambda_{\tau(e)}\right) \circ \varphi_u^{-1} = \Lambda_u$$

Proof. Let $n \in \mathbb{N}$ and

$$E := \prod_{\substack{\alpha\in E_u^{(*)}\\ |\alpha|\le n}} E_\alpha \times \prod_{\substack{\alpha\in E_u^{(*)}\\ n<|\alpha|}} \Xi_{\tau(\alpha_{|\alpha|})} \subseteq \Omega_u$$

where $E_\varnothing$ is a Borel subset of Ξ_u and E_α is a Borel subset of $\Xi_{\tau(\alpha_{|\alpha|})}$. Now

$$\begin{aligned}
&\left(\lambda_u \times \prod_{e\in E_u} \Lambda_{\tau(e)}\right) \circ \varphi_u^{-1}(E)\\
=&\left(\lambda_u \times \prod_{e\in E_u} \Lambda_{\tau(e)}\right)\\
&\qquad\Bigg\{\Big(((S_e)_{e\in E_u},(p_e)_{e\in E_u}),(\omega_{\tau(e)})_{e\in E_u}\Big) \in \Xi_u \times \prod_{e\in E_u} \Omega_{\tau(e)}\Bigg|\\
&\qquad\quad ((S_e)_{e\in E_u},(p_e)_{e\in E_u}) \in E_\varnothing,\ \pi_{\tau(e),\alpha}(\omega_{\tau(e)}) \in E_{e\alpha} \quad \text{for} \quad |\alpha| \le n-1\Bigg\}\\
=&\lambda_u(E_\varnothing)\left(\prod_{e\in E_u} \Lambda_{\tau(e)}\right)\left(E_\varnothing \times \prod_{e\in E_u}\left(\left(\prod_{\substack{\alpha\in E_{\tau(e)}^{(*)}\\ |\alpha|\le n-1}} E_{e\alpha}\right) \times \left(\prod_{\substack{\alpha\in E_{\tau(e)}^{(*)}\\ n-1<|\alpha|}} \Xi_{\tau(\alpha_{|\alpha|})}\right)\right)\right)\\
=&\lambda_u(E_\varnothing) \prod_{e\in E_u} \Lambda_{\tau(e)}\left(\left(\prod_{\substack{\alpha\in E_{\tau(e)}^{(*)}\\ |\alpha|\le n-1}} E_{e\alpha}\right) \times \left(\prod_{\substack{\alpha\in E_{\tau(e)}^{(*)}\\ n-1<|\alpha|}} \Xi_{\tau(\alpha_{|\alpha|})}\right)\right)\\
=&\lambda_u(E_\varnothing) \prod_{e\in E_u}\left(\lambda_{\tau(e)}(E_e) \times \prod_{\substack{\alpha\in E_{\tau(e)}^{(*)}\setminus\{\varnothing\}\\ |\alpha|\le n-1}} \lambda_{\tau(\alpha_{|\alpha|})}(E_{e\alpha})\right)
\end{aligned}$$

$$=\left(\lambda_u \times \prod_{\alpha\in E_u^{(*)}\setminus\{\varnothing\}} \lambda_{\tau(\alpha_{|\alpha|})}\right)\left(\prod_{\substack{\alpha\in E_u^{(*)}\\|\alpha|\le n}} E_\alpha \times \prod_{\substack{\alpha\in E_u^{(*)}\\n<|\alpha|}} \Xi_{\tau(\alpha_{|\alpha|})}\right)$$

$$=\Lambda_u(E)$$

whence $\lambda_u \times \prod_{e\in E_u} \Lambda_{\tau(e)}) \circ \varphi_u^{-1} = \Lambda_u$. □

Lemma 4.3.2. *For $u \in V$ and $\alpha \in E_u^{(*)} \setminus \{\varnothing\}$,*

$$\Lambda_u \circ \mathcal{S}_\alpha^{-1} = \Lambda_{\tau(\alpha_{|\alpha|})}.$$

Proof. Let $n \in \mathbb{N}$ and

$$E = \prod_{\substack{\sigma\in E^{(*)}_{\tau(\alpha_{|\alpha|})}\\|\sigma|\le n}} E_\sigma \times \prod_{\substack{\sigma\in E^{(*)}_{\tau(\alpha_{|\alpha|})}\\n<|\sigma|}} \Xi_{\tau(\sigma_{|\sigma|})} \subseteq \Omega_{\tau(\alpha_{|\alpha|})}$$

where $E_\varnothing$ is a Borel subset of $\Xi_{\tau(\alpha_{|\alpha|})}$ and E_σ is a Borel subset of $\Xi_{\tau(\sigma_{|\sigma|})}$. Now

$$\begin{aligned}
&\Lambda_u \circ \mathcal{S}_\alpha^{-1}(E)\\
&=\Lambda_u(\{(\xi_\beta)_{\beta\in E_u^{(*)}} \in \Omega_u \mid \mathcal{S}_\alpha((\xi_\beta)_{\beta\in E_u^{(*)}}) \in E\})\\
&=\Lambda_u\left(\left\{(\xi_\beta)_{\beta\in E_u^{(*)}} \in \Omega_u \,\middle|\, (\xi_{\alpha\beta})_{\beta\in E^{(*)}_{\tau(\alpha_{|\alpha|})}} \in \right.\right.\\
&\qquad\qquad \left.\left. \prod_{\substack{\sigma\in E^{(*)}_{\tau(\alpha_{|\alpha|})}\\|\sigma|\le n}} E_\sigma \times \prod_{\substack{\sigma\in E^{(*)}_{\tau(\alpha_{|\alpha|})}\\n<|\sigma|}} \Xi_{\tau(\sigma_{|\sigma|})}\right\}\right)\\
&=\Lambda_u\left(\left\{(\xi_\beta)_{\beta\in E_u^{(*)}} \in \Omega_u \,\middle|\, (\xi_{\alpha\beta})_{\beta\in E^{(*)}_{\tau(\alpha_{|\alpha|})},|\beta|\le n} \in \prod_{\substack{\sigma\in E^{(*)}_{\tau(\alpha_{|\alpha|})}\\|\sigma|\le n}} E_\beta\right\}\right)\\
&=\left(\lambda_u \times \prod_{\beta\in E_u^{(*)}\setminus\{\varnothing\}} \lambda_{\tau(\beta_{|\beta|})}\right)\\
&\qquad\left(\left\{(\xi_\beta)_{\beta\in E_u^{(*)}} \in \Omega_u \,\middle|\, (\xi_{\alpha\beta})_{\beta\in E^{(*)}_{\tau(\alpha_{|\alpha|})},|\beta|\le n} \in \prod_{\substack{\sigma\in E^{(*)}_{\tau(\alpha_{|\alpha|})}\\|\sigma|\le n}} E_\beta\right\}\right)
\end{aligned}$$

$$= \prod_{\substack{\sigma \in E^{(*)}_{\tau(\alpha_{|\alpha|})} \\ |\sigma| \le n}} \lambda_{\tau(\sigma_{|\sigma|})}(E_\beta)$$

$$= \left(\lambda_{\tau(\alpha_{|\alpha|})} \times \prod_{\sigma \in E^{(*)}_{\tau(\alpha_{|\alpha|})} \setminus \{\varnothing\}} \lambda_{\tau(\sigma_{|\sigma|})} \right) \left(\prod_{\substack{\sigma \in E^{(*)}_{\tau(\alpha_{|\alpha|})} \\ |\sigma| \le n}} E_\sigma \times \prod_{\substack{\sigma \in E^{(*)}_{\tau(\alpha_{|\alpha|})} \\ n < |\sigma|}} \Xi_{\tau(\sigma_{|\sigma|})} \right)$$

$$= \Lambda_{\tau(\alpha_{|\alpha|})}(E)$$

whence $\Lambda_u \circ \mathcal{S}_\alpha^{-1} = \Lambda_{\tau(\alpha_{|\alpha|})}$. □

Lemma 4.3.3. *We have*

$$\Lambda_u((\Omega_u)_0) = 1\,.$$

In particular $\Lambda(\Omega) = 1$.

Proof. For $w \in V$ and $a > 0$ write

$$T_w(a) = \{\omega \in \Omega_w \mid \exists \sigma \in E_u^{\mathbb{N}} : \prod_{n=1}^{\infty} \operatorname{Lip}(S_{\sigma|n}) \ge a\}\,.$$

Define $p :]0,1[\to [0,1]$ by

$$p(a) = \max_w \Lambda_w(T_w(a))\,.$$

The function p is clearly decreasing. We claim that

$$p = 0\,. \tag{4.3.1}$$

If (4.3.1) is satisfied then

$$\Lambda_u((\Omega_u)_0) = \Lambda_u\left(\Omega_u \setminus \bigcup_{n \in \mathbb{N}} T_u(\tfrac{1}{n})\right) = \Lambda_u(\Omega_u) = 1\,.$$

We will now prove (4.3.1). Since $\operatorname{Lip}(S) < 1$ for all contractions there exists $0 < a_0 < 1$ such that

$$\lambda_w(\{((S_e)_{e \in E_v}, (p_e)_{e \in E_v}) \in \Xi_w | \max_e \operatorname{Lip}(S_e) \ge a_0\}) < \frac{1}{\operatorname{card} E_w} \tag{4.3.2}$$

for all $w \in V$.

Now choose $v \in V$ such that $p(a_0) = \Lambda_v(T_v(a_0))$. Lemma 4.3.1 implies that

$$\begin{aligned}
p(a_0) &= \Lambda_v(T_v(a_0)) \\
&= \left(\left(\lambda_v \times \prod_{e \in E_v} \Lambda_{\tau(e)}\right) \circ \varphi_v^{-1}\right)(T_v(a_0)) \\
&= \left(\lambda_v \times \prod_{e \in E_v} \Lambda_{\tau(e)}\right) \\
&\qquad \left\{ \left(((S_e)_{e \in E_v}, (p_e)_{e \in E_v}), (\omega_{\tau(e)})_{e \in E_v}\right) \in \Xi_v \times \prod_{e \in E_v} \Omega_{\tau(e)} \right| \\
&\qquad\quad \left. \exists \sigma \in E_v^{\mathbb{N}} : \operatorname{Lip}(S_{\sigma_1}) \prod_{n=2}^{\infty} \operatorname{Lip}(S_{\sigma_2 \ldots \sigma_n}(\omega_{\tau(e)})) \geq a_0 \right\} \\
&= \left(\lambda_v \times \prod_{e \in E_v} \Lambda_{\tau(e)}\right) \\
&\qquad \left\{ \left(((S_e)_{e \in E_v}, (p_e)_{e \in E_v}), (\omega_{\tau(e)})_{e \in E_v}\right) \in \Xi_v \times \prod_{e \in E_v} \Omega_{\tau(e)} \right| \\
&\qquad\quad \left. \exists \varepsilon \in E_v : \exists \tau \in E_{\tau(\varepsilon)}^{\mathbb{N}} : \operatorname{Lip}(S_\varepsilon) \prod_{n=1}^{\infty} \operatorname{Lip}(S_{\tau|n}(\omega_{\tau(\varepsilon)})) \geq a_0 \right\} \\
&\leq \sum_{\varepsilon \in E_v} \left(\lambda_v \times \prod_{e \in E_v} \Lambda_{\tau(e)}\right) \\
&\qquad \left\{ \left(((S_e)_{e \in E_v}, (p_e)_{e \in E_v}), (\omega_{\tau(e)})_{e \in E_v}\right) \in \Xi_v \times \prod_{e \in E_v} \Omega_{\tau(e)} \right| \\
&\qquad\quad \left. \exists \tau \in E_{\tau(\varepsilon)}^{\mathbb{N}} : \operatorname{Lip}(S_\varepsilon) \prod_{n=1}^{\infty} \operatorname{Lip}(S_{\tau|n}(\omega_{\tau(\varepsilon)})) \geq a_0 \right\} \\
&\leq \sum_{\varepsilon \in E_v} \int \left(\left(\prod_{e \in E_v} \Lambda_{\tau(e)}\right) \left\{ (\omega_{\tau(e)})_{e \in E_v} \in \prod_{e \in E_v} \Omega_{\tau(e)} \right| \right. \\
&\qquad\quad \left. \left. \exists \tau \in E_{\tau(\varepsilon)}^{\mathbb{N}} : \operatorname{Lip}(S_\varepsilon) \prod_{n=1}^{\infty} \operatorname{Lip}(S_{\tau|n}(\omega_{\tau(\varepsilon)})) \geq a_0 \right\}\right) \\
&\qquad\qquad d\lambda_v\left((S_e)_{e \in E_v}, (p_e)_{e \in E_v}\right) \qquad (4.3.3)
\end{aligned}$$

$$= \sum_{\varepsilon \in E_v} \int_{\{((S_e)_{e\in E_v},(p_e)_{e\in E_v}) \mid \mathrm{Lip}(S_\varepsilon)\geq a_0\}} \left(\left(\prod_{e\in E_v} \Lambda_{\tau(e)} \right) \left\{ (\omega_{\tau(e)})_{e\in E_v} \in \prod_{e\in E_v} \Omega_{\tau(e)} \middle| \exists \tau \in E^{\mathbb{N}}_{\tau(\varepsilon)} : \mathrm{Lip}(S_\varepsilon) \prod_{n=1}^{\infty} \mathrm{Lip}(S_{\tau|n}(\omega_{\tau(\varepsilon)})) \geq a_0 \right\} \right) d\lambda_v\big((S_e)_{e\in E_v}, (p_e)_{e\in E_v}\big)$$

$$= \sum_{\varepsilon \in E_v} \int_{\{((S_e)_{e\in E_v},(p_e)_{e\in E_v}) \mid \mathrm{Lip}(S_\varepsilon)\geq a_0\}} \Lambda_{\tau(\varepsilon)} \left(\left\{ \omega_{\tau(\varepsilon)} \in \Omega_{\tau(\varepsilon)} \middle| \exists \tau \in E^{\mathbb{N}}_{\tau(\varepsilon)} : \mathrm{Lip}(S_\varepsilon) \prod_{n=1}^{\infty} \mathrm{Lip}(S_{\tau|n}(\omega_{\tau(\varepsilon)})) \geq a_0 \right\} \right) d\lambda_v\big((S_e)_{e\in E_v}, (p_e)_{e\in E_v}\big)$$

$$= \sum_{\varepsilon \in E_v} \int_{\{((S_e)_{e\in E_v},(p_e)_{e\in E_v}) \mid \mathrm{Lip}(S_\varepsilon)\geq a_0\}} \Lambda_{\tau(\varepsilon)} \left\{ \omega_{\tau(\varepsilon)} \in \Omega_{\tau(\varepsilon)} \middle| \exists \tau \in E^{\mathbb{N}}_{\tau(\varepsilon)} : \prod_{n=1}^{\infty} \mathrm{Lip}(S_{\tau|n}(\omega_{\tau(\varepsilon)})) \geq a_0 \right\} d\lambda_v\big((S_e)_{e\in E_v}, (p_e)_{e\in E_v}\big)$$

$$= \sum_{\varepsilon \in E_v} \int_{\{((S_e)_{e\in E_v},(p_e)_{e\in E_v}) \mid \mathrm{Lip}(S_\varepsilon)\geq a_0\}} \Lambda_{\tau(\varepsilon)}(T_{\tau(\varepsilon)}(a_0)) \, d\lambda_v\big((S_e)_{e\in E_v}, (p_e)_{e\in E_v}\big)$$

$$\leq \sum_{\varepsilon \in E_v} \int_{\{((S_e)_{e\in E_v},(p_e)_{e\in E_v}) \mid \max_e \mathrm{Lip}(S_e)\geq a_0\}} p(a_0) \, d\lambda_v\big((S_e)_{e\in E_v}, (p_e)_{e\in E_v}\big)$$

$$\leq p(a_0) \max_w \Big(\mathrm{card}(E_w) \lambda_w(\{((S_e)_{e\in E_w}, (p_e)_{e\in E_w}) \mid \max_e \mathrm{Lip}(S_e) \geq a_0\}) \Big). \tag{4.3.4}$$

It follows from (4.3.2) and (4.3.4) that $p(a_0) = 0$. Hence $A := \{a \in]0,1[\mid p(a) = 0\} \neq \varnothing$ and $a := \inf A \leq a_0$. We claim that

$$a = 0.$$

Otherwise $0 < a < \frac{a}{a_0}$ and we can thus choose $\alpha \in]a, \frac{a}{a_0}[$. Since $\alpha a_0 < a := \inf A$ and p is decreasing,

$$p(\alpha a_0) > 0. \tag{4.3.5}$$

Pick $v \in V$ such that $p(\alpha a_0) = \Lambda_v(T_v(\alpha a_0))$. Then (by arguments similar to the proof of (4.3.3)),

$$\begin{aligned} p(\alpha a_0) &= \Lambda_v(T_v(\alpha a_0)) \\ &= \sum_{\varepsilon \in E_v} \int \left(\left(\prod_{e \in E_v} \Lambda_{\tau(e)} \right) \left\{ (\omega_{\tau(e)})_{e \in E_v} \in \prod_{e \in E_v} \Omega_{\tau(e)} \,\middle|\, \right. \right. \\ &\qquad \left. \left. \exists \tau \in E^{\mathbb{N}}_{\tau(\varepsilon)} : \mathrm{Lip}(S_\varepsilon) \prod_{n=1}^{\infty} \mathrm{Lip}(S_{\tau|n}(\omega_{\tau(\varepsilon)})) \geq \alpha a_0 \right\} \right) \\ &\qquad d\lambda_v((S_e)_{e \in E_v}, (p_e)_{e \in E_v}) \\ &= \sum_{\varepsilon \in E_v} \int \Lambda_{\tau(\varepsilon)} \left(\left\{ \omega_{\tau(\varepsilon)} \in \Omega_{\tau(\varepsilon)} \,\middle|\, \right. \right. \\ &\qquad \left. \left. \exists \tau \in E^{\mathbb{N}}_{\tau(\varepsilon)} : \mathrm{Lip}(S_\varepsilon) \prod_{n=1}^{\infty} \mathrm{Lip}(S_{\tau|n}(\omega_{\tau(\varepsilon)})) \geq \alpha a_0 \right\} \right) \\ &\qquad d\lambda_v((S_e)_{e \in E_v}, (p_e)_{e \in E_v}) . \end{aligned} \tag{4.3.6}$$

Since $a < \alpha$, $0 = p(\alpha) = \max_w \Lambda_w(T_w(\alpha))$. Hence $\Lambda_{\tau(\varepsilon)}(T_{\tau(\varepsilon)}(\alpha)) = 0$ for all $\varepsilon \in E_v$, whence

$$\begin{aligned} 1 &= \Lambda_{\tau(\varepsilon)}(\Omega_{\tau(\varepsilon)} \setminus T_{\tau(\varepsilon)}(\alpha)) \\ &= \Lambda_{\tau(\varepsilon)} \left(\left\{ \omega_{\tau(\varepsilon)} \in \Omega_{\tau(\varepsilon)} \,\middle|\, \forall \sigma \in E^{\mathbb{N}}_{\tau(\varepsilon)} : \prod_{n=1}^{\infty} \mathrm{Lip}(S_{\sigma|n}(\omega_{\tau(\varepsilon)})) < \alpha \right\} \right) \end{aligned} \tag{4.3.7}$$

for all $\varepsilon \in E_v$. It follows from (4.3.6) and (4.3.7) that

$$\begin{aligned} p(\alpha a_0) &= \sum_{\varepsilon \in E_v} \int_{\{((S_e)_{e \in E_v}, (p_e)_{e \in E_v}) | \mathrm{Lip}(S_\varepsilon) \geq a_0\}} \Lambda_{\tau(\varepsilon)} \left(\left\{ \omega_{\tau(\varepsilon)} \in \Omega_{\tau(\varepsilon)} \,\middle|\, \right. \right. \\ &\qquad \left. \left. \exists \tau \in E^{\mathbb{N}}_{\tau(\varepsilon)} : \mathrm{Lip}(S_\varepsilon) \prod_{n=1}^{\infty} \mathrm{Lip}(S_{\tau|n}(\omega_{\tau(\varepsilon)})) \geq \alpha a_0 \right\} \right) \\ &\qquad d\lambda_v((S_e)_{e \in E_v}, (p_e)_{e \in E_v}) \\ &\leq \sum_{\varepsilon \in E_v} \int_{\{((S_e)_{e \in E_v}, (p_e)_{e \in E_v}) | \mathrm{Lip}(S_\varepsilon) \geq a_0\}} \Lambda_{\tau(\varepsilon)} \left\{ \omega_{\tau(\varepsilon)} \in \Omega_{\tau(\varepsilon)} \,\middle|\, \right. \\ &\qquad \left. \exists \tau \in E^{\mathbb{N}}_{\tau(\varepsilon)} : \prod_{n=1}^{\infty} \mathrm{Lip}(S_{\tau|n}(\omega_{\tau(\varepsilon)})) \geq \alpha a_0 \right\}) \\ &\qquad d\lambda_v((S_e)_{e \in E_v}, (p_e)_{e \in E_v}) \end{aligned}$$

$$= \sum_{\varepsilon\in E_v} \int_{\{((S_e)_{e\in E_v},(p_e)_{e\in E_v}) \mid \mathrm{Lip}(S_\varepsilon)\geq a_0\}} \Lambda_{\tau(\varepsilon)}(T_{\tau(\varepsilon)}(\alpha a_0)) \, d\lambda_v((S_e)_{e\in E_v}, (p_e)_{e\in E_v})$$

$$\leq \sum_{\varepsilon\in E_v} \int_{\{((S_e)_{e\in E_v},(p_e)_{e\in E_v}) \mid \max_e \mathrm{Lip}(S_e)\geq a_0\}} p(\alpha a_0)\, d\lambda_v((S_e)_{e\in E_v}, (p_e)_{e\in E_v})$$

$$\leq p(\alpha a_0) \max_w \Big(\mathrm{card}(E_w) \lambda_w(\{((S_e)_{e\in E_w}, (p_e)_{e\in E_w}) \mid \max_e \mathrm{Lip}(S_e) \geq a_0\}) \Big). \tag{4.3.8}$$

By combining (4.3.2) and (4.3.8),

$$p(\alpha a_0) = 0$$

which contradicts (4.3.5). Hence $a = 0$, and so $p \equiv 0$. □

We now turn toward the proof of Theorem 2.4.1.

Proof of Theorem 2.4.1

i) First observe that φ_u actually maps $\Xi_u \times \prod_{e\in E_u} (\Omega_{\tau(e)})_0$ into $(\Omega_u)_0$. Now let

$$\zeta = \Big(((S_e)_{e\in E_u}, (p_e)_{e\in E_u}), (\omega_{\tau(e)}) \Big) \in \Xi_u \times \prod_{e\in E_u} (\Omega_{\tau(e)})_0$$

and $\gamma_u = ((\gamma_\varnothing^{(u)}, (\gamma_\alpha^{(u)})_{\alpha\in E_u^{(*)}\setminus\{\varnothing\}}) \in \mathcal{P}(X_u) \times \prod_{u\in V} \mathcal{P}(X_{\tau(\alpha_{|\alpha|})})$ for $u \in V$. Write $S_\gamma(\omega_{\tau(e)}) = S_\gamma^{(\tau(e))}$ and $p_\gamma(\omega_{\tau(e)}) = p_\gamma^{(\tau(e))}$ for $\gamma \in E_{\tau(e)}^{(*)}$. Then

$$\Phi_u\Big((1_{\Xi_u} \times \prod_{e\in E_u} \mu_{\tau(e)})(\zeta)\Big)$$

$$= \Phi_u\Big(((S_e)_{e\in E_u}, (p_e)_{e\in E_u})), (\mu_{\tau(e)}(\omega_{\tau(e)}))_{e\in E_u} \Big)$$

$$= \sum_{e\in E_u} p_e \mu_{\tau(e)}(\omega_{\tau(e)}) \circ S_e^{-1}$$

$$= \sum_{e\in E_u} p_e \Big(\lim_n \sum_{\alpha\in E_{\tau(e)}^{(n)}} p_{\alpha|1}^{(\tau(e))} \dots p_{\alpha|n}^{(\tau(e))} \gamma_\alpha^{(\tau(e))} \circ (S_{\alpha|1}^{(\tau(e))} \circ \dots \circ S_{\alpha|n}^{(\tau(e))})^{-1} \Big) \circ S_e^{-1}$$

[by Lemma 4.2.3]

$$= \lim_n \sum_{e\in E_u} \sum_{\alpha\in E_{\tau(e)}^{(n)}} p_e p_{\alpha|1}^{(\tau(e))} \dots p_{\alpha|n}^{(\tau(e))} \gamma_\alpha^{(\tau(e))} \circ (S_e \circ S_{\alpha|1}^{(\tau(e))} \circ \dots \circ S_{\alpha|n}^{(\tau(e))})^{-1}$$

$$= \mu_u(\varphi_u(\zeta)).$$

This proves i).

iii) Let $M \subseteq \mathcal{P}(X_v)$ be closed w.r.t. the weak topology. It is easily seen by induction that the following holds

$$\begin{aligned}
&\left(T_\lambda^n((Q_u)_u)\right)_v(M) \\
&= \left(\Lambda_v \times \prod_{\alpha \in E_v^{(*)}} Q_{\tau(\alpha_{|\alpha|})}\right) \\
&\qquad \left\{\left(\omega, \left(\nu_\varnothing, (\nu_\alpha)_{\alpha \in E_v^{(*)} \setminus \{\varnothing\}}\right)\right) \in \Omega_v \times \left(\mathcal{P}(X_v) \times \prod_{\alpha \in E_v^{(*)} \setminus \{\varnothing\}} \mathcal{P}(X_{\tau(\alpha_{|\alpha|})})\right) \right| \\
&\qquad\quad \left. \sum_{\alpha \in E_v^{(n)}} p_{\alpha|1}(\omega) \ldots p_{\alpha||\alpha|}(\omega) \nu_\alpha \circ \left(S_{\alpha|1}(\omega) \circ \cdots \circ S_{\alpha||\alpha|}(\omega)\right)^{-1} \in M \right\}.
\end{aligned}$$

Hence (by Fatou's Lemma)

$$\begin{aligned}
&\limsup_n \left(T_\lambda^n((Q_u)_u)\right)_v(M) \\
&= \limsup_n \left(\left(\Lambda_v \times \prod_{\alpha \in E_v^{(*)}} Q_{\tau(\alpha_{|\alpha|})}\right)\right. \\
&\qquad \left\{\left(\omega, \left(\nu_\varnothing, (\nu_\alpha)_{\alpha \in E_v^{(*)} \setminus \{\varnothing\}}\right)\right) \in \Omega_v \times \left(\mathcal{P}(X_v) \times \prod_{\alpha \in E_v^{(*)} \setminus \{\varnothing\}} \mathcal{P}(X_{\tau(\alpha_{|\alpha|})})\right) \right| \\
&\qquad\quad \left. \left. \sum_{\alpha \in E_v^{(n)}} p_{\alpha|1}(\omega) \ldots p_{\alpha||\alpha|}(\omega) \nu_\alpha \circ \left(S_{\alpha|1}(\omega) \circ \cdots \circ S_{\alpha||\alpha|}(\omega)\right)^{-1} \in M \right\}\right) \\
&\leq \left(\Lambda_v \times \prod_{\alpha \in E_v^{(*)}} Q_{\tau(\alpha_{|\alpha|})}\right) \\
&\qquad \left(\limsup_n \left\{\left(\omega, \left(\nu_\varnothing, (\nu_\alpha)_{\alpha \in E_v^{(*)} \setminus \{\varnothing\}}\right)\right) \in \Omega_v \times \left(\mathcal{P}(X_v) \times \prod_{\alpha \in E_v^{(*)} \setminus \{\varnothing\}} \mathcal{P}(X_{\tau(\alpha_{|\alpha|})})\right) \right| \right. \\
&\qquad\quad \left. \left. \sum_{\alpha \in E_v^{(n)}} p_{\alpha|1}(\omega) \ldots p_{\alpha||\alpha|}(\omega) \nu_\alpha \circ \left(S_{\alpha|1}(\omega) \circ \cdots \circ S_{\alpha||\alpha|}(\omega)\right)^{-1} \in M \right\}\right).
\end{aligned}$$

Now Theorem 2.3.1 and the fact that M is closed yield

$$
\begin{aligned}
&\limsup_n \left(T_\lambda^n((Q_u)_u)\right)_v(M) \\
&\leq \left(\Lambda_v \times \prod_{\alpha\in E_v^{(*)}} Q_{\tau(\alpha_{|\alpha|})}\right) \\
&\qquad \left(\left\{\left(\omega,\left(\nu_\varnothing,(\nu_\alpha)_{\alpha\in E_v^{(*)}\setminus\{\varnothing\}}\right)\right)\in\Omega_v\times\left(\mathcal{P}(X_v)\times\prod_{\alpha\in E_v^{(*)}\setminus\{\varnothing\}}\mathcal{P}(X_{\tau(\alpha_{|\alpha|})})\right)\right|\right. \\
&\qquad\qquad \left.\left.\sum_{\alpha\in E_v^{(n)}} p_{\alpha|1}(\omega)\dots p_{\alpha||\alpha|}(\omega)\nu_\alpha\circ\left(S_{\alpha|1}(\omega)\circ\dots\circ S_{\alpha||\alpha|}(\omega)\right)^{-1}\in M\right\}\right)
\end{aligned}
$$

[by Lemma 4.3.3]

$$
\begin{aligned}
&= \left(\Lambda_v \times \prod_{\alpha\in E_v^{(*)}} Q_{\tau(\alpha_{|\alpha|})}\right) \\
&\qquad \left(\left\{\left(\omega,\left(\nu_\varnothing,(\nu_\alpha)_{\alpha\in E_v^{(*)}\setminus\{\varnothing\}}\right)\right)\in(\Omega_v)_0\times\left(\mathcal{P}(X_v)\times\prod_{\alpha\in E_v^{(*)}}\mathcal{P}(X_{\tau(\alpha_{|\alpha|})})\right)\right|\right. \\
&\qquad\qquad \left.\left.\mu_v(\omega)\in M\right\}\right) \\
&=\Lambda_v(\{\omega\in(\Omega_v)_0 \mid \mu_v(\omega)\in M\}) \\
&=(\lambda_v\circ\mu_v^{-1})(M) = P_{\lambda,v}(M).
\end{aligned}
$$

As M was arbitrary the above inequality implies that

$$
T_\lambda^n((Q_u)_u)_v \xrightarrow{\text{w}} P_{\lambda,v}.
$$

ii) We have

$$
\begin{aligned}
T_\lambda((P_{\lambda,u})_u)_v &= \left(\lambda_u\times\prod_{e\in E_v} P_{\lambda,\tau(e)}\right)\circ\Phi_v^{-1} \\
&= \left(\lambda_v\times\prod_{e\in E_v}\Lambda_{\tau(e)}\circ\mu_{\tau(e)}^{-1}\right)\circ\Phi_v^{-1} \\
&= \left(\lambda_v\times\prod_{e\in E_v}\Lambda_{\tau(e)}\right)\circ\left(\lambda_v\times\prod_{e\in E_v}\mu_{\tau(e)}\right)^{-1}\circ\Phi_v^{-1}
\end{aligned}
$$

$$
\begin{aligned}
&[\text{by i)}]\\
&= \left(\lambda_v \times \prod_{e\in E_v} \Lambda_{\tau(e)}\right) \circ \varphi_v^{-1} \circ \mu_v^{-1}\\
&[\text{by Lemma 4.3.1}]\\
&= \Lambda_v \circ \mu_v^{-1}\\
&= P_{\lambda,v}
\end{aligned}
$$

and iii) clearly implies that $(P_{\lambda,u})_u$ is the only fixed point of T_λ. □

Proof of Theorem 2.4.2

Follows easily from Theorem 2.4.1 and Lemma 4.3.1. □

Lemma 4.3.4. *Let X_v be compact for all v and assume that there exist numbers $0 < T < 1$ and $0 < c$ such that*

$$p_e \leq T \quad \Lambda_v\text{-a.s.}$$

for all $v \in V$ and $e \in E_v$, and

$$\operatorname{dist}\left(S_{\alpha e}(X_{\tau(e)}), S_{\alpha\varepsilon}(X_{\tau(\varepsilon)})\right) \geq c \quad \Lambda_v\text{-a.s.}$$

for all $v \in V$, $\alpha \in E_v$ and $e, \varepsilon \in E_{\tau(\alpha_{|\alpha|})}$ with $e \neq \varepsilon$. Then the following assertion holds for all $v \in V$ and Λ_v-a.a. $\omega \in \Omega_v$:

$$\mu_v(\omega)(\{x\}) = 0 \quad \text{for } x \in X_v.$$

Proof. Let $v \in V$ and write

$$
\begin{aligned}
\Omega_v^\circ = (\Omega_v)_0 \bigcup \{\omega \in \Omega_v \mid &1)\ p_e(\omega) \leq T \text{ for } w \in V \text{ and } e \in E_w;\\
&2)\quad \operatorname{dist}\left(S_{\alpha e}(\omega)(X_{\tau(e)}), S_{\alpha\varepsilon}(\omega)(X_{\tau(\varepsilon)})\right) \geq c\\
&\qquad \text{for all } w \in V,\ \alpha \in E_w \text{ and } e, \varepsilon \in E_{\tau(\alpha_{|\alpha|})} \text{ with } e \neq \varepsilon\}.
\end{aligned}
$$

Clearly $\Lambda_v(\Omega_v^\circ) = 1$. We claim that $\mu_v(\omega)(\{x\}) = 0$ for $\omega \in \Omega_v^\circ$ and $x \in X_v$. Let $\omega \in \Omega_v^\circ$ and $x \in X_v$. If $x \in X_v \setminus C_v(\omega)$, $\mu_v(\omega)(\{x\}) = 0$. Now assume that $x = \pi_v(\omega)(\sigma) \in C_v(\omega)$. As $C_{\sigma|n}(\omega) \searrow \{x\}$, $\mu_v(\omega)(\{x\}) = \inf_n \mu_v(\omega)(C_{\sigma|n}(\omega)) = \inf_n \prod_{k=1}^n p_{\sigma|k}(\omega) \leq \inf_n T^n = 0$ (since $\mu_v(\omega)(C_{\sigma|n}(\omega)) = \prod_{k=1}^n p_{\sigma|k}(\omega)$ for all n by Theorem 2.3.3). □

4.4 Proofs of the Theorems in Section 2.5

Proof of Proposition 2.5.1

i) The matrix $A(q,t)$ is clearly non-negative. Now fix $u, v \in V$. Since (E, V) is strongly connected there exists a path $\alpha \in E^{(*)}_{uv}$ from u to v, whence

$$\left(A(q,t)^{|\alpha|}\right)_{u,v} = \sum_{\beta \in E^{|\alpha|}_{uv}} \prod_{k=1}^{|\alpha|} A_{\iota(\beta_k),\tau(\beta_k)}(q,,t) \geq \prod_{k=1}^{|\alpha|} A_{\iota(\alpha_k),\tau(\alpha_k)}(q,,t) > 0\,.$$

This shows that $A(q,t)$ is irreducible.

ii) Fix $q, t \in \mathbb{R}$. It follows easily from condition (I) that each entry $A_{u,v}(q,t)$ in $A(q,t)$ is a real analytic function of q and t. Now, the largest zero of a polynomial is an analytic function of the coefficients of the polynomial in the region where that zero is a simple zero. The matrix $A(q,t)$ is irreducible so the spectral radius $\Phi(q,t)$ is a simple zero of the characteristic polynomial.

iii) Since $A(p,t)$ is irreducible there exists a positive Perron-Frobenius eigenvector $(\rho_u)_{u\in V}$ with eigenvalue $\Phi(p,t)$, i.e.

$$\sum_v A_{u,v}(p,t)\rho_v = \Phi(p,t)\rho_u \qquad \text{for all} \quad u \in V\,.$$

Since $0 < p_e < 1$ for λ_v-a.e. for all $v \in V$,

$$\begin{aligned}
\sum_v A_{u,v}(q,t)\rho_v &= \sum_v \mathbf{E}_{\lambda_v}\left[\sum_{e\in E_{uv}} p_e^q \operatorname{Lip}(S_e)^t\right]\rho_v \\
&= \sum_v \int \left(\sum_{e\in E_{uv}} p_e^q \operatorname{Lip}(S_e)^t\right) d\lambda_u\ \rho_v \\
&< \sum_v \int \left(\sum_{e\in E_{uv}} p_e^p \operatorname{Lip}(S_e)^t\right) d\lambda_u\ \rho_v \\
&= \sum_v A_{u,v}(q,t)\rho_v = \rho_u\,.
\end{aligned}$$

The Perron-Frobenius theorem therefore implies that

$$\Phi(p,t) > \Phi(q,t)\,.$$

iv) Similar to the proof of iii).

v) Proof of $\lim_{t\to-\infty}\Phi(q,t) = \infty$: First observe that

$$A_{u,v}(q,t) \to \infty \quad \text{as} \quad t \to -\infty \tag{4.4.1}$$

for all $u, v \in V$ with $E_{u,v} \neq \varnothing$. Indeed, $\sum_{e\in E_{u,v}} p_e^q \operatorname{Lip}(S_e)^t \nearrow \infty$ as $t \to -\infty$ for all $((S_e)_{e\in E_u}, (p_e)_{e\in E_u}) \in \Xi_u$, and the monotone convergence theorem therefore implies that $A_{u,v}(q,t) = \mathbf{E}_{\lambda_u}[\sum_{e\in E_{uv}} p_e^q \operatorname{Lip}(S_e)^t] \nearrow \infty$ as $t \to -\infty$.

Let $K \in \mathbb{R}$. Choose by (4.4.1) a number t_0 such that $A_{u,v}(q,t) \geq K$ for $t \leq t_0$ and $u, v \in V$ with $E_{u,v} \neq \varnothing$. Fix $t \leq t_0$. Since each row in $A(q,t)$ has at least one non-zero entry, $\sum_v A_{u,v}(q,t) \geq K$ for each u, and so

$$A(q,t)\begin{pmatrix}1\\ \vdots\\ 1\end{pmatrix} \geq K\begin{pmatrix}1\\ \vdots\\ 1\end{pmatrix}.$$

The Perron-Frobenius theorem therefore implies that $\Phi(q,t) \geq K$ for $t \leq t_0$.

Proof of $\lim_{t\to\infty} \Phi(q,t) = 0$: First observe that

$$A_{u,v}(q,t) \to 0 \quad \text{as} \quad t \to \infty \tag{4.4.2}$$

Indeed, $\sum_{e\in E_{u,v}} p_e^q \operatorname{Lip}(S_e)^t \to 0$ as $t \to \infty$ and (by (I)) $\sum_{e\in E_{u,v}} p_e^q \operatorname{Lip}(S_e)^t \leq \operatorname{card}(E_{uv})e^{|q\log\Delta|}$ for all $((S_e)_{e\in E_u}, (p_e)_{e\in E_u}) \in \Xi_u$. The dominated convergence theorem therefore implies that $A_{u,v}(q,t) = \mathbf{E}_{\lambda_u}[\sum_{e\in E_{uv}} p_e^q \operatorname{Lip}(S_e)^t] \to 0$ as $t \to \infty$.

Let $\varepsilon > 0$. Choose by (4.4.2) a number t_0 such that $A_{u,v}(q,t) \leq \varepsilon/\operatorname{card}(V)$ for $t \geq t_0$ and $u, v \in V$. Fix $t \geq t_0$. Then clearly $\sum_v A_{uv}(q,t) \leq \operatorname{card}(V)\varepsilon/\operatorname{card}(V) = \varepsilon$ for each u, whence

$$A(q,t)\begin{pmatrix}1\\ \vdots\\ 1\end{pmatrix} \leq \varepsilon\begin{pmatrix}1\\ \vdots\\ 1\end{pmatrix}.$$

The Perron-Frobenius theorem therefore implies that $\Phi(q,t) \leq \varepsilon$ for $t \geq t_0$.

vi) Similar to the proof of v).

vii) Since $A(q_i,t_i)$ is irreducible there exists a positive Perron-Frobenius eigenvector $(\rho_{iu})_{u\in V}$ for $A(q_i,t_i)$ with eigenvalue $\Phi(q_i,t_i)$, i.e.

$$\sum_v A_{u,v}(q_i,t_i)\rho_{iv} = \Phi(q_i,t_i)\rho_{iu} \qquad \text{for all} \quad u \in V.$$

Hence

$$\sum_v A_{u,v}(\alpha_1 q_1 + \alpha_2 q_2, \alpha_1 t_1 + \alpha_2 t_2)\rho_{1v}^{\alpha_1}\rho_{2v}^{\alpha_2}$$

$$= \sum_v \int \left(\sum_{e\in E_{u,v}} p_e^{\alpha_1 q_1 + \alpha_2 q_2} \operatorname{Lip}(S_e)^{\alpha_1 t_1 + \alpha_2 t_2} \right) d\lambda_u \ \rho_{1v}^{\alpha_1}\rho_{2v}^{\alpha_2}$$

$$= \sum_v \int \left(\sum_{e\in E_{u,v}} \left(p_e^{q_1} \operatorname{Lip}(S_e)^{t_1}\rho_{1v}\right)^{\alpha_1} \left(p_e^{q_2} \operatorname{Lip}(S_e)^{t_1}\rho_{2v}\right)^{\alpha_2} \right) d\lambda_u$$

[by Hölder's inequality]

$$\leq \left(\sum_v \int \sum_{e\in E_{u,v}} p_e^{q_1} \operatorname{Lip}(S_e)^{t_1}\rho_{1v}\, d\lambda_u \right)^{\alpha_1} \cdot$$

$$\left(\sum_v \int \sum_{e\in E_{u,v}} p_e^{q_2} \operatorname{Lip}(S_e)^{t_2}\rho_{1v}\, d\lambda_u \right)^{\alpha_2}$$

$$= (A_{u,v}(q_1,t_1)\rho_{1v})^{\alpha_1} (A_{u,v}(q_2,t_2)\rho_{1v})^{\alpha_2}$$

$$= \Phi(q_1,t_1)^{\alpha_1}\Phi(q_2,t_2)^{\alpha_2}\rho_{1u}^{\alpha_1}\rho_{2u}^{\alpha_2}$$

and the Perron-Frobenius theorem therefore implies that $\Phi(\alpha_1 q_1 + \alpha_2 q_2, \alpha_1 t_1 + \alpha_2 t_2) \leq \Phi(q_1,t_1)^{\alpha_1}\Phi(q_2,t_2)^{\alpha_2}$. □

Proof of Proposition 2.5.2

i) The function Φ is real analytic and neither of the partial derivatives vanishes. The implicit function theorem therefore implies that β is analytic.
ii) Let $q < p$ and assume that $\beta(q) \leq \beta(p)$. Since Φ is strictly decreasing in each variable,

$$1 = \Phi(q, \beta(q)) > \Phi(p, \beta(q)) \geq \Phi(p, \beta(p)) = 1$$

which is a contradiction.
iii) Follows from Proposition 2.5.1.
iv) Let $q, p \in \mathbb{R}$ and $\alpha \in [0,1]$. Then

$$\begin{aligned}\Phi(\alpha q + (1-\alpha)p, \alpha\beta(q) + (1-\alpha)\beta(p)) &\leq \Phi(q,\beta(q))^{\alpha}\Phi(p,\beta(p))^{1-\alpha}\\ &= 1^{\alpha}1^{1-\alpha} = 1\\ &= \Phi(\alpha q + (1-\alpha)p, \beta(\alpha q + (1-\alpha)p))\end{aligned}$$

whence $\alpha\beta(q) + (1-\alpha)\beta(p) \geq \beta(\alpha q + (1-\alpha)p)$.

v) We have

$$\sum_v A_{u,v}(1,0) = \sum_v \int \sum_{e\in E_{uv}} p_e \, d\lambda_u = \int \left(\sum_v \sum_{e\in E_{uv}} p_e \right) d\lambda_u = \int d\lambda_u = 1$$

for all $u \in V$, i.e.

$$A(1,0) \begin{pmatrix} 1 \\ \vdots \\ 1 \end{pmatrix} = \begin{pmatrix} 1 \\ \vdots \\ 1 \end{pmatrix} .$$

Perron-Frobenius theorem therefore implies that $\Phi(1,0) = 1$, whence $\beta(1) = 0$. □

Observe that if $\alpha \in E_u^{(*)}$, then independence implies that

$$\mathbf{E}_{\lambda_u}[p_{\alpha_1}^q \operatorname{Lip}(S_{\alpha_1})^t] \left(\prod_{i=2}^{|\alpha|} \mathbf{E}_{\lambda_{\tau(\alpha_{i-1})}}[p_{\alpha_i}^q \operatorname{Lip}(S_{\alpha_i})^t] \right) = \mathbf{E}_{\Lambda_u} \left[\prod_{i=1}^{|\alpha|} p_{\alpha|i}^q \operatorname{Lip}(S_{\alpha|i})^t \right] . \tag{4.4.3}$$

This fact will be used tactically throughout the rest of the exposition. Let $n \in \mathbb{N}$. It follows from (2.5.7) and (2.5.8) that

$$\sum_v \sum_{\alpha\in E_{uv}^{(n)}} \rho_u^{-1} \mathbf{E}_{\lambda_u}[p_{\alpha_1}^q \operatorname{Lip}(S_{\alpha_1})^{\beta(q)}] \left(\prod_{i=2}^{n} \mathbf{E}_{\lambda_{\tau(\alpha_{i-1})}}[p_{\alpha_i}^q \operatorname{Lip}(S_{\alpha_i})^{\beta(q)}] \right) \rho_v = 1 \tag{4.4.4}$$

$$\sum_u \sum_{\alpha\in E_{uv}^{(n)}} \sigma_u \mathbf{E}_{\lambda_u}[p_{\alpha_1}^q \operatorname{Lip}(S_{\alpha_1})^{\beta(q)}] \left(\prod_{i=2}^{n} \mathbf{E}_{\lambda_{\tau(\alpha_{i-1})}}[p_{\alpha_i}^q \operatorname{Lip}(S_{\alpha_i})^{\beta(q)}] \right) \sigma_v^{-1} = 1 \tag{4.4.5}$$

for all $u, v \in V$. Equations (2.5.10) and (4.4.4) imply that

$$\begin{aligned} 1 &= \sum_u \sigma_u \rho_u \\ &= \sum_u \sum_v \sum_{\gamma\in E_{uv}^{(n)}} \sigma_u \mathbf{E}_{\lambda_u}[p_{\gamma_1}^q \operatorname{Lip}(S_{\gamma_1})^{\beta(q)}] \left(\prod_{i=2}^{n} \mathbf{E}_{\lambda_{\tau(\gamma_{i-1})}}[p_{\gamma_i}^q \operatorname{Lip}(S_{\gamma_i})^{\beta(q)}] \right) \rho_v . \end{aligned}$$

Hence, by differentiation w.r.t. q (here prime $'$ denotes differentiation w.r.t. q),

$$
\begin{aligned}
0 =& 1' \\
=& \left(\sum_u \sum_v \sum_{\gamma \in E_{uv}^{(n)}} \sigma_u \left(\mathbf{E}_{\lambda_u}[p_{\gamma_1}^q \operatorname{Lip}(S_{\gamma_1})^{\beta(q)}] \left(\prod_{i=2}^n \mathbf{E}_{\lambda_{\tau(\gamma_{i-1})}}[p_{\gamma_i}^q \operatorname{Lip}(S_{\gamma_i})^{\beta(q)}] \right) \right) \rho_v \right) \\
=& \sum_u \sum_v \sum_{\gamma \in E_{uv}^{(n)}} \sigma_u' \left(\mathbf{E}_{\lambda_u}[p_{\gamma_1}^q \operatorname{Lip}(S_{\gamma_1})^{\beta(q)}] \left(\prod_{i=2}^n \mathbf{E}_{\lambda_{\tau(\gamma_{i-1})}}[p_{\gamma_i}^q \operatorname{Lip}(S_{\gamma_i})^{\beta(q)}] \right) \right) \rho_v + \\
& \sum_u \sum_v \sum_{\gamma \in E_{uv}^{(n)}} \sigma_u \left(\mathbf{E}_{\lambda_u}[p_{\gamma_1}^q \operatorname{Lip}(S_{\gamma_1})^{\beta(q)}] \left(\prod_{i=2}^n \mathbf{E}_{\lambda_{\tau(\gamma_{i-1})}}[p_{\gamma_i}^q \operatorname{Lip}(S_{\gamma_i})^{\beta(q)}] \right) \right)' \rho_v + \\
& \sum_u \sum_v \sum_{\gamma \in E_{uv}^{(n)}} \sigma_u \left(\mathbf{E}_{\lambda_u}[p_{\gamma_1}^q \operatorname{Lip}(S_{\gamma_1})^{\beta(q)}] \left(\prod_{i=2}^n \mathbf{E}_{\lambda_{\tau(\gamma_{i-1})}}[p_{\gamma_i}^q \operatorname{Lip}(S_{\gamma_i})^{\beta(q)}] \right) \right) \rho_v' \\
& \text{[by (4.4.4) and (4.4.5)]} \\
=& \sum_u \sigma_u' \rho_u + \\
& \sum_u \sum_v \sum_{\gamma \in E_{uv}^{(n)}} \sigma_u \left(\mathbf{E}_{\lambda_u}[p_{\gamma_1}^q \operatorname{Lip}(S_{\gamma_1})^{\beta(q)}] \left(\prod_{i=2}^n \mathbf{E}_{\lambda_{\tau(\gamma_{i-1})}}[p_{\gamma_i}^q \operatorname{Lip}(S_{\gamma_i})^{\beta(q)}] \right) \right)' \rho_v + \\
& \sum_v \sigma_v \rho_v' \\
=& \left(\sum_u \sigma_u \rho_u \right)' + \\
& \sum_u \sum_v \sum_{\gamma \in E_{uv}^{(n)}} \sigma_u \left(\mathbf{E}_{\lambda_u}[p_{\gamma_1}^q \operatorname{Lip}(S_{\gamma_1})^{\beta(q)}] \left(\prod_{i=2}^n \mathbf{E}_{\lambda_{\tau(\gamma_{i-1})}}[p_{\gamma_i}^q \operatorname{Lip}(S_{\gamma_i})^{\beta(q)}] \right) \right)' \rho_v \\
& \text{[by (2.5.10)]} \\
=& \sum_u \sum_v \sum_{\gamma \in E_{uv}^{(n)}} \sigma_u \left(\mathbf{E}_{\lambda_u}[p_{\gamma_1}^q \operatorname{Lip}(S_{\gamma_1})^{\beta(q)}] \left(\prod_{i=2}^n \mathbf{E}_{\lambda_{\tau(\gamma_{i-1})}}[p_{\gamma_i}^q \operatorname{Lip}(S_{\gamma_i})^{\beta(q)}] \right) \right)' \rho_v \\
=& \sum_u \sum_v \sum_{\gamma \in E_{uv}^{(n)}} \sigma_u \mathbf{E}_{\Lambda_u} \left[\prod_{i=1}^n p_{\gamma_i}^q \operatorname{Lip}(S_{\gamma_i})^{\beta(q)} \right]' \rho_v \,. \qquad (4.4.6)
\end{aligned}
$$

Since $|\frac{d}{dq}(p_e^q \operatorname{Lip}(S_e)^{\beta(q)})| = \left|\log(p_e) p_e^q \beta'(q) \operatorname{Lip}(S_e)^\beta(q)\right| \leq |\beta'(q)||\log\Delta|(1 \vee$

$\Delta^q)(1 \vee \Delta^{\beta(q)})$ and β and β' are bounded on compact intervals,

$$\mathbf{E}_{\Lambda_u}\left[\prod_{i=1}^{n} p_{\gamma_i}^q \operatorname{Lip}(S_{\gamma_i})^{\beta(q)}\right]'$$
$$= \int \frac{d}{dq}\left(\prod_{i=1}^{n} p_{\gamma|i}^q \operatorname{Lip}(S_{\gamma|i})^{\beta(q)}\right) d\Lambda_u$$
$$= \int \log\left(\prod_{i=1}^{n} p_{\gamma|i}\right)\left(\prod_{i=1}^{n} p_{\gamma|i}\right)^q \left(\prod_{i=1}^{n} \operatorname{Lip}(S_{\gamma|i})\right)^{\beta(q)} d\Lambda_u +$$
$$\beta'(q)\int \log\left(\prod_{i=1}^{n} \operatorname{Lip}(S_{\gamma|i})\right)\left(\prod_{i=1}^{n} p_{\gamma|i}\right)^q \left(\prod_{i=1}^{n} \operatorname{Lip}(S_{\gamma|i})\right)^{\beta(q)} d\Lambda_u\,. \tag{4.4.7}$$

Substituting (4.4.6) into (4.4.7) and solving for $\beta'(q)$ yield

$$\beta'(q) = -\frac{\sum_u \sum_v \sum_{\gamma\in E_{uv}^{(n)}} \sigma_u(q) \int \log\left(\prod_{i=1}^n p_{\gamma|i}\right)\left(\prod_{i=1}^n p_{\gamma|i}\right)^q \left(\prod_{i=1}^n \operatorname{Lip}(S_{\gamma|i})\right)^{\beta(q)} d\Lambda_u \rho_v(q)}{\sum_u \sum_v \sum_{\gamma\in E_{uv}^{(n)}} \sigma_u(q) \int \log\left(\prod_{i=1}^n \operatorname{Lip}(S_{\gamma|i})\right)\left(\prod_{i=1}^n p_{\gamma|i}\right)^q \left(\prod_{i=1}^n \operatorname{Lip}(S_{\gamma|i})\right)^{\beta(q)} d\Lambda_u \rho_v(q)} \tag{4.4.8}$$

Define for $n \in \mathbb{N}$, $\overline{a}_n$ and $\underline{a}_n$ by

$$\overline{a}_n = \sup_{\gamma\in E^{(n)}} \|\ell_\gamma\|_\infty\,, \quad \underline{a}_n = \inf_{\gamma\in E^{(n)}} \|\ell_\gamma\|_{-\infty}\,.$$

Observe that for $\alpha_1, \ldots, \alpha_n \in E^{(*)}$ and $\tau(\alpha_i) = \iota(\alpha_{i+1})$,

$$\kappa_{\alpha_1\ldots\alpha_n} = \log\left(\prod_{i=1}^{|\alpha_1|} p_{\alpha_1|i}\right) + \log\left(\prod_{i=1}^{|\alpha_2|} p_{\alpha_1\alpha_2||\alpha_1|+i}\right) + \ldots$$
$$+ \log\left(\prod_{i=1}^{|\alpha_n|} p_{\alpha_1\ldots\alpha_n||\alpha_1\ldots\alpha_{n-1}|+i}\right)$$
$$= \kappa_{\alpha_1} + \kappa_{\alpha_2}(\mathcal{S}_{\alpha_1}) + \cdots + \kappa_{\alpha_n}(\mathcal{S}_{\alpha_1\ldots\alpha_{n-1}}) \tag{4.4.9}$$

and similarly

$$\chi_{\alpha_1\ldots\alpha_n} = \chi_{\alpha_1} + \chi_{\alpha_2}(\mathcal{S}_{\alpha_1}) + \cdots + \chi_{\alpha_n}(\mathcal{S}_{\alpha_1\ldots\alpha_{n-1}})\,. \tag{4.4.10}$$

Lemma 4.4.1. *We have for $u \in V$ and $n \in \mathbb{N}$*

i) $\Lambda_u(\{\omega \in \Omega_u \mid \underline{a} \leq \ell_\gamma(\mathcal{S}_\alpha(\omega)) \leq \overline{a}$ *for all* $\alpha \in E_u^{(*)}$ *and* $\gamma \in E_{\tau(\alpha_{|\alpha|})}^{(*)}\}) = 1$.

ii) $\Lambda_u(\{\omega \in \Omega_u \mid \underline{a}_n \leq \ell_\gamma(\mathcal{S}_\alpha(\omega)) \leq \overline{a}_n$ *for all* $\alpha \in E_u^{(*)}$ *and* $\gamma \in E_{\tau(\alpha_{|\alpha|})}^{(n)}\}) = 1$.

Proof. i) Since $\Lambda_u \circ \mathcal{S}_\alpha^{-1} = \Lambda_{\tau(\alpha_{|\alpha|})}$ (cf. Lemma 4.3.2),

$$\begin{aligned}
&\Lambda_u(\{\omega \in \Omega_u \mid \underline{a} \leq \ell_\gamma(\mathcal{S}_\alpha(\omega)) \leq \overline{a}\}) \\
=&\Lambda_u(\mathcal{S}_\alpha^{-1}\{\omega \in \Omega_{\tau(\alpha_{|\alpha|})} \mid \underline{a} \leq \ell_\gamma(\omega) \leq \overline{a}\}) \\
=&\Lambda_{\tau(\alpha_{|\alpha|})}(\{\omega \in \Omega_{\tau(\alpha_{|\alpha|})} \mid \underline{a} \leq \ell_\gamma(\omega) \leq \overline{a}\}) = 1
\end{aligned} \tag{4.4.11}$$

The result in i) now follows from (4.4.11) and the fact that Z is finite and $E_u^{(*)}$ is countable.
ii) Similar to the proof of i). □

Lemma 4.4.2. *Assume conditions (I) and (III). Then*

i) $\lim_n \overline{a}_n = \overline{a}$.

ii) $\lim_n \underline{a}_n = \underline{a}$.

Proof. i) Proof of $\limsup_n \overline{a}_n \leq \overline{a}$: If $\gamma \in E_u^{(n)}$ is a path of length n, it may be partitioned into simple cycles $\gamma_1, \ldots, \gamma_k$ plus at most card(E) additional edges $e_1, \ldots, e_m$ such that $\alpha = \gamma_1 \ldots \gamma_k e_1 \ldots e_m$. Hence

$$\begin{aligned}
\kappa_\alpha =& \kappa_{\gamma_1} + \kappa_{\gamma_2}(\mathcal{S}_{\gamma_1}) + \cdots + \kappa_{\gamma_k}(\mathcal{S}_{\gamma_1 \ldots \gamma_{k-1}}) + \\
&\quad \kappa_{e_1 \ldots e_m}(\mathcal{S}_{\gamma_1 \ldots \gamma_k}) \\
\geq& \kappa_{\gamma_1} + \kappa_{\gamma_2}(\mathcal{S}_{\gamma_1}) + \cdots + \kappa_{\gamma_k}(\mathcal{S}_{\gamma_1 \ldots \gamma_{k-1}}) + \\
&\quad \log\left(\mathrm{Lip}(S_{\gamma_1 \ldots \gamma_k e_1})^{\overline{a}} \cdot \ldots \cdot \mathrm{Lip}(S_{\gamma_1 \ldots \gamma_k e_1 \ldots e_m})^{\overline{a}} \left(\frac{\Delta}{T^{\overline{a}}}\right)^m\right) \\
\geq& \kappa_{\gamma_1} + \kappa_{\gamma_2}(\mathcal{S}_{\gamma_1}) + \cdots + \kappa_{\gamma_k}(\mathcal{S}_{\gamma_1 \ldots \gamma_{k-1}}) + \\
&\quad \overline{a}\chi_{e_1 \ldots e_m}(\mathcal{S}_{\gamma_1 \ldots \gamma_k e_1}) + \mathrm{card}(E)(0 \wedge \log(\Delta/T^{\overline{a}})) \\
&\text{[by Lemma 4.4.1]} \\
=& \overline{a}\chi_{\gamma_1} + \overline{a}\chi_{\gamma_2}(\mathcal{S}_{\gamma_1}) + \cdots + \overline{a}\chi_{\gamma_k}(\mathcal{S}_{\gamma_1 \ldots \gamma_{k-1}}) + \\
&\quad \overline{a}\chi_{e_1 \ldots e_m}(\mathcal{S}_{\gamma_1 \ldots \gamma_k e_1}) + \mathrm{card}(E)(0 \wedge \log(\Delta/T^{\overline{a}})) \\
=& \overline{a}\chi_\alpha(\omega) + \overline{c} \quad \Lambda_u\text{-a.s.}
\end{aligned}$$

where $\overline{c} = \mathrm{card}(E)(0 \wedge \log(\Delta/T^{\overline{a}}))$. Consequently

$$\ell_\alpha = \frac{\kappa_\alpha}{\chi_\alpha} \leq \overline{a} + \frac{\overline{c}}{\chi_\alpha} \leq \overline{a} + \frac{\overline{c}}{n \log T} \quad \Lambda_u\text{-a.s.}$$

Now, since $\alpha \in E_u^{(n)}$ was arbitrary the above inequality implies that $\overline{a}_n \leq \overline{a} + \frac{\overline{c}}{n \log T}$ for $n \in \mathbb{N}$, and so $\limsup_n \overline{a}_n \leq \overline{a}$.
Proof of $\overline{a} \leq \liminf_n \overline{a}_n$: Let $n \in \mathbb{N}$ and $\varepsilon > 0$. Since Z is finite there exists $\zeta \in Z$ such that $\overline{a} = \|\zeta\|_\infty$. For $k \in \mathbb{N}$ write $\zeta^k = \overbrace{\zeta \ldots \zeta}^{k \text{ times}}$ and $\zeta^0 = \varnothing$. Now write $n = |\zeta| m + r$ where $m \in \mathbb{N} \cup \{0\}$ and $r \in \{0, \ldots, |\zeta| - 1\}$ and put $\gamma = \zeta^m \zeta_1 \ldots \zeta_r$. Let

$$M := \{\omega \in \Omega_{\iota(\zeta_1)} \mid \overline{a} - \varepsilon \leq \ell_\zeta(\omega)\}.$$

Then clearly

$$\Lambda_u(M) > 0.$$

Put

$$\begin{aligned}
A_0 = & \Bigg\{ \Big(((S_e)_{e \in E_u}, (p_e)_{e \in E_u}), ((S_{(\zeta|i)e})_{e \in E_{\tau(\zeta_i)}}, (p_{(\zeta|i)e})_{e \in E_{\tau(\zeta_i)}})_{i=1,\ldots,|\zeta|-1} \Big) \in \\
& \qquad \Xi_u \times \prod_{i=1}^{|\zeta|-1} \Xi_{\tau(\zeta_i)} \Bigg| \\
& \qquad \overline{a} - \varepsilon \leq \frac{\log(p_{\zeta|1} \cdot \ldots \cdot p_{\zeta||\zeta|})}{\log(\mathrm{Lip}(S_{\zeta|1}) \cdot \ldots \cdot \mathrm{Lip}(S_{\zeta||\zeta|}))} \Bigg\} \\
A_k = & \Bigg\{ \Big(((S_{\zeta^k(\zeta|i)e})_{e \in E_{\tau(\zeta^k(\zeta|i)_{|\zeta^k|+i})}}, (p_{(\zeta|i)e})_{e \in E_{\tau(\zeta^k(\zeta|i)_{|\zeta^k|+i})}})_{i=0,\ldots,|\zeta|-1} \Big) \in \\
& \qquad \prod_{i=0}^{|\zeta|-1} \Xi_{\tau(\zeta^k(\zeta|i)_{|\zeta^k|+i})} \Bigg| \\
& \qquad \overline{a} - \varepsilon \leq \frac{\log(p_{\zeta^k(\zeta|1)} \cdot \ldots \cdot p_{\zeta^k(\zeta||\zeta|)})}{\log(\mathrm{Lip}(S_{\zeta^k(\zeta|1)}) \cdot \ldots \cdot \mathrm{Lip}(S_{\zeta^k(\zeta||\zeta|)}))} \Bigg\} \\
& \qquad \text{for} \quad k = 1, \ldots, m-1 \\
A = & A_0 \times A_1 \times \cdots \times A_{m-1} \times \prod_{\substack{\alpha \in E_u^{(*)} \setminus \{\varnothing\} \\ \alpha \neq \varnothing, \zeta^m|1, \ldots, \zeta^m||\zeta^m|-1}} \Xi_{\tau(\alpha_{|\alpha|})}.
\end{aligned}$$

Then

$$\begin{aligned}
\Lambda_u(A) &= \left(\lambda_u \times \prod_{i=1}^{|\zeta|-1} \lambda_{\tau(\zeta_i)} \right)(A_0) \cdot \prod_{k=1}^{m-1} \left(\left(\prod_{i=0}^{|\zeta|-1} \lambda_{\tau(\zeta^k(\zeta|i)_{|\zeta^k|+i})} \right)(A_k) \right) \\
&= \left(\left(\lambda_u \times \prod_{i=1}^{|\zeta|-1} \lambda_{\tau(\zeta_i)} \right)(A_0) \right)^m
\end{aligned}$$

$$
\begin{aligned}
&=\Lambda_u\left(A_0 \times \prod_{\substack{\alpha\in E_u^{(*)}\setminus\{\varnothing\}\\ \alpha\neq\varnothing,\varsigma|1,\dots,\varsigma||\varsigma|-1}} \Xi_{\tau(\alpha_{|\alpha|})}\right)^m\\
&=\Lambda_u\left(\left\{\omega\in\Omega_{\iota(\varsigma_1)}\,\middle|\,\overline{a}-\varepsilon\le\frac{\log(p_{\varsigma|1}(\omega)\cdot\ldots\cdot p_{\varsigma||\varsigma|}(\omega))}{\log(\mathrm{Lip}(S_{\varsigma|1}(\omega))\cdot\ldots\cdot\mathrm{Lip}(S_{\varsigma||\varsigma|}(\omega)))}\right\}\right)^m\\
&=\Lambda_u(M)^m>0\,.
\end{aligned}
$$

For $\omega\in M$,

$$
\begin{aligned}
\kappa_\gamma(\omega)=&\kappa_\varsigma(\omega)+\kappa_\varsigma(S_\varsigma(\omega))+\cdots+\kappa_\varsigma(S_{\varsigma^{m-1}}(\omega))+\\
&\qquad\kappa_{\varsigma_1,\dots,\varsigma_r}(S_{\varsigma^m}(\omega))\\
\le&\kappa_\varsigma(\omega)+\kappa_\varsigma(S_\varsigma(\omega))+\cdots+\kappa_\varsigma(S_{\varsigma^{m-1}}(\omega))+\\
&\qquad\log\left(\mathrm{Lip}(S_{\varsigma^m\varsigma_1})^{\overline{a}-\varepsilon}\cdot\ldots\cdot\mathrm{Lip}(S_{\varsigma^m\varsigma_1\dots\varsigma_r})^{\overline{a}-\varepsilon}\left(\frac{T}{\Delta^{\overline{a}-\varepsilon}}\right)\right)\\
\le&\kappa_\varsigma(\omega)+\kappa_\varsigma(S_\varsigma(\omega))+\cdots+\kappa_\varsigma(S_{\varsigma^{m-1}}(\omega))+\\
&\qquad(\overline{a}-\varepsilon)\chi_{\varsigma_1\dots\varsigma_r}(S_{\varsigma^m}(\omega))+\mathrm{card}(E)(0\vee\log(T/\Delta^{\overline{a}-\varepsilon}))\\
&\left[\text{since }\omega\in M\right]\\
\le&(\overline{a}-\varepsilon)\chi_\varsigma(\omega)+(\overline{a}-\varepsilon)\chi_\varsigma(S_\varsigma(\omega))+\cdots+(\overline{a}-\varepsilon)\chi_\varsigma(S_{\varsigma^{m-1}}(\omega))+\\
&\qquad(\overline{a}-\varepsilon)\chi_{\varsigma_1\dots\varsigma_r}(S_{\varsigma^m}(\omega))+\mathrm{card}(E)(0\vee\log(T/\Delta^{\overline{a}-\varepsilon}))\\
\le&(\overline{a}-\varepsilon)\chi_\gamma(\omega)+\mathrm{card}(E)(0\vee\log(T/\Delta^{\overline{a}-\varepsilon}))=(\overline{a}-\varepsilon)\chi_\gamma(\omega)+\underline{c}_\varepsilon
\end{aligned}
$$

where $\underline{c}_\varepsilon=\mathrm{card}(E)(0\vee\log(T/\Delta^{\overline{a}-\varepsilon}))$. Hence

$$
\ell_\gamma(\omega)=\frac{\kappa_\gamma(\omega)}{\chi_\gamma(\omega)}\ge\overline{a}-\varepsilon+\frac{\underline{c}_\varepsilon}{\chi_\gamma(\omega)}\ge\overline{a}-\varepsilon+\frac{\underline{c}_\varepsilon}{n\log T} \tag{4.4.12}
$$

for $\omega\in M$. Now, since $\Lambda_u(M)>0$, inequality (4.4.12) implies that

$$
\overline{a}_n\ge\|\ell_\gamma\|_\infty\ge\overline{a}-\varepsilon+\frac{\underline{c}_\varepsilon}{n\log T}
$$

for $n\in\mathbb{N}$. Letting $n\to\infty$ finally yields $\liminf_n\overline{a}_n\ge\overline{a}-\varepsilon$ for all $\varepsilon>0$.
ii) Similar to the proof of i). □

We are now ready to prove Proposition 2.5.3.

Proof of Proposition 2.5.3

i) By (4.4.8),

$$\beta'(q)+\overline{a}_n$$
$$=\frac{\sum_u\sum_v\sum_{\gamma\in E^{(n)}_{uv}}\sigma_u(q)\int\kappa_\gamma\left(\prod_{i=1}^n p_{\gamma|i}\right)^q\left(\prod_{i=1}^n \mathrm{Lip}(S_{\gamma|i})\right)^{\beta(q)}d\Lambda_u\rho_v(q)}{\sum_u\sum_v\sum_{\gamma\in E^{(n)}_{uv}}\sigma_u(q)\int\chi_\gamma\left(\prod_{i=1}^n p_{\gamma|i}\right)^q\left(\prod_{i=1}^n \mathrm{Lip}(S_{\gamma|i})\right)^{\beta(q)}d\Lambda_u\rho_v(q)}+\overline{a}_n$$
$$=-\frac{\sum_u\sum_v\sum_{\gamma\in E^{(n)}_{uv}}\sigma_u(q)\int\overbrace{(\kappa_\gamma-\overline{a}_n\chi_\gamma)}^{\geq 0}\left(\prod_{i=1}^n p_{\gamma|i}\right)^q\left(\prod_{i=1}^n \mathrm{Lip}(S_{\gamma|i})\right)^{\beta(q)}d\Lambda_u\rho_v(q)}{\sum_u\sum_v\sum_{\gamma\in E^{(n)}_{uv}}\sigma_u(q)\int\chi_\gamma\left(\prod_{i=1}^n p_{\gamma|i}\right)^q\left(\prod_{i=1}^n \mathrm{Lip}(S_{\gamma|i})\right)^{\beta(q)}d\Lambda_u\rho_v(q)}$$
$$\geq 0$$

for all $n\in\mathbb{N}$. Hence (by Lemma 4.4.2)

$$\beta'(q)+\overline{a}=\beta'(q)+\lim_n\overline{a}_n\geq 0$$

and so $\frac{d}{dq}(\beta(q)+\overline{a}q)\geq 0$ for all $q\in\mathbb{R}$, i.e. $q\to\beta(q)+\overline{a}q$ is increasing.

ii) By the convexity of β it suffices to prove,

ii1) $aq+\beta(q)\not\to-\infty$ as $q\to-\infty$ for $a<\overline{a}$.
ii2) $aq+\beta(q)\to-\infty$ as $q\to-\infty$ for $\overline{a}<a$.

Proof of ii1). Let $a<\overline{a}$ and assume that

$$\beta(q)+aq\to-\infty\quad\text{as}\quad q\to-\infty. \tag{4.4.13}$$

Choose $\zeta\in Z$ such that $\overline{a}=\|\ell_\zeta\|_\infty$. Let $\iota(\zeta_1)=u$ and $|\zeta|=n$. It follows from (4.4.3) and (4.4.4) that

$$\rho_u(q)=\sum_v\sum_{\alpha\in E^{(n)}_{uv}}\int\prod_{i=1}^n p^q_{\alpha|i}\,\mathrm{Lip}(S_{\alpha|i})^{\beta(q)}\,d\Lambda_u\rho_v(q)$$
$$>\int\prod_{i=1}^n p^q_{\zeta|i}\,\mathrm{Lip}(S_{\zeta|i})^{\beta(q)}\,d\Lambda_u\rho_u(q)$$

whence (by Fatou's Lemma)

$$
\begin{aligned}
1 &\geq \liminf_{q\to-\infty} \int \prod_{i=1}^{n} p_{\zeta|i}^{q}\,\mathrm{Lip}(S_{\zeta|i})^{\beta(q)}\,d\Lambda_u \\
&= \liminf_{q\to-\infty} \int \left(\prod_{i=1}^{n} \mathrm{Lip}(S_{\zeta|i})\right)^{q\ell_\zeta+\beta(q)} d\Lambda_u \\
&\geq \liminf_{q\to-\infty} \int_{\{a\leq\ell_\zeta\leq\overline{a}\}} \left(\prod_{i=1}^{n} \mathrm{Lip}(S_{\zeta|i})\right)^{q\ell_\zeta+\beta(q)} d\Lambda_u \\
&\geq \int_{\{a\leq\ell_\zeta\leq\overline{a}\}} \liminf_{q\to-\infty} \left(\prod_{i=1}^{n} \mathrm{Lip}(S_{\zeta|i})\right)^{q\ell_\zeta+\beta(q)} d\Lambda_u\,. \qquad (4.4.14)
\end{aligned}
$$

Now, if $\omega \in \Omega_u$ and $a \leq \ell_\zeta(\omega)$, $aq + \beta(q) \geq \ell_\zeta(\omega)q + \beta(q)$ for $q < 0$, and (4.4.13) therefore implies that

$$\ell_\zeta(\omega)q + \beta(q) \to -\infty \quad \text{as} \quad q \to -\infty \qquad \text{for} \quad a \leq \ell_\zeta(\omega)\,. \qquad (4.4.15)$$

By combining (4.4.14) and (4.4.15),

$$1 \geq \int_{\{a\leq\ell_\zeta\leq\overline{a}\}} \infty\,d\Lambda_u = \infty\cdot\Lambda_u(\{a\leq\ell_\zeta\leq\overline{a}\}) = \infty \qquad (4.4.16)$$

since $\Lambda_u(\{a \leq \ell_\zeta \leq \overline{a}\}) > 0$ because $a < \overline{a} = \|\ell_\zeta\|_\infty$. Equation (4.4.16) yields the desired contradiction. This proves ii1).

Proof of ii2). Let $\overline{a} < a$. Since $q \to \overline{a}q + \beta(q)$ is increasing, $\overline{a}q + \beta(q) \leq \overline{a}0 + \beta(0) = \beta(0)$ for $q < 0$, whence

$$aq + \beta(q) = (a-\overline{a})q + (\overline{a}q + \beta(q)) \leq (a-\overline{a})q + \beta(0) \qquad \text{for} \quad q < 0\,.$$

Hence

$$\limsup_{q\to-\infty}(aq + \beta(q)) \leq \limsup_{q\to-\infty}((a-\overline{a})q + \beta(q)) = -\infty\,.$$

iii)–iv) Similar to the proofs of i) and ii).
v) Since $q \to \overline{a}q + \beta(q)$ is increasing, $\overline{e} := \lim_{q\to-\infty}(\overline{a}q + \beta(q))$ exists and

$$\overline{e} := \lim_{q\to-\infty}(\overline{a}q + \beta(q)) \in [-\infty, \infty[\,.$$

Now assume, in order to get a contradiction, that

$$\overline{e} := \lim_{q\to-\infty}(\overline{a}q + \beta(q)) = -\infty. \qquad (4.4.17)$$

Let $u := \iota(\zeta_1)$ and $n := |\zeta|$. Arguments similar to the proof of (4.4.14) shows that

$$1 \geq \int_{\{\ell_\zeta = \overline{a}\}} \liminf_{q \to -\infty} \left(\prod_{i=1}^{n} \mathrm{Lip}(S_{\zeta|i}) \right)^{q\ell_\zeta + \beta(q)} d\Lambda_u \,. \tag{4.4.18}$$

Now, if $\omega \in \{\ell_\zeta = \overline{a}\}$ then $\ell_\zeta(\omega) = \overline{a}$, and (4.4.17) therefore implies that

$$\ell_\zeta(\omega)q + \beta(q) \to -\infty \quad \text{as} \quad q \to \infty \qquad \text{for} \quad \omega \in \{\ell_\zeta = \overline{a}\} \tag{4.4.19}$$

By combining (4.4.18) and (4.4.19),

$$1 \geq \int_{\{\ell_\zeta = \overline{a}\}} \infty \, d\Lambda_u = \infty \cdot \Lambda_u(\{\ell_\zeta = \overline{a}\}) = \infty \,. \tag{4.4.20}$$

since $\Lambda_u(\{\ell_\zeta = \overline{a}\}) > 0$. Equation (4.4.20) yields the desired contradiction.
vi) Similar to the proof of v). □

We now turn toward the proof of Proposition 2.5.4. In order to prove this proposition we need the following result.

Theorem 4.4.3. *Let $n \in \mathbb{N}$ and A be an irreducible non-negative matrix with index of imprimitivity $h \in \mathbb{N}$, i.e.*

$$h := \mathrm{card}\{\lambda \in \mathbb{C} \mid \lambda \textit{ is an eigenvalue of } A \textit{ with } |\lambda| = r(A)\,\}$$

(recall that $r(\cdot)$ denotes spectral radius). Then the following are equivalent

i) *A^n is irreducible.*
ii) *n and h are relatively prime.*

Proof. See [Min, p. 66, Problem 8]. □

Proof of Proposition 2.5.4

i) ⇒ ii). Assume that $\underline{a} = \overline{a}$. For each $\varepsilon > 0$ then there exists an integer $N_\varepsilon \in \mathbb{N}$ such that $\underline{a}_n, \overline{a}_n \in]a - \varepsilon, a + \varepsilon[$ for $n \geq N_\varepsilon$. For $n \geq N_\varepsilon$ and $\gamma \in E^{(n)}$,

$$\frac{\log(p_{\gamma|1} \cdot \ldots \cdot p_{\gamma|n})}{\log(\mathrm{Lip}(S_{\gamma|1}) \cdot \ldots \cdot \mathrm{Lip}(S_{\gamma|n}))} = \ell_\gamma \in [\underline{a}_n, \overline{a}_n] \subseteq]a - \varepsilon, a + \varepsilon[\quad \Lambda_{\iota(\gamma_1)}\text{-a.s.}$$

Hence, for $\gamma \in E^{(n)}$, then we get $\Lambda_{\iota(\gamma_1)}$-a.s.,

$$\begin{aligned} \mathrm{Lip}(S_{\gamma|1})^{a+\varepsilon} \cdot \ldots \cdot \mathrm{Lip}(S_{\gamma|n})^{a+\varepsilon} &\leq p_{\gamma|1} \cdot \ldots \cdot p_{\gamma|n} \\ &\leq \mathrm{Lip}(S_{\gamma|1})^{a-\varepsilon} \cdot \ldots \cdot \mathrm{Lip}(S_{\gamma|n})^{a-\varepsilon} \,. \end{aligned} \tag{4.4.21}$$

Proof of $\beta(q) \le -\beta(0)q + \beta(0)$ for $0 \le q$: Let $\varepsilon > 0$ and $0 \le q$. Let h denote the index of imprimitivity of $A(q, -(a-\varepsilon)q + \beta(0))$. Now choose an integer $n \ge N_\varepsilon$ such that n and h are relatively prime. Since $0 \le q$, (4.4.21) implies that

$$\sum_v \sum_{\gamma \in E_{uv}^{(n)}} \mathbf{E}_{\lambda_u}[p_{\gamma_1}^q \operatorname{Lip}(S_{\gamma_1})^{-(a-\varepsilon)q+\beta(q)}].$$

$$\prod_{i=2}^n \mathbf{E}_{\lambda_{\tau(\gamma_{i-1})}}[p_{\gamma_i}^q \operatorname{Lip}(S_{\gamma_i}^{-(a-\varepsilon)q+\beta(q)}]\rho_v(0)$$

$$= \sum_v \sum_{\gamma \in E_{uv}^{(n)}} \int \left(\prod_{i=1}^n p_{\gamma|i}^q \operatorname{Lip}(S_{\gamma|i})^{-(a-\varepsilon)q+\beta(0)} \right) d\Lambda_u \ \rho_v(0)$$

$$\le \sum_v \sum_{\gamma \in E_{uv}^{(n)}} \int \left(\prod_{i=1}^n \operatorname{Lip}(S_{\gamma|i})^{q(a-\varepsilon)-(a-\varepsilon)q+\beta(0)} \right) d\Lambda_u \ \rho_v(0)$$

$$= \sum_v \sum_{\gamma \in E_{uv}^{(n)}} \int \left(\prod_{i=1}^n p_{\gamma|i}^0 \operatorname{Lip}(S_{\gamma|i})^{\beta(0)} \right) d\Lambda_u \ \rho_v(0)$$

[by (4.4.4)]

$$= \rho_u(0)$$

for $u \in V$. Hence

$$A(q, -(a-\varepsilon)q + \beta(0))^n \rho(0) \le \rho(0). \tag{4.4.22}$$

Now, since n and h are relatively prime, Theorem 4.4.3 implies that $A(q, -(a-\varepsilon)q + \beta(0))^n$ is irreducible. Equation (4.4.22) together with the Perron-Frobenius theorem therefore imply that

$$r\big(A(q, -(a-\varepsilon)q + \beta(0))^n\big) \le 1$$

(where $r(\cdot)$ denotes spectral radius), whence

$$\Phi(q, -(a-\varepsilon)q + \beta(0)) = r\big(A(q, -(a-\varepsilon)q + \beta(0))\big) \le 1,$$

and so $\beta(q) \le -(a-\varepsilon)q + \beta(0)$ for all $\varepsilon > 0$. Hence

$$\beta(q) \le -aq + \beta(0) \quad \text{for} \quad 0 \le q. \tag{4.4.23}$$

Proof of $\beta(q) \le -\beta(0)q + \beta(0)$ for $q < 0$: For $q < 0$ similar arguments (using the first inequality in (4.4.21)) show that $\beta(q) \le -(a+\varepsilon)q + \beta(0)$ for all $\varepsilon > 0$. Hence

$$\beta(q) \le -aq + \beta(0) \quad \text{for} \quad q < 0. \tag{4.4.24}$$

Proof of $\beta(q) \geq -\beta(0)q + \beta(0)$ for $0 \leq q$: Let $\varepsilon > 0$ and $0 \leq q$. Let h denote the index of imprimitivity of $A(q, -(a+\varepsilon)q + \beta(0))$. Now choose a sequence $(n_k)_k$ of integers such that $n_k \geq N_\varepsilon$, $n_k \to \infty$ as $k \to \infty$ and n_k and h are relatively prime. Since $0 \leq q$, (4.4.21) implies that

$$\sum_v \sum_{\gamma \in E_{uv}^{(n_k)}} \mathbf{E}_{\lambda_u}[p^q_{\gamma_1} \operatorname{Lip}(S_e)^{-(a+\varepsilon)q+\beta(q)}].$$

$$\prod_{i=2}^{n_k} \mathbf{E}_{\lambda_{\tau(\gamma_{i-1})}}[p^q_{\gamma_i} \operatorname{Lip}(S_e)^{-(a+\varepsilon)q+\beta(q)}]\rho_v(0)$$

$$=\sum_v \sum_{\gamma \in E_{uv}^{(n_k)}} \int \left(\prod_{i=1}^{n_k} p^q_{\gamma|i} \operatorname{Lip}(S_{\gamma|i})^{-(a+\varepsilon)q+\beta(0)}\right) d\Lambda_u \ \rho_v(0)$$

$$\geq\sum_v \sum_{\gamma \in E_{uv}^{(n_k)}} \int \left(\prod_{i=1}^{n_k} \operatorname{Lip}(S_{\gamma|i})^{q(a+\varepsilon)-(a+\varepsilon)q+\beta(0)}\right) d\Lambda_u \ \rho_v(0)$$

$$=\sum_v \sum_{\gamma \in E_{uv}^{(n_k)}} \int \left(\prod_{i=1}^{n_k} p^0_{\gamma|i} \operatorname{Lip}(S_{\gamma|i})^{\beta(0)}\right) d\Lambda_u \ \rho_v(0)$$

[by (4.4.4)]

$$=\rho_u(0)$$

for $u \in V$. Hence

$$A(q, -(a+\varepsilon)q + \beta(0))^{n_k} \rho(0) \geq \rho(0). \tag{4.4.25}$$

Now, since n_k and h are relatively prime, Theorem 4.4.3 implies that $A(q, -(a-\varepsilon)q + \beta(0))^{n_k}$ is irreducible. Equation (4.4.25) together with the Perron-Frobenius theorem therefore imply that

$$r(A^{n_k}) \geq 1$$

for $k \in \mathbb{N}$ where $A := A(q, -(a+\varepsilon)q + \beta(0))$. Hence

$$\|A^{n_k}\|^{\frac{1}{n_k}} \geq (r(A))^{\frac{1}{n_k}} \geq 1 \quad \text{for} \quad k \in \mathbb{N},$$

where $\|A^{n_k}\| := \sup_{\|x\| \leq 1} \|A^{n_k} x\|$ denotes the operator norm of A^{n_k} w.r.t. an arbitrary norm $\|\cdot\|$ on $\mathbb{R}^{\operatorname{card}(V)}$. Consequently (since $n_k \to \infty$)

$$\Phi(q, -(a+\varepsilon)q + \beta(q)) = r(A) = \lim_n \|A^n\|^{\frac{1}{n}} \geq 1$$

and so $\beta(q) \geq -(a+\varepsilon)q + \beta(0)$ for all $\varepsilon > 0$. Hence

$$\beta(q) \geq -aq + \beta(0) \quad \text{for} \quad 0 \leq q. \tag{4.4.26}$$

Proof of $\beta(q) \geq -\beta(0)q + \beta(0)$ for $q < 0$: For $q < 0$ similar arguments (using the second inequality in (4.4.21)) show that $\beta(q) \geq -(a-\varepsilon)q + \beta(0)$ for all $\varepsilon > 0$. Hence

$$\beta(q) \geq -aq + \beta(0) \quad \text{for} \quad q < 0. \tag{4.4.27}$$

The desired result now follows from (4.4.23), (4.4.24), (4.4.26) and (4.4.27).

ii) $\Rightarrow$ i). We claim that

$$\beta(0) \leq \underline{a}. \tag{4.4.28}$$

Otherwise there exists an $\varepsilon > 0$ and a simple cycle $\zeta \in Z$ such that

$$\Lambda_u(\ell_\zeta \leq \beta(0) - \varepsilon) > 0$$

where $u := \iota(\zeta_1)$. For $q > 0$,

$$\begin{aligned}
\rho_u(q) &= \sum_v \sum_{\gamma \in E_{uv}^{(|\zeta|)}} \int \left(\prod_{i=1}^{|\zeta|} p_{\gamma|i}^q \operatorname{Lip}(S_{\gamma|i})^{\beta(q)} \right) d\Lambda_u \, \rho_v(q) \\
&\geq \int \left(\prod_{i=1}^{|\zeta|} p_{\zeta|i}^q \operatorname{Lip}(S_{\zeta|i})^{\beta(q)} \right) d\Lambda_u \, \rho_u(q)
\end{aligned}$$

whence (since $\beta(q) = -\beta(0)q + \beta(0)$)

$$\begin{aligned}
1 &\geq \int \left(\prod_{i=1}^{|\zeta|} p_{\zeta|i}^q \operatorname{Lip}(S_{\zeta|i})^{-\beta(0)q+\beta(0)} \right) d\Lambda_u \\
&\geq \int_{\{\ell_\zeta \leq \beta(0)-\varepsilon\}} \left(\prod_{i=1}^{|\zeta|} p_{\zeta|i}^q \operatorname{Lip}(S_{\zeta|i})^{-\beta(0)q+\beta(0)} \right) d\Lambda_u \\
&= \int_{\{\ell_\zeta \leq \beta(0)-\varepsilon\}} \left(\prod_{i=1}^{|\zeta|} \operatorname{Lip}(S_{\zeta|i})^{q(\ell_\zeta(\omega)-\beta(0))+\beta(0)} \right) d\Lambda_u \\
&\geq \int_{\{\ell_\zeta \leq \beta(0)-\varepsilon\}} \left(\prod_{i=1}^{|\zeta|} \operatorname{Lip}(S_{\zeta|i})^{-\varepsilon q+\beta(0)} \right) d\Lambda_u
\end{aligned}$$

for $q > 0$. Hence (by Fatou's lemma)

$$\begin{aligned}
1 &\geq \liminf_{q\to\infty} \int_{\{\ell_\zeta \leq \beta(0)-\varepsilon\}} \left(\prod_{i=1}^{|\zeta|} \operatorname{Lip}(S_{\zeta|i})^{-\varepsilon q+\beta(0)} \right) d\Lambda_u(\omega) \\
&\geq \int_{\{\ell_\zeta \leq \beta(0)-\varepsilon\}} \liminf_{q\to\infty} \left(\prod_{i=1}^{|\zeta|} \operatorname{Lip}(S_{\zeta|i})^{-\varepsilon q+\beta(0)} \right) d\Lambda_u(\omega) \\
&= \int_{\{\ell_\zeta \leq \beta(0)-\varepsilon\}} \infty \, d\Lambda_u = \infty \cdot \Lambda_u(\{\ell_\zeta \leq \beta(0) - \varepsilon\}) = \infty
\end{aligned}$$

which is a contradiction. This proves (4.4.28).

Similar arguments show that

$$\overline{a} \leq \beta(0). \tag{4.4.29}$$

By (4.4.28) and (4.4.29), $\underline{a} = \beta(0) = \overline{a}$. □

Proof of Proposition 2.5.5

Case 1. Follows immediately from Proposition 2.5.4.

Case 2. i) Suppose that β is not strictly convex. Then β is affine on some non-degenerate interval and hence (since β is real-analytic) affine on $\mathbb{R}$, i.e. $\beta(q) = \beta(0) - \beta(0)q$ (since $\beta(1) = 0$.) The previous proposition therefore implies that $\underline{a} = \overline{a}$ which is a contradiction.

ii) The function β is strictly convex by i), and $\alpha := -\beta'$ is therefore strictly decreasing (cf. [Rob, Theorem B, p. 10].)

iii) Let $n \in \mathbb{N}$. It follows from the definition of $\overline{a}_n$ that

$$\log\left(\prod_{i=1}^{n} p_{\gamma|i}\right) \geq \overline{a}_n \log\left(\prod_{i=1}^{n} \mathrm{Lip}(S_{\gamma|i})\right) \quad \Lambda_{\iota(\gamma_1)}\text{-a.s.} \tag{4.4.30}$$

By (4.4.30) and (4.4.8),

$$\begin{aligned}
&\alpha(q)\\
&= -\beta'(q)\\
&= \frac{\sum_u \sum_v \sum_{\gamma \in E_{uv}^{(n)}} \sigma_u(q) \int \log\left(\prod_{i=1}^n p_{\gamma|i}\right) \left(\prod_{i=1}^n p_{\gamma|i}\right)^q \left(\prod_{i=1}^n \mathrm{Lip}(S_{\gamma|i})\right)^{\beta(q)} d\Lambda_u \rho_v(q)}{\sum_u \sum_v \sum_{\gamma \in E_{uv}^{(n)}} \sigma_u(q) \int \log\left(\prod_{i=1}^n \mathrm{Lip}(S_{\gamma|i})\right) \left(\prod_{i=1}^n p_{\gamma|i}\right)^q \left(\prod_{i=1}^n \mathrm{Lip}(S_{\gamma|i})\right)^{\beta(q)} d\Lambda_u \rho_v(q)}\\
&\leq \frac{\sum_u \sum_v \sum_{\gamma \in E_{uv}^{(n)}} \sigma_u(q) \int \overline{a}_n \log\left(\prod_{i=1}^n \mathrm{Lip}(S_{\gamma|i})\right) \left(\prod_{i=1}^n p_{\gamma|i}\right)^q \left(\prod_{i=1}^n \mathrm{Lip}(S_{\gamma|i})\right)^{\beta(q)} d\Lambda_u \rho_v(q)}{\sum_u \sum_v \sum_{\gamma \in E_{uv}^{(n)}} \sigma_u(q) \int \log\left(\prod_{i=1}^n \mathrm{Lip}(S_{\gamma|i})\right) \left(\prod_{i=1}^n p_{\gamma|i}\right)^q \left(\prod_{i=1}^n \mathrm{Lip}(S_{\gamma|i})\right)^{\beta(q)} d\Lambda_u \rho_v(q)}\\
&= \overline{a}_n
\end{aligned}$$

for all $n \in \mathbb{N}$. Hence

$$\alpha(q) \leq \lim_n \overline{a}_n = \overline{a} \quad \text{for} \quad q \in \mathbb{R}. \tag{4.4.31}$$

It follows by a similar argument that

$$\underline{a} \leq \alpha(q) \quad \text{for} \quad q \in \mathbb{R}.$$

We will now prove that $\alpha(q) \to \overline{a}$ as $q \to -\infty$. Let $a < \overline{a}$. Proposition 2.5.3 shows that $aq + \beta(q) \to \infty$ as $q \to -\infty$, and we can thus choose a number $q_0 < 0$ such that $\beta(q) > -aq$ for $q < q_0$. Hence $-\frac{\beta(q)}{q} \geq a$ for $q < q_0$, and so

$$\liminf_{q \to -\infty} -\frac{\beta(q)}{q} \geq a. \tag{4.4.32}$$

Since $\alpha = -\beta'$ is decreasing, (4.4.32) implies that $\lim_{q\to-\infty} \alpha(q) \geq a$ for all $a < \bar{a}$, whence $\lim_{q\to-\infty} \alpha(q) = \bar{a}$.
It follows by a similar argument that $\lim_{q\to\infty} \alpha(q) = \underline{a}$. □

Proof of Proposition 2.5.6

Case 1. Follows immediately from the fact that $\beta(q) = a - aq$ in Case 1.
Case 2. First note that [Rob, Theorem D, p. 34] implies that

$$\beta^*(\alpha(q)) = q\alpha(q) + \beta(q) \quad \text{for } q \in \mathbb{R} \tag{4.4.33}$$

i)-ii) Obvious.
iv) Existence: Let $f(q) := -a_{\max}q$ for $q \in \mathbb{R}$. For $q < 0$, $-\frac{\beta(q)}{q} \geq a_{\max}$ whence

$$\beta(q) \geq -qa_{\max} = f(q) \quad \text{for} \quad q < 0\,. \tag{4.4.34}$$

Choose $\varepsilon > 0$ such that $a_{\max} + \varepsilon < \bar{a}$. Next choose a sequence $(q_n)_n$ in $]-\infty, 0[$ satisfying

$$-\frac{\beta(q_n)}{q_n} \to a_{\max}\,. \tag{4.4.35}$$

We claim that $(q_n)_n$ is bounded. Indeed, pick $N \in \mathbb{N}$ such that $-\frac{\beta(q_n)}{q_n} \leq a_{\max} + \varepsilon$ for $n \geq N$. We have $0 \geq q_n(a_{\max}+\varepsilon)+\beta(q_n)$ for $n \geq N$, whence $\limsup_n(q_n(a_{\max}+\varepsilon)+\beta(q_n)) \leq 0$ and Proposition 2.5.3 therefore implies that $(q_n)_n$ is bounded. Since $(q_n)_n$ is bounded there exists a subsequence $(q_{n_k})_k$ and a $q_{\min} \leq 0$ such that

$$q_{n_k} \to q_{\min}\,. \tag{4.4.36}$$

We claim that $q_{\min} \neq 0$. Otherwise the continuity of β implies that $\infty \geq \bar{a} > a_{\max} = \lim_k -\frac{\beta(q_{n_k})}{q_{n_k}} = \infty$ (since $\beta(q_{n_k}) \geq \beta(0) > 0$) which is a contradiction. Since $q_{\min} \neq 0$, (4.4.35) and the continuity of β show that $a_{\max} = \lim_k -\frac{\beta(q_{n_k})}{q_{n_k}} = -\frac{\beta(q_{\min})}{q_{\min}}$ i.e.

$$\beta(q_{\min}) = -a_{\max}q_{\min} = f(q_{\min})\,. \tag{4.4.37}$$

Now equations (4.4.34) and (4.4.35) imply that $\beta'(q_{\min}) = -a_{\max}$, i.e. $\alpha(q_{\min}) = -\beta'(q_{\min}) = a_{\max}$. Also by (4.4.37),

$$\beta^*(a_{\max}) = \beta^*(\alpha(q_{\min})) = q_{\min}\alpha(q_{\min}) + \beta(q_{\min}) = q_{\min}a_{\max} + \beta(q_{\min}) = 0\,.$$

Uniqueness: The uniqueness is immediate since $\alpha(\mathbb{R}) =]\underline{a}, \bar{a}[$ and α is strictly decreasing and continuous.
Finally we must prove that the graph of β has the line that passes through the origin with slope $\frac{\beta(q_{\min})}{q_{\min}}$ as a tangent at the point $(q_{\min}, \beta(q_{\min}))$. We have $0 = \beta^*(\alpha(q_{\min})) = q_{\min}\alpha(q_{\min}) + \beta(q_{\min})$ whence

$$\beta'(q_{\min}) = -\alpha(q_{\min}) = \frac{\beta(q_{\min})}{q_{\min}}$$

which clearly proves the assertion.
v) Similar to the proof of iv).
iii) Proof of $\underline{a} \leq a_{\min}$: If $a_{\min} < \underline{a}$ then

$$\sup_{0<q} \left(-\frac{\beta(q)}{q} \right) = a_{\min} < \underline{a}$$

and we can thus choose an $\varepsilon > 0$ such that $(\underline{a} - \varepsilon)q + \beta(q) \geq 0$ for $q > 0$. This in turn implies that

$$\liminf_{q \to \infty} ((\underline{a} - \varepsilon)q + \beta(q)) \geq 0$$

which contradicts Proposition 2.5.3.
Proof of $\overline{a} \leq a_{\max}$: Similar to the proof of $\underline{a} \leq a_{\min}$.
Proof of $a_{\min} \leq a_{\max}$: Follows from iv) and v) since α is decreasing and $\underline{a} \leq a_{\min}$ and $a_{\max} \leq \overline{a}$.
vi) Follows from iv) and v).
vii) We have $\beta^*(\alpha(1)) = 1 \cdot \alpha(1) + \beta(1) = \alpha(1)$.
viii) Clearly $\beta^*(\alpha) \leq 1 \cdot \alpha + \beta(\alpha) = \alpha$.
ix) We have $\beta^*(\alpha(q)) = q\alpha(q) + \beta(q)$ for $q \in \mathbb{R}$. Hence

$$(\beta^*)'(\alpha(q))\alpha'(q) = q\alpha'(q) + \alpha(q) + \beta'(q) = q\alpha'(q).$$

Thus $(\beta^*)'(\alpha(1))\alpha'(1) = \alpha'(1)$, whence

$$(\beta^*)'(\alpha(1)) = 1. \tag{4.4.38}$$

The assertion in ix) follows from (4.4.38) and vii).
x) The function β^* is strictly concave and it is thus sufficient to prove that $(\beta^*)'(\alpha(0)) = 0$. It follows from the proof of part ix) that $(\beta^*)'(\alpha(q))\alpha'(q) = q\alpha'(q)$, whence $(\beta^*)'(\alpha(0))\alpha'(0) = 0$, which implies that $(\beta^*)'(\alpha(0)) = 0$. Also $\beta^*(\alpha(q)) = q\alpha(q) + \beta(q)$, whence

$$\max_{\alpha \geq 0} \beta^*(\alpha) = \beta^*(\alpha(0)) = 0\alpha(0) + \beta(0) = \beta(0).$$

xi) Let $V = \{v\}$ and $E = \{1, \ldots, N\}$. Put $\Lambda_v = \Lambda$. Clearly $\rho_v(q) = 1 = \sigma_v(q)$ for all $q \in \mathbb{R}$. It follows from (4.4.6) and (4.4.7) that

$$0 = \sum_{i=1}^{N} \int \log(p_i) p_i^q \operatorname{Lip}(S_i)^{\beta(q)} \, d\Lambda +$$
$$\beta'(q) \sum_{i=1}^{N} \int \log(\operatorname{Lip}(S_i)) p_i^q \operatorname{Lip}(S_i)^{\beta(q)} \, d\Lambda \tag{4.4.39}$$

Differentiation of (4.4.39) yields

$$0 = \sum_{i=1}^{N} \int \left(\log(p_i) + \beta'(q)\log(\mathrm{Lip}(S_i))\right)^2 p_i^q \,\mathrm{Lip}(S_i)^{\beta(q)}\, d\Lambda +$$

$$\beta''(q) \sum_{i=1}^{N} \int \log(\mathrm{Lip}(S_i)) p_i^q \,\mathrm{Lip}(S_i)^{\beta(q)}\, d\Lambda$$

whence

$$\alpha''(q) = -\beta''(q) = \frac{\sum_{i=1}^{N} \int \left(\log(p_i) + \beta'(q)\log(\mathrm{Lip}(S_i))\right)^2 p_i^q \,\mathrm{Lip}(S_i)^{\beta(q)}\, d\Lambda}{\sum_{i=1}^{N} \int \log(\mathrm{Lip}(S_i)) p_i^q \,\mathrm{Lip}(S_i)^{\beta(q)}\, d\Lambda} \tag{4.4.40}$$

Finally note that for each fixed $q \in \mathbb{R}$ there exists an $i = 1, \ldots, N$ satisfying

$$\Lambda\Big(\log(p_i) + \beta'(q)\log(\mathrm{Lip}(S_i)) \neq 0 \Big) > 0 \,. \tag{4.4.41}$$

Otherwise $\ell_i = \frac{\kappa_i}{\chi_i} = \frac{\log p_i}{\log(\mathrm{Lip}(S_i))} = -\beta'(q)$ Λ-a.s. for all simple cycles i, whence $\underline{a} = \overline{a}$, contradicting the fact that $\underline{a} < \overline{a}$.

It follows from (4.4.40) and (4.4.41) that $\alpha'(q) < 0$. □

The next lemma plays a crucial role in the estimates in the remaining parts of the exposition.

Lemma 4.4.4. *Assume conditions (I) and (III). For each* $q \in]q_{\min}, q_{\max}[$ *there exists a number* $\varepsilon_q \in]0, \infty]$ *such that*

$$\beta(\lambda q) < \lambda\beta(q) \quad \text{for} \quad \lambda \in]1, 1 + \varepsilon_q[\,.$$

Proof. Case 1. In this case $\varepsilon_q = \infty$. Indeed, $\lambda\beta(q) = \lambda(\beta(0) - \beta(0)q) = \lambda\beta(0) - \lambda\beta(0)q > \beta(0) - \beta(0)\lambda q = \beta(\lambda q)$ for $\lambda > 1$ since $\beta(0) > 0$.
Case 2. The assertion is an immediate consequence of the geometry underlying the construction of $q_{\min}$ and $q_{\max}$ (cf. figure 2.5.2.) □

Chapter 5
The Random Variable $X_{u,q}$

5.1 The Random Variable $X_{u,q}$

For $u \in V$ let $\mathbf{Max}\,(u)$ denote the family of all finite maximal antichains in $E_u^{\mathbb{N}}$, i.e. $\Gamma \in \mathbf{Max}\,(u)$ if and only if Γ is a finite subset of $E_u^{(*)}$ and for each $\sigma \in E_u^{\mathbb{N}}$ there is a unique $\gamma \in \Gamma$ such that $\sigma \in [\gamma]$. For $\Gamma \in \mathbf{Max}(u)$ write

$$|\Gamma| = \max_{\gamma \in \Gamma} |\gamma| .$$

If $\Gamma \in \mathbf{Max}(u)$ and $\gamma \in \Gamma$ write $\Gamma(\gamma) = \{\alpha \in E^{(*)}_{\tau(\gamma_{|\gamma|})} \mid \gamma\alpha \in \Gamma\}$. Clearly $\Gamma(\gamma) \in \mathbf{Max}(\tau(\gamma_{|\gamma|}))$. We define an order $\prec$ in $\mathbf{Max}\,(u)$ by

$$\Sigma \prec \Gamma \qquad \Leftrightarrow \qquad \left\{ \begin{array}{c} \text{For each } \gamma \in \Gamma \text{ there exists} \\ \text{a (unique) } \sigma \in \Sigma \text{ such that} \\ \sigma \prec \gamma \end{array} \right\}$$

It is easy to see that $(\mathbf{Max}\,(u), \prec)$ is a directed set.

Define for $q \in \mathbb{R}, u \in V$ and $\Gamma \in \mathbf{Max}\,(u)$ a random variable

$$X_{u,q,\Gamma} : \Omega_u \to \mathbb{R}$$

by

$$X_{u,q,\Gamma} = \sum_{\alpha \in \Gamma} \rho_u(q)^{-1} \left(\prod_{i=1}^{|\alpha|} p^q_{\alpha|i} \operatorname{Lip}(S_{\alpha|i})^{\beta(q)} \right) \rho_{\tau(\alpha_{|\alpha|})}(q) .$$

For $n \in \mathbb{N}$ write $X_{u,q,n} = X_{u,q,E_u^{(n)}}$. We will now define some auxiliary random variables related to $X_{u,q,\Gamma}$. For $u \in V$, $\alpha \in E_u^{(*)}$ and $\Gamma \in \mathbf{Max}(\tau(\alpha_{|\alpha|}))$ define

$$X_{u,\alpha,q,\Gamma} : \Omega_u \to \mathbb{R}$$

by

$$X_{u,\alpha,q,\Gamma} = \sum_{\gamma\in\Gamma} \rho^{-1}_{\tau(\alpha_{|\alpha|})} \left(\prod_{i=1}^{|\gamma|} p^q_{\alpha(\gamma|i)} \operatorname{Lip}(S_{\alpha(\gamma|i)})^{\beta(q)} \right) \rho_{\tau(\gamma_{|\gamma|})}(q).$$

Put $X_{u,\alpha,q,E^{(n)}_{\tau(\alpha_{|\alpha|})}} = X_{u,\alpha,q,n}$. Define for $\Gamma \in \mathbf{Max}(u)$ the σ-algebra $\mathcal{A}_{u,\Gamma}$ by

$$\mathcal{A}_{u,\Gamma} = \sigma((\pi_{u,\alpha||\alpha|-1})_{\alpha\in\Gamma})$$

and put $\mathcal{A}_{u,E^{(n)}_u} = \mathcal{A}_{u,n}$.

Proposition 5.1.1. *Assume condition (I) is satisfied. Then the following hold*

i) $X_{u,q,\Gamma} \geq 0$ *for* $\Gamma \in \mathbf{Max}(u)$.

ii) *The family* $(X_{u,q,\Gamma})_{\Gamma\in\mathbf{Max}(u)}$ *is a martingale adapted to* $(\mathcal{A}_{u,\Gamma})_{\Gamma\in\mathbf{Max}(u)}$.

Proof. i) Obvious.

ii) The random variable $X_{u,q,\Gamma}$ is clearly $\mathcal{A}_{u,\Gamma}$-measurable and condition (I) implies that $X_{u,q,\Gamma}$ is bounded and hence integrable.

Let Σ, $\Gamma \in \mathbf{Max}(u)$ with $\Sigma \prec \Gamma$. Now

$$\begin{aligned}
&\mathbf{E}_{\Lambda_u}[X_{u,q,\Gamma} \mid \mathcal{A}_{u,\Sigma}] \\
=&\mathbf{E}_{\Lambda_u}\left[\sum_{\alpha\in\Gamma} \rho_u^{-1} \left(\prod_{i=1}^{|\alpha|} p^q_{\alpha|i} \operatorname{Lip}(S_{\alpha|i})^{\beta(q)} \right) \rho_{\tau(\alpha_{|\alpha|})} \middle| \mathcal{A}_{u,\Sigma} \right] \\
=&\mathbf{E}_{\Lambda_u}\left[\sum_{\alpha\in\Sigma} \sum_{\gamma\in\Gamma(\alpha)} \rho_u^{-1} \left(\prod_{i=1}^{|\alpha|} p^q_{\alpha|i} \operatorname{Lip}(S_{\alpha|i})^{\beta(q)} \right) \cdot \right. \\
&\qquad \left. \left(\prod_{j=1}^{|\gamma|} p^q_{\alpha(\gamma|j)} \operatorname{Lip}(S_{\alpha(\gamma|j)})^{\beta(q)} \right) \rho_{\tau(\gamma_{|\gamma|})} \middle| \mathcal{A}_{u,\Sigma} \right] \\
=&\sum_{\alpha\in\Sigma} \sum_{\gamma\in\Gamma(\alpha)} \rho_u^{-1} \left(\prod_{i=1}^{|\alpha|} p^q_{\alpha|i} \operatorname{Lip}(S_{\alpha|i})^{\beta(q)} \right) \cdot \\
&\qquad \mathbf{E}_{\Lambda_u}\left[\prod_{j=1}^{|\gamma|} p^q_{\alpha(\gamma|j)} \operatorname{Lip}(S_{\alpha(\gamma|j)})^{\beta(q)} \middle| \mathcal{A}_{u,\Sigma} \right] \rho_{\tau(\gamma_{|\gamma|})} \\
=&\sum_{\alpha\in\Sigma} \left(\rho_u^{-1} \left(\prod_{i=1}^{|\alpha|} p^q_{\alpha|i} \operatorname{Lip}(S_{\alpha|i})^{\beta(q)} \right) \sum_{\gamma\in\Gamma(\alpha)} \cdot \right. \\
&\qquad \left. \mathbf{E}_{\Lambda_u}\left[\prod_{j=1}^{|\gamma|} p^q_{\alpha(\gamma|j)} \operatorname{Lip}(S_{\alpha(\gamma|j)})^{\beta(q)} \right] \rho_{\tau(\gamma_{|\gamma|})} \right)
\end{aligned} \tag{5.1.1}$$

since $\prod_{i=1}^{|\alpha|} p^q_{\alpha|i} \operatorname{Lip}(S_{\alpha|i})^{\beta(q)}$ is $\mathcal{A}_{u,\Sigma}$-measurable for $\alpha \in \Sigma$, and the random variable $\prod_{j=1}^{|\gamma|} p^q_{\alpha(\gamma|j)} \operatorname{Lip}(S_{\alpha(\gamma|j)})^{\beta(q)}$ is $\mathcal{A}_{u,\Sigma}$-independent for $\alpha \in \Sigma$.

Since $(p^q_{\alpha(\gamma|j)} \operatorname{Lip}(S_{\alpha(\gamma|j)})^{\beta(q)})_{j=1,\dots,|\gamma|}$ are independent for $\gamma \in E^{(*)}_{\tau(\alpha_{|\alpha|})}$,

$$\begin{aligned}
&\mathbf{E}_{\Lambda_u}\left[\prod_{j=1}^{|\gamma|} p^q_{\alpha(\gamma|j)} \operatorname{Lip}(S_{\alpha(\gamma|j)})^{\beta(q)}\right] \\
&= \prod_{j=1}^{|\gamma|} \mathbf{E}_{\Lambda_u}\left[p^q_{\alpha(\gamma|j)} \operatorname{Lip}(S_{\alpha(\gamma|j)})^{\beta(q)}\right] \\
&= \mathbf{E}_{\lambda_{\tau(\alpha_{|\alpha|})}}[p^q_{\gamma_1} \operatorname{Lip}(S_{\gamma_1})^{\beta(q)}] \left(\prod_{j=2}^{|\gamma|} \mathbf{E}_{\lambda_{\tau(\gamma_{j-1})}}[p^q_{\gamma_j} \operatorname{Lip}(S_{\gamma_j})^{\beta(q)}]\right)
\end{aligned} \tag{5.1.2}$$

for $\gamma \in E^{(*)}_{\tau(\alpha_{|\alpha|})}$.

It follows from (5.1.1) and (5.1.2) that

$$\begin{aligned}
&\mathbf{E}_{\Lambda_u}[X_{u,q,\Gamma} \mid \mathcal{A}_{u,\Sigma}] \\
&= \sum_{\alpha \in \Sigma} \left(\rho_u^{-1} \left(\prod_{i=1}^{|\alpha|} p^q_{\alpha|i} \operatorname{Lip}(S_{\alpha|i})^{\beta(q)}\right) \cdot \rho_{\tau(\alpha_{|\alpha|})} \cdot \right. \\
&\left. \sum_{\gamma \in \Gamma(\alpha)} \rho^{-1}_{\tau(\alpha_{|\alpha|})} \mathbf{E}_{\lambda_{\tau(\alpha_{|\alpha|})}}[p^q_{\gamma_1} \operatorname{Lip}(S_{\gamma_1})^{\beta(q)}] \left(\prod_{j=2}^{|\gamma|} \mathbf{E}_{\lambda_{\tau(\gamma_{j-1})}}[p^q_{\gamma_j} \operatorname{Lip}(S_{\gamma_j})^{\beta(q)}]\right) \rho_{\tau(\gamma_{|\gamma|})} \right)
\end{aligned} \tag{5.1.3}$$

However, since $\Gamma(\alpha) \in \mathbf{Max}(\tau(\alpha_{|\alpha|}))$ and $\sum_{v \in V} \sum_{e \in E_{uv}} \rho_u^{-1} \mathbf{E}_{\lambda_u}[p^q_e \operatorname{Lip}(S_e)^{\beta(q)}] \rho_v = 1$ for all $u \in V$,

$$\sum_{\gamma \in \Gamma(\alpha)} \rho^{-1}_{\tau(\alpha_{|\alpha|})} \mathbf{E}_{\lambda_{\tau(\alpha_{|\alpha|})}}[p^q_{\gamma_1} \operatorname{Lip}(S_{\gamma_1})^{\beta(q)}] \left(\prod_{j=2}^{|\gamma|} \mathbf{E}_{\lambda_{\tau(\gamma_{j-1})}}[p^q_{\gamma_j} \operatorname{Lip}(S_{\gamma_j})^{\beta(q)}]\right) \rho_{\tau(\gamma_{|\gamma|})} = 1$$

Hence

$$\mathbf{E}_{\Lambda_u}[X_{u,q,\Gamma} \mid \mathcal{A}_{u,\Sigma}] = \sum_{\alpha \in \Sigma} \rho_u^{-1} \left(\prod_{i=1}^{|\alpha|} p^q_{\alpha|i} \operatorname{Lip}(S_{\alpha|i})^{\beta(q)}\right) \rho_{\tau(\alpha_{|\alpha|})} = X_{u,q,\Sigma}. \qquad \square$$

It follows from Proposition 5.1.1 that $(X_{u,q,n})_{n\in\mathbb{N}}$ is a non-negative martingale adapted to the filtration $(\mathcal{A}_{u,n})_{n\in\mathbb{N}}$. The martingale convergence theorem therefore implies that there exists a random variable $X_{u,q} \geq 0$ defined on Ω_u satisfying

$$\mathbf{E}_{\Lambda_u}[X_{u,q}] \leq \liminf_n \mathbf{E}_{\Lambda_u}[X_{u,q,n}] = 1$$
$$X_{u,q,n} \to X_{u,q} \quad \Lambda_u\text{-a.s.} \tag{5.1.4}$$

5.2 The Positive Moments of $X_{u,q}$

We will later (cf. Chapter 6) define a random measure

$$\mathcal{M}_{u,q} : \Omega_u \to \mathcal{P}(X_u)$$

supported Λ_u-a.s. on $\Delta_{\mu_u}(\alpha(q))$ and satisfying $\mathcal{M}_{u,q}(C_u) = X_{u,q}$ Λ_u-a.s. The random measure $\mathcal{M}_{u,q}$ will be used in the proof of

$$\mathcal{H}_{\mu}^{q,\beta(q)-\delta}(\Delta_\mu(\alpha(q)) = \infty \tag{5.2.1}$$

for all $\delta > 0$ and P_u-a.a. μ. The proof of (5.2.1) is based on the following observation: for Λ_u-a.a. $\omega \in \Omega_u$ the statement below holds,

$$\forall E \subseteq \operatorname{supp} \mu_u(\omega) : \mathcal{M}_{u,q}(\omega)(E) > 0 \;\Rightarrow\; \mathcal{H}_{\mu}^{q,\beta(q)-\delta}(E) = \infty \,.$$

Thus, we need to know that $X_{u,q} = \mathcal{M}_{u,q}(\Delta_{\mu_u}(\alpha(q)) > 0$ Λ_u-a.s. An easy argument (given in Proposition 5.2.3) shows that in order for $X_{u,q}$ to be positive Λ_u-a.s., i.e. $\Lambda_u(X_{u,q} > 0) = 1$, it suffices to prove that $\Lambda_u(X_{u,q} > 0) > 0$. The proof of the latter inequality is based on the fact that the almost sure convergence asserted in (5.1.4) can be strengthened to L^1-convergence for $q \in]q_{\min}, q_{\max}[$, i.e.

$$X_{u,q,n} \to X_{u,q} \quad \text{in } L^1 \tag{5.2.2}$$

for $q \in]q_{\min}, q_{\max}[$. In order to prove (5.2.2) it clearly suffices to show that $(X_{u,q,n})_n$ is an $L^{1+\varepsilon}$ bounded martingale for some $\varepsilon > 0$. Hence, we need to show that $(X_{u,q,n})_n$ has uniformly bounded moments of order $1+\varepsilon$. This is one of a number of reasons for studying the moments of $X_{u,q,n}$. A detailed study of the moments of the auxiliary random variables $X_{u,q,n}$ seems to be an unavoidable problem in the theory of random self-similar sets (and measures), cf. the papers by Mauldin & Williams [Mau1], Graf [Gra1], Arbeiter [Ar], Patzschke & Zähle [Pat] and Holley and Waymire [Ho] which all spend considerable time and effort in order to prove that various higher order moments are bounded. However, Falconer [Fa4] seems to manage without all these tedious and elaborate considerations, cf. in particular [Fa4, equation (24)] which Falconer establishes without any reference to the behaviour of the higher order moments.

We now state and prove the main theorem in this section.

Theorem 5.2.1. *Assume condition (I). Let $u \in V$ and $q \in]q_{\min}, q_{\max}[$. Then*

$$(X_{u,q,\Gamma})_{\Gamma \in \mathbf{Max}(u)}$$

is an L^p-bounded martingale for $p \in [1, 1+\varepsilon_q[$.

Proof. It is clearly sufficient to prove that

$$\max_{v \in V} \sup_{\Gamma \in \mathbf{Max}(v)} \|X_{v,q,\Gamma}\|_{1+p}^{1+p} < \infty$$

for $p \in \mathbb{Q} \cap [0, \varepsilon_q[$. Let $l \in \mathbb{N}$. For $\varepsilon_q < \infty$ choose $m_l \in \mathbb{N} \cup \{0\}$ such that

$$\frac{m_l}{l} < \varepsilon_q \leq \frac{m_l + 1}{l}.$$

For $\varepsilon_q = \infty$ put $m_l = \infty$. Next, write $\mathbb{N}_l = \{0, \ldots, m_l\}$ for $m_l < \infty$, and write $\mathbb{N}_l = \{0, 1, \ldots\}$ for $m_l = \infty$. It suffices to prove that

$$\max_{v \in V} \sup_{\Gamma \in \mathbf{Max}(v)} \|X_{v,q,\Gamma}\|_{1+\frac{p}{l}}^{1+\frac{p}{l}} < \infty$$

for $l \in \mathbb{N}$ and $p \in \mathbb{N}_l$. For $l \in \mathbb{N}$ and $p \in \mathbb{N}_l$, let $\mathbf{S}(l,p)$ denote the statement:

$$\max_{v \in V} \sup_{k \in \{0,\ldots,p\}} \sup_{\Gamma \in \mathbf{Max}(v)} \|X_{v,q,\Gamma}\|_{1+\frac{k}{l}}^{1+\frac{k}{l}} < \infty.$$

We must prove that $\mathbf{S}(l,p)$ is a true statement for all $l \in \mathbb{N}$ and $p \in \mathbb{N}_l$. Fix an $l \in \mathbb{N}$. Below we prove that $\mathbf{S}(l,p)$ is a true statement for all $p \in \mathbb{N}_l$ by induction on p.

1. The start of the induction: We must prove that $\mathbf{S}(l,0)$ is true. The martingale property implies that

$$\mathbf{E}_{\Lambda_u}[X_{u,q,\Gamma} \mid \mathcal{A}_{u,E_u}] = X_{u,q,E_u}$$

for all $\Gamma \in \mathbf{Max}(u)$, whence (by taking expectation)

$$\|X_{u,q,\Gamma}\|_{1+\frac{0}{l}}^{1+\frac{0}{l}} = \mathbf{E}_{\lambda_u}[X_{u,q,\Gamma}] = \mathbf{E}_{\lambda_u}[X_{u,q,E_u}] = \max_v \mathbf{E}_{\lambda_v}[X_{v,q,E_v}] < \infty$$

for all $\Gamma \in \mathbf{Max}(u)$. This proves that $\mathbf{S}(l,0)$ is true.

2. The inductive step: Assume that $\mathbf{S}(l,p)$ is true for some $p \in \{0, \ldots, m_l - 1\}$ if $\varepsilon_q < \infty$, or that $\mathbf{S}(l,p)$ is true for some $p \in \{0, 1, \ldots\}$ if $\varepsilon_q = \infty$; i.e. we are assuming that

$$M := \max_{v \in V} \sup_{k \in \{0,\ldots,p\}} \sup_{\Gamma \in \mathbf{Max}(v)} \|X_{v,q,\Gamma}\|_{1+\frac{k}{l}}^{1+\frac{k}{l}} < \infty.$$

We must now prove that $\mathbf{S}(l,p+1)$ is true.

To simplify notation write

$$H_e^{(n)} = \rho_u^{-\frac{n}{l}} \mathbf{E}_{\lambda_u}\left[\left(\sum_{e \in E_u} p_e^q \operatorname{Lip}(S_e)^{\beta(q)}\right)^{\frac{n}{l}}\right] \rho_{\tau(e)}^{\frac{n}{l}}$$

$$L_e^{(n)} = \int \prod_{e \in E_u} \left(\rho_u^{-1} p_e^q \operatorname{Lip}(S_e)^{\beta(q)} \rho_{\tau(e)}\right)^{\frac{n}{l}} \, d\lambda_u$$

for $u \in V$ and $e \in E_u$. Now observe that

$$
\begin{aligned}
&\forall u \in V : \forall e \in E_u : \forall \Gamma \in \mathbf{Max}(u) : \forall n \in \mathbb{N} : \\
&\|X_{u,q,\Gamma}^{\frac{1}{t}}\|_n^n \\
&= \int \Bigg(\Bigg(\sum_{e \in E_u} \rho_u^{-1} p_e^q \operatorname{Lip}(S_e)^{\beta(q)} \rho_{\tau(e)} \cdot \\
&\qquad \sum_{\alpha \in \Gamma(e)} \rho_{\tau(e)}^{-1} \left(\prod_{i=1}^{|\alpha|} p_{e(\alpha|i)}^q \operatorname{Lip}(S_{e(\alpha|i)})^{\beta(q)}\right) \rho_{\tau(\alpha_{|\alpha|})}\Bigg)^{\frac{1}{t}}\Bigg)^n d\Lambda_u \\
&= \int \left(\left(\sum_{e \in E_u} \rho_u^{-1} p_e^q \operatorname{Lip}(S_e)^{\beta(q)} \rho_{\tau(e)} X_{\tau(e),q,\Gamma(e)}(\mathcal{S}_e)\right)^{\frac{1}{t}}\right)^n d\Lambda_u \\
&\text{[by Lemma 4.3.1]} \\
&= \int \left(\left(\sum_{e \in E_u} \rho_u^{-1} p_e^q \operatorname{Lip}(S_e)^{\beta(q)} \rho_{\tau(e)} X_{\tau(e),q,\Gamma(e)}(\mathcal{S}_e)\right)^{\frac{1}{t}}\right)^n \\
&\qquad d\left(\left(\lambda_u \times \prod_{e \in E_u} \Lambda_{\tau(e)}\right) \circ \varphi_u^{-1}\right) \\
&\le \int_{\Xi_u \times \prod_{e \in E_u} \Omega_{\tau(e)}} \left(\sum_{e \in E_u} \left(\rho_u^{-1} p_e^q \operatorname{Lip}(S_e)^{\beta(q)} \rho_{\tau(e)}\right)^{\frac{1}{t}} X_{\tau(e),q,\Gamma(e)}(\omega_{\tau(e)})^{\frac{1}{t}}\right)^n \\
&\qquad d\left(\lambda_u \times \prod_{e \in E_u} \Lambda_{\tau(e)}\right)\left(((S_e)_{e \in E_u}, (p_e)_{e \in E_u}), (\omega_{\tau(e)})_{e \in E_u}\right) \\
&= \int_{\Xi_u} \int_{\prod_{e \in E_u} \Omega_{\tau(e)}} \sum_{\substack{\sum_{e \in E_u} n_e = n \\ (n_e)_e \subseteq \mathbb{N}_0}} \frac{n!}{\prod_{e \in E_u} n_e!} \left(\prod_{e \in E_u} \left(\rho_u^{-1} p_e^q \operatorname{Lip}(S_e)^{\beta(q)} \rho_{\tau(e)}\right)^{\frac{n_e}{t}}\right) \cdot \\
&\qquad \left(\prod_{e \in E_u} X_{\tau(e),q,\Gamma(e)}(\omega_{\tau(e)})^{\frac{n_e}{t}}\right) \cdot \\
&\qquad d\left(\prod_{e \in E_u} \Lambda_{\tau(e)}\right)\left((\omega_{\tau(e)})_{e \in E_u}\right) d\lambda_u\left((S_e)_{e \in E_u}, (p_e)_{e \in E_u}\right)
\end{aligned}
$$

$$= \sum_{\substack{\sum_{e\in E_u} n_e = n \\ (n_e)_e \subseteq \mathbb{N}_0}} \frac{n!}{\prod_{e\in E_u} n_e!} \int \left(\prod_{e\in E_u} \left(\rho_u^{-1} p_e^q \operatorname{Lip}(S_e)^{\beta(q)} \rho_{\tau(e)} \right)^{\frac{n_e}{l}} \right) d\lambda_u \cdot$$

$$\prod_{e\in E_u} \|X_{\tau(e),q,\Gamma(e)}\|_{\frac{n_e}{l}}^{\frac{n_e}{l}}$$

$$= \sum_{e\in E_u} \int \left(\rho_u^{-1} p_e^q \operatorname{Lip}(S_e)^{\beta(q)} \rho_{\tau(e)} \right)^{\frac{n}{l}} d\lambda_u \, \|X^{\frac{1}{l}}_{\tau(e),q,\Gamma(e)}\|_n^n +$$

$$\sum_{\substack{\sum_{e\in E_u} n_e = n \\ (n_e)_e \subseteq \{0,\ldots,n-1\}}} \frac{n!}{\prod_{e\in E_u} n_e!} \int \left(\prod_{e\in E_u} \left(\rho_u^{-1} p_e^q \operatorname{Lip}(S_e)^{\beta(q)} \rho_{\tau(e)} \right)^{\frac{n_e}{l}} \right) d\lambda_u \cdot$$

$$\prod_{e\in E_u} \|X^{\frac{1}{l}}_{\tau(e),q,\Gamma(e)}\|_{n_e}^{n_e}$$

$$= \sum_{e\in E_u} H_e^{(n)} \|X^{\frac{1}{l}}_{\tau(e),q,\Gamma(e)}\|_n^n +$$

$$\sum_{\substack{\sum_{e\in E_u} n_e = n \\ (n_e)_e \subseteq \{0,\ldots,n-1\}}} \frac{n!}{\prod_{e\in E_u} n_e!} L_u^{(n_e)} \prod_{e\in E_u} \|X^{\frac{1}{l}}_{\tau(e),q,\Gamma(e)}\|_{n_e}^{n_e} \tag{5.2.3}$$

Let $u \in V$ and $\Gamma \in \mathbf{Max}(u)$. Then (5.2.3) implies that

$$\|X_{u,q,\Gamma}\|_{1+\frac{p+1}{l}}^{1+\frac{p+1}{l}}$$

$$= \|X^{\frac{1}{l}}_{u,q,\Gamma}\|_{1+p+l}^{1+p+l}$$

[by (5.2.3) applied to $\|X^{\frac{1}{l}}_{u,q,\Gamma}\|_{1+p+l}^{1+p+l}$]

$$= \sum_{e_1\in E_u} H_{e_1}^{(l+p+1)} \|X^{\frac{1}{l}}_{\tau(e_1),q,\Gamma(e_1)}\|_{l+p+1}^{l+p+1} +$$

$$\sum_{\substack{\sum_{e_1\in E_u} n_{e_1} = l+p+1 \\ (n_{e_1})_{e_1} \subseteq \{0,\ldots,l+p\}}} \frac{(l+p+1)!}{\prod_{e_1\in E_u} n_{e_1}!} L_u^{(n_{e_1})} \prod_{e_1\in E_u} \|X^{\frac{1}{l}}_{\tau(e_1),q,\Gamma(e_1)}\|_{n_{e_1}}^{n_{e_1}}$$

[by (5.2.3) applied to $\|X^{\frac{1}{l}}_{\tau(e_1),q,\Gamma(e_1)}\|^{1+p+l}_{1+p+l}$]

$$= \sum_{e_1 \in E_u} \sum_{e_2 \in E_{\tau(e_1)}} H^{(l+p+1)}_{e_1} H^{(l+p+1)}_{e_2} \|X^{\frac{1}{l}}_{\tau(e_1 e_2),q,\Gamma(e_1 e_2)}\|^{l+p+1}_{l+p+1} +$$

$$\sum_{\substack{\sum_{e_1 \in E_u} n_{e_1} = l+p+1 \\ (n_{e_1})_{e_1} \subseteq \{0,\dots,l+p\}}} \frac{(l+p+1)!}{\prod_{e_1 \in E_u} n_{e_1}!} L^{(n_{e_1})}_u \prod_{e_1 \in E_u} \|X^{\frac{1}{l}}_{\tau(e_1),q,\Gamma(e_1)}\|^{n_{e_1}}_{n_{e_1}} +$$

$$\sum_{e_1 \in E_u} H^{(l+p+1)}_{e_1} .$$

$$\left(\sum_{\substack{\sum_{e_2 \in E_{\tau(e_1)}} n_{e_2} = l+p+1 \\ (n_{e_2})_{e_2} \subseteq \{0,\dots,l+p\}}} \frac{(l+p+1)!}{\prod_{e_2 \in E_{\tau(e_1)}} n_{e_2}!} L^{(n_{e_1})}_u \prod_{e_2 \in E_{\tau(e_1)}} \|X^{\frac{1}{l}}_{\tau(e_2),q,\Gamma(e_2)}\|^{n_{e_2}}_{n_{e_2}} \right)$$

$\vdots$

[by repeated application of (5.2.3)]

$$= \sum_{\gamma \in E^{(|\Gamma|)}_u} H^{(l+p+1)}_{\gamma_1} \cdot \ldots \cdot H^{(l+p+1)}_{\gamma_{|\gamma|}} \|X^{\frac{1}{l}}_{\tau(\gamma_{|\gamma|}),q,\Gamma(\gamma)}\|^{l+p+1}_{l+p+1} +$$

$$\sum_{\substack{\sum_{e \in E_u} n_e = l+p+1 \\ (n_e)_e \subseteq \{0,\dots,l+p\}}} \frac{(l+p+1)!}{\prod_{e \in E_u} n_e!} L^{(n_e)}_u \prod_{e \in E_u} \|X^{\frac{1}{l}}_{\tau(e),q,\Gamma(e)}\|^{n_e}_{n_e} +$$

$$\sum_{n=1}^{|\Gamma|} \sum_{\gamma \in E^{(n)}_u} H^{(l+p+1)}_{\gamma_1} \cdot \ldots \cdot H^{(l+p+1)}_{\gamma_n} .$$

$$\left(\sum_{\substack{\sum_{e \in E_{\tau(\gamma_n)}} n_e = l+p+1 \\ (n_e)_e \subseteq \{0,\dots,l+p\}}} \frac{(l+p+1)!}{\prod_{e \in E_{\tau(\gamma_n)}} n_e!} L^{(n_e)}_{\tau(\gamma_n)} \prod_{e \in E_{\tau(\gamma_n)}} \|X^{\frac{1}{l}}_{\tau(e),q,\Gamma(\gamma e)}\|^{n_e}_{n_e} \right)$$

$$\leq \sum_{\gamma\in E_u^{(|\Gamma|)}} H_{\gamma_1}^{(l+p+1)}\cdot\ldots\cdot H_{\gamma_{|\gamma|}}^{(l+p+1)}\|X_{\tau(\gamma_{|\gamma|}),q,\Gamma(\gamma)}^{\frac{1}{l}}\|_{l+p+1}^{l+p+1}+$$

$$\sum_{\substack{\sum_{e\in E_u} n_e=l+p+1\\(n_e)_e\subseteq\{0,\ldots,l+p\}}} \frac{(l+p+1)!}{\prod_{e\in E_u} n_e!}\left(\max_{\substack{v\in V\\k=0,\ldots,l+p}} L_v^{(k)}\right)\left(\prod_{e\in E_u}(2\vee M)^{n_e}\right)+$$

$$\sum_{n=1}^{|\Gamma|}\sum_{\gamma\in E_u^{(n)}} H_{\gamma_1}^{(l+p+1)}\cdot\ldots\cdot H_{\gamma_n}^{(l+p+1)}\cdot$$

$$\left(\sum_{\substack{\sum_{e\in E_{\tau(\gamma_n)}} n_e=l+p+1\\(n_e)_e\subseteq\{0,\ldots,l+p\}}} \frac{(l+p+1)!}{\prod_{e\in E_{\tau(\gamma_n)}} n_e!}\left(\max_{\substack{v\in V\\k=0,\ldots,l+p}} L_v^{(k)}\right)\left(\prod_{e\in E_{\tau(\gamma_n)}}(2\vee M)^{n_e}\right)\right)$$

$$\leq \sum_{\gamma\in E_u^{(|\Gamma|)}} H_{\gamma_1}^{(l+p+1)}\cdot\ldots\cdot H_{\gamma_{|\gamma|}}^{(l+p+1)}\|X_{\tau(\gamma_{|\gamma|}),q,\Gamma(\gamma)}^{\frac{1}{l}}\|_{l+p+1}^{l+p+1}+$$

$$\left(\max_{\substack{v\in V\\k=0,\ldots,l+p}} L_v^{(k)}\right)(2\vee M)^{l+p+1}\,\mathrm{card}(E_u)^{p+l+1}+$$

$$\sum_{n=1}^{|\Gamma|}\sum_{\gamma\in E_u^{(n)}} H_{\gamma_1}^{(l+p+1)}\cdot\ldots\cdot H_{\gamma_n}^{(l+p+1)}\cdot$$

$$\left(\max_{\substack{v\in V\\k=0,\ldots,l+p}} L_v^{(k)}\right)(2\vee M)^{l+p+1}\,\mathrm{card}(E_{\tau(\gamma_n)})^{p+l+1}$$

$$\leq(1+c_0)\sum_{n=1}^{\infty}\sum_{\gamma\in E_u^{(n)}} H_{\gamma_1}^{(l+p+1)}\cdot\ldots\cdot H_{\gamma_n}^{(l+p+1)}+c_0 \qquad (5.2.4)$$

since

$$\|X_{\tau(e),q,\Gamma(\gamma e)}^{\frac{1}{l}}\|_{n_e}^{n_e}=\|X_{\tau(e),q,\Gamma(\gamma e)}\|_{\frac{n_e}{l}}^{\frac{n_e}{l}}$$

$$\leq\begin{cases}1+\|X_{\tau(e),q,\Gamma(\gamma e)}\|_1=2\leq 2\vee M & \text{for}\quad n_e\leq l\\ \|X_{\tau(e),q,\Gamma(\gamma e)}\|_{1+\frac{n_e-l}{l}}^{1+\frac{n_e-1}{l}}\leq M\leq 2\vee M & \text{for}\quad l<n_e\leq l+p\end{cases}$$

(here $c_0=2+\max_v(\mathrm{card}\,E_v)^{l+p+1}\left(\max_{v\in V;k=0,\ldots,l+p} L_v^{(k)}\right)(2\vee M)^{l+p+1}<\infty$.)

The matrix

$$A:=A((1+\tfrac{p+1}{l})q,(1+\tfrac{p+1}{l})\beta(q))=\left(\sum_{e\in E_{uv}}\rho_u^{l+p+1}H_e^{(l+p+1)}\rho_v^{-(l+p+1)}\right)_{u,v\in V}$$

is irreducible with spectral radius $\Phi := \Phi((1+\frac{p+1}{l})q, (1+\frac{p+1}{l})\beta(q)) > 0$. Perron-Frobenius theorem therefore implies that there exists a positive eigenvector $z = (z_u)_{u\in V}$ of A with eigenvalue Φ, i.e.

$$\sum_v \sum_{e\in E_{uv}} \rho_u^{l+p+1} H_e^{(l+p+1)} \rho_v^{-(l+p+1)} z_v = \Phi z_u$$

for $u \in V$, whence

$$\sum_{\gamma\in E_u^{(n)}} \rho_u^{l+p+1} \left(\prod_{i=1}^{|\gamma|} H_{\gamma|i}^{(l+p+1)} \right) \rho_v^{-(l+p+1)} z_v = \Phi^n z_u$$

for $n \in \mathbb{N}$ and $\gamma \in E_u^{(n)}$. The above equation implies that

$$\sum_{\gamma\in E_u^{(n)}} H_{\gamma|1}^{(l+p+1)} \cdot \ldots \cdot H_{\gamma|n}^{(l+p+1)} \le c_1 \Phi^n \tag{5.2.5}$$

where $c_1 = (\overline{\rho}/\underline{\rho})^{l+p+1}(\max_v z_v / \min_v z_v)$. By combining (5.2.4) and (5.2.5),

$$\|X_{u,q,\Gamma}\|_{1+\frac{p+1}{l}}^{1+\frac{p+1}{l}} \le (1+c_0)c_1 \sum_{n=1}^{\infty} \Phi^n + c_0 \,. \tag{5.2.6}$$

Finally note that $\Phi := \Phi((1+\frac{p+1}{l})q, (1+\frac{p+1}{l})\beta(q)) < \Phi((1+\frac{p+1}{l})q, \beta((1+\frac{p+1}{l})q)) = 1$ since $1+\frac{p+1}{l} \le 1+\frac{m_l}{l} < 1+\varepsilon_q$. Hence $\sum_{n=1}^{\infty} \Phi^n = \frac{1}{1-\Phi}$, and equation (5.2.6) therefore implies that

$$\|X_{u,q,\Gamma}\|_{1+\frac{p+1}{l}}^{1+\frac{p+1}{l}} \le (1+c_0)c_1 \tfrac{1}{1-\Phi} + c_0 \tag{5.2.7}$$

for all $\Gamma \in \mathbf{Max}(u)$. This proves the inductive step since the right hand side of equation (5.2.7) is independent of Γ. □

If in addition $q \in]q_{\min}, q_{\max}[$, Theorem 5.2.1 and the L^p martingale convergence theorem imply that $X_{u,q} \in \bigcap_{1\le p<1+\varepsilon_q} L^p(\Omega_u, \Lambda_u)$ and

$$\|X_{u,q,n} - X_{u,q}\|_p \to 0 \quad \text{for} \quad 1 \le p < 1+\varepsilon_q \,. \tag{5.2.8}$$

The next proposition summarizes most of the properties of $X_{u,q}$ and $X_{u,\alpha,q,\Gamma}$.

Proposition 5.2.2. *Assume that condition (I) is satisfied. Let $q \in \mathbb{R}$, $u \in V$ and $\alpha \in E_u^{(*)}$. Let $\Gamma, \Sigma \in \mathbf{Max}(u)$ with $\Sigma \prec \Gamma$, and $\Delta, \Pi \in \mathbf{Max}(\tau(\alpha_{|\alpha|}))$ with $\Pi \prec \Delta$. Then the following hold.*

i)

$$\mathbf{E}_{\Lambda_u}[X_{u,q,\Gamma}] = 1\,.$$

If in addition $q \in]q_{\min}, q_{\max}[$ then

$$\mathbf{E}_{\Lambda_u}[X_{u,q}] = 1\,.$$

ii) $X_{u,q,\Gamma} = \sum_{\sigma\in\Sigma} \rho_u^{-1} \left(\prod_{i=1}^{|\sigma|} p_{\sigma|i}^q \operatorname{Lip}(S_{\sigma|i})^{\beta(q)}\right) \rho_{\tau(\sigma_{|\sigma|})} X_{u,\sigma,q,\Gamma(\sigma)}\,.$

iii) $X_{u,\alpha,q,\Delta}$

$$= \sum_{\pi\in\Pi} \rho_{\tau(\alpha_{|\alpha|})}^{-1} \left(\prod_{i=1}^{|\pi|} p_{\alpha(\pi|i)}^q \operatorname{Lip}(S_{\alpha(\pi|i)})^{\beta(q)}\right) \rho_{\tau(\alpha\pi_{|\alpha\pi|})} X_{u,\alpha\pi,q,\Delta(\pi)}\,.$$

iv) $X_{u,\alpha,q,\Gamma} = X_{\tau(\alpha_{|\alpha|}),q,\Gamma}(\mathcal{S}_\alpha)$.

v) *The limit*

$$\lim_n X_{u,\alpha,q,n} := X_{u,\alpha,q}$$

exists Λ_u-a.e.

vi) $X_{u,\alpha,q} = X_{\tau(\alpha_{|\alpha|}),q}(\mathcal{S}_\alpha)$

vii) $X_{u,q} = \sum_{\sigma\in\Sigma} \rho_u^{-1} \left(\prod_{i=1}^{|\sigma|} p_{\sigma|i}^q \operatorname{Lip}(S_{\sigma|i})^{\beta(q)}\right) \rho_{\tau(\sigma_{|\sigma|})} X_{u,\sigma,q}$ Λ_u-a.s.

viii) $X_{u,\alpha,q}$

$$= \sum_{\pi\in\Pi} \rho_{\tau(\alpha_{|\alpha|})}^{-1} \left(\prod_{i=1}^{|\pi|} p_{\alpha(\pi|i)}^q \operatorname{Lip}(S_{\alpha(\pi|i)})^{\beta(q)}\right) \rho_{\tau(\alpha\pi_{|\alpha\pi|})} X_{u,\alpha\pi,q} \quad \Lambda_u\text{-a.s.}$$

ix) *The family $(X_{u,\gamma,q})_{\gamma\in\Gamma}$ is independent.*

x) *The random variables $(X_{u,q}, X_{u,\gamma,q})_{\gamma\in E_{uu}^{(*)}}$ are identically distributed.*

Proof. i) It follows immediately from the martingale property that

$$\mathbf{E}_{\Lambda_u}[X_{u,q,\Gamma}] = \mathbf{E}_{\Lambda_u}\left[\mathbf{E}_{\Lambda_u}[X_{u,q,\Gamma}|\mathcal{A}_{u,E_u}]\right] = \mathbf{E}_{\Lambda_u}[X_{u,q,1}] = 1\,.$$

If in addition $q \in]q_{\min}, q_{\max}[$, (5.2.8) implies that

$$|\mathbf{E}_{\Lambda_u}[X_{u,q}] - 1| = |\mathbf{E}_{\Lambda_u}[X_{u,q}] - \mathbf{E}_{\Lambda_u}[X_{u,q,n}]| \le \|X_{u,q} - X_{u,q,n}\|_1 \to 0\,.$$

ii) We have

$$X_{u,q,\Gamma} = \sum_{\gamma\in\Gamma} \rho_u^{-1} \left(\prod_{i=1}^{|\gamma|} p_{\gamma|i}^q \operatorname{Lip}(S_{\gamma|i})^{\beta(q)}\right) \rho_{\tau(\gamma_{|\gamma|})}$$

$$= \sum_{\sigma\in\Sigma} \sum_{\alpha\in\Gamma(\sigma)} \rho_u^{-1} \left(\prod_{i=1}^{|\sigma|} p_{\sigma|i}^q \operatorname{Lip}(S_{\sigma|i})^{\beta(q)}\right) \cdot$$

$$\left(\prod_{i=1}^{|\alpha|} p_{\sigma(\alpha|i)}^q \operatorname{Lip}(S_{\sigma(\alpha|i)})^{\beta(q)}\right) \rho_{\tau(\sigma\alpha_{|\sigma\alpha|})}$$

$$= \sum_{\sigma\in\Sigma} \left(\rho_u^{-1} \left(\prod_{i=1}^{|\sigma|} p_{\sigma|i}^q \operatorname{Lip}(S_{\sigma|i})^{\beta(q)} \right) \rho_{\tau(\sigma_{|\sigma|})} \cdot \right.$$

$$\left. \sum_{\alpha\in\Gamma(\sigma)} \rho_{\tau(\sigma_{|\sigma|})}^{-1} \left(\prod_{i=1}^{|\alpha|} p_{\sigma(\alpha|i)}^q \operatorname{Lip}(S_{\sigma(\alpha|i)})^{\beta(q)} \right) \rho_{\tau(\sigma\alpha_{|\sigma\alpha|})} \right)$$

$$= \sum_{\sigma\in\Sigma} \rho_u^{-1} \left(\prod_{i=1}^{|\sigma|} p_{\sigma|i}^q \operatorname{Lip}(S_{\sigma|i})^{\beta(q)} \right) \rho_{\tau(\sigma_{|\sigma|})} X_{u,\sigma,q,\Gamma(\sigma)} .$$

iii) Follows by a calculation similar to the one in the proof of ii).

iv) Obvious.

v)-vi) Follows from iv).

vii) Follows from ii).

viii) Follows from iii).

ix) Let $\Gamma_0 \subseteq \Gamma$ and let $(B_\gamma)_{\gamma\in\Gamma_0}$ be a family of Borel subsets of $\mathbb{R}$. Then

$$\Lambda_u \left(X_{u,\gamma,q} \in B_\gamma \quad \text{for all} \quad \gamma \in \Gamma_0 \right)$$

$$= \Lambda_u \left(X_{\tau(\gamma_{|\gamma|}),q}(S_\gamma) \in B_\gamma \quad \text{for all} \quad \gamma \in \Gamma_0 \right)$$

$$= \left(\lambda_u \times \prod_{\alpha\in E_u^{(*)}\setminus\{\varnothing\}} \lambda_\alpha \right) \left(S_\gamma \in X_{\tau(\gamma_{|\gamma|}),q}^{-1}(B_\gamma) \quad \text{for all} \quad \gamma\in\Gamma_0 \right)$$

$$= \left(\lambda_u \times \left(\prod_{\substack{\alpha\in E_u^{(*)}\setminus\{\varnothing\} \\ \forall\gamma\in\Gamma_0:\gamma\not\prec\alpha}} \lambda_\alpha \right) \times \prod_{\gamma\in\Gamma_0} \left(\prod_{\substack{\alpha\in E_u^{(*)} \\ \gamma\prec\alpha}} \lambda_\alpha \right) \right)$$

$$\left\{ \left(\xi_\varnothing, (\xi_\alpha)_{\alpha\in E_u^{(*)}\setminus\{\varnothing\}} \right) \in \Xi_u \times \prod_{\alpha\in E_u^{(*)}\setminus\{\varnothing\}} \Xi_{\tau(\alpha_{|\alpha|})} \right|$$

$$\left. (\xi_{\gamma\rho})_{\rho\in E_{\tau(\gamma_{|\gamma|})}^{(*)}} \in X_{\tau(\gamma_{|\gamma|}),q}^{-1}(B_\gamma) \quad \text{for all} \quad \gamma\in\Gamma_0 \right\}$$

$$= \left(\lambda_u \times \left(\prod_{\substack{\alpha \in E_u^{(*)} \setminus \{\varnothing\} \\ \forall \gamma \in \Gamma_0 : \gamma \not\prec \alpha}} \lambda_\alpha \right) \times \prod_{\gamma \in \Gamma_0} \left(\prod_{\substack{\alpha \in E_u^{(*)} \\ \gamma \prec \alpha}} \lambda_\alpha \right) \right)$$

$$\left(\Xi_u \times \prod_{\substack{\alpha \in E_u^{(*)} \setminus \{\varnothing\} \\ \forall \gamma \in \Gamma_0 : \gamma \not\prec \alpha}} \Xi_{\tau(\alpha_{|\alpha|})} \times \right.$$

$$\left. \prod_{\gamma \in \Gamma_0} \left\{ (\xi_{\gamma\rho})_{\rho \in E^{(*)}_{\tau(\gamma_{|\gamma|})}} \in \prod_{\rho \in E^{(*)}_{\tau(\gamma_{|\gamma|})}} \Xi_{\tau(\gamma\rho_{|\gamma\rho|})} \middle| (\xi_{\gamma\rho})_{\rho \in E^{(*)}_{\tau(\gamma_{|\gamma|})}} \in X^{-1}_{\tau(\gamma_{|\gamma|}),q}(B_\gamma) \right\} \right)$$

$$= \prod_{\gamma \in \Gamma_0} \left(\left(\prod_{\substack{\alpha \in E_u^{(*)} \\ \gamma \prec \alpha}} \lambda_\alpha \right) \right.$$

$$\left. \left\{ (\xi_{\gamma\rho})_{\rho \in E^{(*)}_{\tau(\gamma_{|\gamma|})}} \in \prod_{\rho \in E^{(*)}_{\tau(\gamma_{|\gamma|})}} \Xi_{\tau(\gamma\rho_{|\gamma\rho|})} \middle| (\xi_{\gamma\rho})_{\rho \in E^{(*)}_{\tau(\gamma_{|\gamma|})}} \in X^{-1}_{\tau(\gamma_{|\gamma|}),q}(B_\gamma) \right\} \right)$$

$$= \prod_{\gamma \in \Gamma_0} \Lambda_u \left(\mathcal{S}_\gamma \in X^{-1}_{\tau(\gamma_{|\gamma|}),q}(B_\gamma) \right)$$

$$= \prod_{\gamma \in \Gamma_0} \Lambda_u \left(X_{\tau(\gamma_{|\gamma|}),q}(\mathcal{S}_\gamma) \in (B_\gamma) \right)$$

$$= \prod_{\gamma \in \Gamma_0} \Lambda_u \left(X_{u,\gamma,q} \in (B_\gamma) \right) .$$

x) Let $\gamma \in E_{uu}^{(*)}$ and let B be a Borel subset of $\mathbb{R}$. Lemma 4.3.2 implies that

$$\begin{aligned} \Lambda_u(X_{u,\gamma,q} \in B) &= \Lambda_u(X_{\tau(\gamma_{|\gamma|}),q}(\mathcal{S}_\gamma) \in B) = \Lambda_u(\mathcal{S}_\gamma^{-1}(X_{\tau(\gamma_{|\gamma|}),q} \in B)) \\ &= \Lambda_{\tau(\gamma_{|\gamma|})}(X_{\tau(\gamma_{|\gamma|}),q} \in B) = \Lambda_u(X_u \in B) \quad \square \end{aligned}$$

Proposition 5.2.3. *Assume that conditions (I) and (II) are satisfied and let* $q \in]q_{\min}, q_{\max}[$. *Then*

$$X_{u,q} > 0 \quad \Lambda_u\text{-a.s.}$$

for all $u \in V$.

Proof. We have by Lemma 4.3.1 for $u \in V$,

$$\Lambda_u(\{\omega \in \Omega_u \mid X_{u,q}(\omega) = 0\})$$

$$= \left(\left(\lambda_u \times \prod_{e \in E_u} \Lambda_{\tau(e)}\right) \circ \varphi_u^{-1}\right) (\{\omega \in \Omega_u \mid X_{u,q}(\omega) = 0\})$$

$$= \left(\lambda_u \times \prod_{e \in E_u} \Lambda_{\tau(e)}\right)$$

$$\left(\left\{\left(((S_e)_{e \in E_u}, (p_e)_{e \in E_u}), (\omega_{\tau(e)})_{e \in E_u}\right) \in \Xi_u \times \prod_{e \in E_u} \Omega_{\tau(e)} \right| \right.$$

$$\left.\left. X_{u,q}\left(\varphi_u\left(((S_e)_{e \in E_u}, (p_e)_{e \in E_u}), (\omega_{\tau(e)})_{e \in E_u}\right)\right) = 0\right\}\right)$$

[by Proposition 5.2.2]

$$= \left(\lambda_u \times \prod_{e \in E_u} \Lambda_{\tau(e)}\right)$$

$$\left(\left\{\left(((S_e)_{e \in E_u}, (p_e)_{e \in E_u}), (\omega_{\tau(e)})_{e \in E_u}\right) \in \Xi_u \times \prod_{e \in E_u} \Omega_{\tau(e)} \right| \right.$$

$$\left.\left. \sum_{e \in E_u} \rho_u^{-1} p_e^q \operatorname{Lip}(S_e)^{\beta(q)} \rho_{\tau(e)} X_{\tau(e),q}(\omega_{\tau(e)}) = 0 \right\}\right)$$

Now, since $\rho_u > 0$ for all $u \in V$ and in addition $\Delta \leq p_e$ and $\Delta \leq \operatorname{Lip}(S_e)$ for all $u \in V$ and λ_u-a.a. $((S_e)_{e \in E_u}, (p_e)_{e \in E_u}) \in \Xi_u$, the above inequality implies that

$$\Lambda_u(\{\omega \in \Omega_u \mid X_{u,q}(\omega) = 0\})$$

$$= \left(\lambda_u \times \prod_{e \in E_u} \Lambda_{\tau(e)}\right)$$

$$\left(\left\{\left(((S_e)_{e \in E_u}, (p_e)_{e \in E_u}), (\omega_{\tau(e)})_{e \in E_u}\right) \in \Xi_u \times \prod_{e \in E_u} \Omega_{\tau(e)} \right| \right.$$

$$\left.\left. X_{\tau(e),q}(\omega_{\tau(e)}) = 0 \quad \text{for all} \quad e \in E_u \right\}\right)$$

$$= \left(\lambda_u \times \prod_{e \in E_u} \Lambda_{\tau(e)}\right) \left(\Xi_u \times \prod_{e \in E_u} \left\{\omega_{\tau(e)} \in \Omega_{\tau(e)} \middle| X_{\tau(e),q}(\omega_{\tau(e)}) = 0\right\}\right)$$

$$= \prod_{e \in E_u} \Lambda_{\tau(e)} \left(\left\{\omega_{\tau(e)} \in \Omega_{\tau(e)} \middle| X_{\tau(e),q}(\omega_{\tau(e)}) = 0\right\}\right) \tag{5.2.9}$$

for all $u \in V$. Put $M := \max_u \Lambda_u(\{\omega \in \Omega_u \mid X_{u,q}(\omega) = 0\})$. Since $M \leq 1$, (5.2.9) implies that

$$M = \max_u \left(\prod_{e \in E_u} \Lambda_{\tau(e)} \left(\{ \omega_{\tau(e)} \in \Omega_{\tau(e)} \big| X_{\tau(e),q}(\omega_{\tau(e)}) = 0 \} \right) \right)$$
$$\leq \max_u \left(M^{\text{card } E_u} \right) = M^{\min_u(\text{card } E_u)} \leq M$$

whence

$$M = M^{\min_u(\text{card } E_u)} . \tag{5.2.10}$$

Since (II) implies that $\min_u(\text{card}\, E_u) \geq 2$, (5.2.10) shows that

$$M = 0 \quad \text{or} \quad M = 1 . \tag{5.2.11}$$

However, since $q \in]q_{\min}, q_{\max}[$, Proposition 5.2.2 shows that $\int X_{u,q}\, d\Lambda_u = 1$ for all u, whence

$$M := \max_u \Lambda_u(\{\omega \in \Omega_u \mid X_{u,q}(\omega) = 0\}) \neq 1 . \tag{5.2.12}$$

It follows immediately from (5.2.11) and (5.2.12) that $\max_u \Lambda_u(\{\omega \in \Omega_u \mid X_{u,q}(\omega) = 0\}) = M = 0$. □

5.3 The negative moments of $X_{u,q}$

Our next goal is to prove that $X_{u,q}$ has finite negative moments of all orders for $q \in]q_{\min}, q_{\max}[$. Our arguments are very similar to [Ho, Corollary 2.5].

Lemma 5.3.1. *Let X be a positive random variable defined on a probability space (Ω, Σ, P). If there exist $M, \varepsilon > 0$ and a number $z \in]0,1[$ such that*

$$\mathbf{E}_P[e^{-tX}] \leq M e^{-\varepsilon t^z} \quad \textit{for} \quad t \geq 0$$

then X has finite negative moments of all orders, i.e.

$$\int X^{-p}\, dP < \infty \quad \textit{for all} \quad p > 0\,.$$

Proof. We have for $p > 0$,

$$\begin{aligned}\int X^{-p}\, dP \leq & e P(X^{-p} \leq e) + \sum_{n=1}^{\infty} e^{n+1} P(e^n \leq X^{-p} \leq e^{n+1}) \\ \leq & e \left(1 + \sum_{n=1}^{\infty} e^n P(X \leq e^{-\frac{n}{p}})\right). \end{aligned} \tag{5.3.1}$$

Now, for each $n \in \mathbb{N}$ and $t > 0$,

$$P(X \leq e^{-\frac{n}{p}}) \leq \int_{\{X \leq e^{-\frac{n}{p}}\}} e^{-tX}\, dP\, e^{te^{-\frac{n}{p}}} \leq \mathbf{E}_P[e^{-tX}] e^{te^{-\frac{n}{p}}} \leq M e^{te^{-\frac{n}{p}} - \varepsilon t^z}\,.$$

Hence

$$P(X \leq e^{-\frac{n}{p}}) \leq M e^{\inf_{t \geq 0}(te^{-\frac{n}{p}} - \varepsilon t^z)}\,.$$

An easy calculus argument shows that

$$\inf_{t \geq 0}(te^{-\frac{n}{p}} - \varepsilon t^z) = -ae^{bn}$$

where $a := \varepsilon^{1/(1-z)} z^{z/(1-z)}(1 - z^{1/(1-z)}) > 0$ and $b := z/(p(1-z)) > 0$. Hence

$$P(X \leq e^{-\frac{n}{p}}) \leq M e^{-ae^{bn}}\,. \tag{5.3.2}$$

By (5.3.1) and (5.3.2),

$$\int X^{-p}\, dP \leq e\left(1 + M \sum_{n=1}^{\infty} e^{-(ae^{bn} - n)}\right) < \infty\,. \quad \square$$

Proposition 5.3.2. *Let $u \in V$ and $q \in]q_{\min}, q_{\max}[$. Assume conditions (I) and (II). Then the following statments hold.*

i) *There exist $M, \varepsilon > 0$ and $z \in]0,1[$ such that*

$$\mathbf{E}_{\Lambda_u}[e^{-tX_{u,q}}] \leq Me^{-\varepsilon t^z} \quad \textit{for} \quad t \geq 0\,.$$

ii) *$X_{u,q}$ has finite moments of all negative orders, i.e.*

$$\int X_{u,q}^{-p} < \infty \quad \textit{for all} \quad p > 0\,.$$

Proof. i) Let $N = \max_v \operatorname{card}(E_v)$, $\delta = (\frac{1}{2} \wedge (\underline{\rho}/\overline{\rho})(\Delta^q \wedge 1)(\Delta^{\beta(q)} \wedge 1))N$ and

$$z = \frac{\log N}{\log N - \log \delta} \in]0,1[\,.$$

By Proposition 5.2.3, $X_{v,q} > 0$ Λ_v-a.s. for all $v \in V$. Hence

$$\varepsilon := -\tfrac{1}{2} \max_v \log \mathbf{E}_{\Lambda_v}[e^{-X_{v,q}}] > 0\,.$$

It follows from (I) and Proposition 5.2.2 that

$$X_{v,q} = \sum_{e \in E_v} \rho_u^{-1} r_e^q \operatorname{Lip}(S_e)^{\beta(q)} \rho_{\tau(e)} X_{v,e,q} \geq \delta N^{-1} \sum_{e \in E_v} X_{v,e,q} \quad \Lambda_v\text{-a.s.}$$

for all $v \in V$. Define $L : \mathbb{R}_+ \to \mathbb{R}$ by

$$L(t) = \max_v \log \mathbf{E}_{\Lambda_v}[e^{-t^{\frac{1}{z}} X_{v,q}}]\,.$$

Since for each $v \in V$, $(X_{v,e,q})_{e \in E_v}$ is an independent family,

$$\begin{aligned}
L(N^n) &= \max_v \log \mathbf{E}_{\Lambda_v}\left[e^{-N^{\frac{n}{z}} X_{v,q}}\right] \\
&\leq \max_v \log \mathbf{E}_{\Lambda_v}\left[e^{-N^{\frac{n}{z}} \delta N^{-1} \sum_{e \in E_v} X_{v,e,q}}\right] \\
&= \max_v \log \mathbf{E}_{\Lambda_v}\left[\prod_{e \in E_v} e^{-N^{\frac{n}{z}-1} \delta X_{v,e,q}}\right] \\
&= \max_v \log \left(\prod_{e \in E_v} \mathbf{E}_{\Lambda_v}\left[e^{-N^{\frac{n}{z}-1} \delta X_{v,e,q}}\right]\right) \\
&= \max_v \sum_{e \in E_v} \log \mathbf{E}_{\Lambda_v}\left[e^{-N^{\frac{n}{z}-1} \delta X_{v,e,q}}\right]
\end{aligned}$$

$$
\begin{aligned}
&= \max_v \sum_{e \in E_v} \log \mathbf{E}_{\Lambda_v} \left[e^{-N^{\frac{n}{2}-1} \delta X_{\tau(e),q}(S_e)} \right] \\
&= \max_v \sum_{e \in E_v} \log \mathbf{E}_{\Lambda_{\tau(e)}} \left[e^{-N^{\frac{n}{2}-1} \delta X_{\tau(e),q}} \right] \\
&= N \max_w \log \mathbf{E}_{\Lambda_w} \left[e^{-N^{\frac{n}{2}-1} \delta X_{w,q}} \right] \\
&= N \max_w \log \mathbf{E}_{\Lambda_w} \left[e^{-N^{\frac{n-1}{2}} X_{w,q}} \right] \\
&= N L(N^{n-1})
\end{aligned}
$$

for $n \in \mathbb{N}$. An induction on n thus yields $L(N^n) \leq N^n L(1) = N^n \max_v \log \mathbf{E}_{\Lambda_v}[e^{-X_{v,q}}]$ $= -2N^n \varepsilon$ for $n \in \mathbb{N}$, whence

$$
\limsup_n \frac{L(N^n)}{N^n} \leq -2\varepsilon \,.
$$

Let $N < t$. Choose $n \in \mathbb{N}$ such that $N^n < t \leq N^{n+1}$. Then we get, since L is decreasing,

$$
\frac{L(t)}{t} \leq \frac{L(N^n)}{N^n}
$$

and so

$$
\limsup_{t \to \infty} \frac{L(t)}{t} \leq \limsup_n \frac{L(N^n)}{N^n} \leq -2\varepsilon \,.
$$

We can thus choose $t_0 > 0$ such that $\frac{L(t)}{t} \leq -\varepsilon$ for $t \geq t_0$. Hence

$$
L(t) \leq \left(\sup_{0 \leq s \leq t_0} L(s) \right) - \varepsilon t := M_0 - \varepsilon t \quad \text{for} \quad t \geq 0 \,.
$$

This clearly implies that

$$
\mathbf{E}_{\Lambda_u} \left[e^{-t X_{u,q}} \right] \leq e^{L(t^2)} \leq e^{M_0 - \varepsilon t^2} = e^{M_0} e^{-\varepsilon t^2} \,.
$$

ii) Follows from i) and Lemma 5.3.1. □

We will now introduce some notation which we will use throughout the rest of the exposition. For $u \in V$ write

$$\begin{aligned}
\Omega'_u = \{\omega \in \Omega_u \mid & 1) \quad \omega \in (\Omega_u)_0\,; \\
& 2) \quad \lim_n X_{u,q,n}(\omega) \quad \text{exists;} \\
& 3) \quad X_{u,q}(\omega) = \\
& \qquad \sum_{\alpha \in E_u^{(n)}} \rho_u^{-1} \left(\prod_{i=1}^{n} p_{\alpha|i}(\omega)^q \operatorname{Lip}(S_{\alpha|i}(\omega))^{\beta(q)} \right) \rho_{\tau(\alpha_{|\alpha|})} X_{u,\alpha,q} \\
& \quad \text{for all } n \in \mathbb{N}; \\
& 4) \quad X_{u,\alpha,q}(\omega) = \\
& \qquad \sum_{\gamma \in E_{\tau(\alpha_{|\alpha|})}} \rho_{\tau(\alpha_{|\alpha|})}^{-1} \left(\prod_{i=1}^{n} p_{\alpha(\gamma|i)}(\omega)^q \operatorname{Lip}(S_{\alpha(\gamma|i)}(\omega))^{\beta(q)} \right) \cdot \\
& \qquad\qquad \rho_{\tau(\alpha\gamma_{|\alpha\gamma|})} X_{u,\alpha\gamma,q} \\
& \quad \text{for all } \alpha \in E_u^{(*)} \text{ and } n \in \mathbb{N} \ \}
\end{aligned} \tag{5.3.3}$$

It follows from Lemma 4.3.3 and Proposition 5.2.2 that

$$\Lambda_u(\Omega'_u) = 1\,. \tag{5.3.4}$$

Chapter 6

The Random Multifractal Construction Measure $\mathcal{M}_{u,q}$ and the $Q_{u,q}$ Measure

6.1 The $\mathcal{M}_{u,q}$ Measure

We begin by stating the main theorem in this section.

Theorem 6.1.1. *Let $q \in \mathbb{R}$. Assume that X_v is compact for all $v \in V$, and that condition (I) holds. Then there exists a random measure*

$$\mathcal{M}_{u,q} : \Omega_u \to \mathcal{M}(X_u)$$

satisfying

i) *$\mathcal{M}_{u,q}$ has total mass $X_{u,q}$ Λ_u-a.s.*

ii) *$\mathcal{M}_{u,q}(C_u) = X_{u,q}$ Λ_u-a.s.*

iii) *The following holds for Λ_u-a.a. $\omega \in \Omega_u$: for all compact subsets K of X_u then*

$$\sum_{\substack{\alpha \in E_u^{(n)} \\ K \cap C_u(\omega) \neq \varnothing}} \rho_u^{-1} \left(\prod_{i=1}^{|\alpha|} p_{\alpha|i}(\omega)^q \operatorname{Lip}(S_{\alpha|i}(\omega))^{\beta(q)} \right) \rho_{\tau(\alpha_{|\alpha|})} X_{u,\alpha,q}(\omega) \searrow \mathcal{M}_{u,q}(\omega)(K)$$

The proof of Theorem 6.1.1 is based on the following lemma.

Lemma 6.1.2. *Let $q \in \mathbb{R}$. Assume that X_v is compact for all $v \in V$, and that condition (I) is satisfied. For $\omega \in \Omega_u$, let*

$$x_\alpha(\omega) \in S_{\alpha|1} \circ \cdots \circ S_{\alpha|\alpha|}(X_{\tau(\alpha_{|\alpha|})}).$$

Then the following statement holds.

i) *There exists a subset $\Omega_u'' \subseteq \Omega_u'$ with $\Lambda_u(\Omega_u'') = 1$ such that the sequence*

$$\left(\sum_{\alpha \in E_u^{(n)}} f(x_\alpha(\omega)) \rho_u^{-1} \left(\prod_{i=1}^{|\alpha|} p_{\alpha|i}(\omega)^q \operatorname{Lip}(S_{\alpha|i}(\omega))^{\beta(q)} \right) \rho_{\tau(\alpha_{|\alpha|})} \right)_{n \in \mathbb{N}}$$

converges for all $\omega \in \Omega_u''$ and all $f \in C^0(X_u)$.

For each $\omega \in \Omega''_u$ *define* $\psi_{u,q}(\omega) : C^0(X_u) \to \mathbb{R}$ *by*

$$\psi_{u,q}(\omega)(f) = \lim_n \sum_{\alpha \in E_u^{(n)}} f(x_\alpha(\omega))\rho_u^{-1} \left(\prod_{i=1}^{|\alpha|} p_{\alpha|i}(\omega)^q \operatorname{Lip}(S_{\alpha|i}(\omega))^{\beta(q)} \right) \rho_{\tau(\alpha_{|\alpha|})} .$$

Then the following statements hold.

ii) $\psi_{u,q}(\omega)$ *is a functional on* $C^0(X_u)$.

iii) $\|\psi_{u,q}(\omega)\| = X_{u,q}(\omega)$.

Proof. i) Let

$$\Omega''_u = \{\omega \in (\Omega_u)_0 \mid \left((S_e(\omega))_{e\in E_u}, (p_e(\omega))_{e\in E_u}\right) \quad \text{satisfies (I)}\}$$

Then clearly $\Lambda_u(\Omega''_u) = 1$. We must now show that

$$\left(\sum_{\alpha \in E_u^{(n)}} f(x_\alpha(\omega))\rho_u^{-1} \left(\prod_{i=1}^{|\alpha|} p_{\alpha|i}(\omega)^q \operatorname{Lip}(S_{\alpha|i}(\omega))^{\beta(q)} \right) \rho_{\tau(\alpha_{|\alpha|})} \right)_{n\in\mathbb{N}}$$

is Cauchy for $\omega \in \Omega''_u$ and $f \in C^0(X_u)$. Fix $\omega \in \Omega''_u$, $f \in C^0(X_u)$ and let $\varepsilon > 0$. Write $p_\alpha(\omega) = p_\alpha$ and $S_\alpha(\omega) = S_\alpha$ for $\alpha \in E_u^{(*)}$. Choose $A > 0$ satisfying

$$2\tfrac{\varepsilon}{A}\left(\tfrac{\varepsilon}{A} + X_{u,q}(\omega) + \|f\|_\infty\right) \le \varepsilon .$$

Now choose $N \in \mathbb{N}$ such that

$$\sup_{\alpha \in E_u^{(n)}} \operatorname{diam} f\left(S_{\alpha|1} \circ \cdots \circ S_{\alpha|\dot{\alpha}|}(X_{\tau(\alpha_{|\alpha|})})\right) \le \varepsilon / A \quad \text{for} \quad n \ge \mathrm{N}. \tag{6.1.1}$$

Next, for each $\alpha \in E_u^{(N)}$ choose an integer $N_\alpha \in \mathbb{N}$ such that

$$|X_{u,\alpha,q,n}(\omega) - X_{u,\alpha,q}(\omega)| \le \varepsilon/(AM) \quad \text{for} \quad n \ge N_\alpha . \tag{6.1.2}$$

where $M := \sum_{\alpha \in E_u^{(N)}} \rho_u^{-1} \left(\prod_{i=1}^{|\alpha|} p_{\alpha|i}^q \operatorname{Lip}(S_{\alpha|i})^{\beta(q)} \right) \rho_{\tau(\alpha_{|\alpha|})}$.

Let $n \ge N + \max_{\alpha \in E_u^{(N)}} N_\alpha$ and $m \in \mathbb{N}$. Then

$$\Bigg| \sum_{\alpha \in E_u^{(m+n)}} f(x_\alpha(\omega))\rho_u^{-1} \left(\prod_{i=1}^{|\alpha|} p_{\alpha|i}^q \operatorname{Lip}(S_{\alpha|i})^{\beta(q)} \right) \rho_{\tau(\alpha_{|\alpha|})} -$$

$$\sum_{\alpha \in E_u^{(n)}} f(x_\alpha(\omega))\rho_u^{-1} \left(\prod_{i=1}^{|\alpha|} p_{\alpha|i}^q \operatorname{Lip}(S_{\alpha|i})^{\beta(q)} \right) \rho_{\tau(\alpha_{|\alpha|})} \Bigg|$$

$$
= \Bigg| \sum_{\alpha \in E_u^{(N)}} \rho_u^{-1} \left(\prod_{i=1}^{|\alpha|} p_{\alpha|i}^q \operatorname{Lip}(S_{\alpha|i})^{\beta(q)} \right) \rho_{\tau(\alpha_{|\alpha|})} \cdot
$$

$$
\left(\sum_{\gamma \in E_{\tau(\alpha_{|\alpha|})}^{(n+m-N)}} f(x_{\alpha\gamma}(\omega)) \rho_{\tau(\alpha_{|\alpha|})}^{-1} \left(\prod_{i=1}^{n+m-N} p_{\alpha(\gamma|i)}^q \operatorname{Lip}(S_{\alpha(\gamma|i)})^{\beta(q)} \right) \rho_{\tau(\gamma_{|\gamma|})} - \right.
$$

$$
\left. \sum_{\gamma \in E_{\tau(\alpha_{|\alpha|})}^{(n-N)}} f(x_{\alpha\gamma}(\omega)) \rho_{\tau(\alpha_{|\alpha|})}^{-1} \left(\prod_{i=1}^{n-N} p_{\alpha(\gamma|i)}^q \operatorname{Lip}(S_{\alpha(\gamma|i)})^{\beta(q)} \right) \rho_{\tau(\gamma_{|\gamma|})} \right) \Bigg|
$$

$$
\leq \sum_{\alpha \in E_u^{(N)}} \rho_u^{-1} \left(\prod_{i=1}^{|\alpha|} p_{\alpha|i}^q \operatorname{Lip}(S_{\alpha|i})^{\beta(q)} \right) \rho_{\tau(\alpha_{|\alpha|})} \cdot
$$

$$
\left(\Bigg| \sum_{\gamma \in E_{\tau(\alpha_{|\alpha|})}^{(n+m-N)}} \left(f(x_{\alpha\gamma}(\omega)) - f(x_\alpha(\omega)) \right) \rho_{\tau(\alpha_{|\alpha|})}^{-1} \cdot \right.
$$

$$
\left(\prod_{i=1}^{n+m-N} p_{\alpha(\gamma|i)}^q \operatorname{Lip}(S_{\alpha(\gamma|i)})^{\beta(q)} \right) \rho_{\tau(\gamma_{|\gamma|})} \Bigg| +
$$

$$
\left(f(x_\alpha(\omega)) X_{u,\alpha,q,n+m-N}(\omega) - f(x_\alpha(\omega)) X_{u,\alpha,q,n-N}(\omega) \right) +
$$

$$
\Bigg| \sum_{\gamma \in E_{\tau(\alpha_{|\alpha|})}^{(n-N)}} \left(f(x_{\alpha\gamma}(\omega)) - f(x_\alpha(\omega)) \right) \rho_{\tau(\alpha_{|\alpha|})}^{-1} \cdot
$$

$$
\left. \left(\prod_{i=1}^{n-N} p_{\alpha(\gamma|i)}^q \operatorname{Lip}(S_{\alpha(\gamma|i)})^{\beta(q)} \right) \rho_{\tau(\gamma_{|\gamma|})} \Bigg| \right)
$$

$$
\leq \sum_{\alpha \in E_u^{(N)}} \rho_u^{-1} \left(\prod_{i=1}^{|\alpha|} p_{\alpha|i}^q \operatorname{Lip}(S_{\alpha|i})^{\beta(q)} \right) \rho_{\tau(\alpha_{|\alpha|})} \cdot
$$

$$
\Big(\operatorname{diam}\left(f(C_\alpha(\omega)) \right) X_{u,\alpha,q,n+m-N}(\omega) +
$$

$$
\|f\|_\infty |X_{u,\alpha,q,n+m-N}(\omega) - X_{u,\alpha,q,n-N}(\omega)| +
$$

$$
\operatorname{diam}\left(f(C_\alpha(\omega)) \right) X_{u,\alpha,q,n-N}(\omega) \Big)
$$

[by (6.1.1) and (6.1.2)]

$$\leq \sum_{\alpha\in E_u^{(N)}} \rho_u^{-1}\left(\prod_{i=1}^{|\alpha|} p_{\alpha|i}^q \operatorname{Lip}(S_{\alpha|i})^{\beta(q)}\right)\rho_{\tau(\alpha_{|\alpha|})}\cdot$$

$$\left(2\frac{\varepsilon}{A}\left(\frac{\varepsilon}{AM}+X_{u,\alpha,q}(\omega)\right)+\|f\|_\infty 2\frac{\varepsilon}{AM}\right)$$

$$=2\frac{\varepsilon}{AM}\left(\frac{\varepsilon}{A}+\|f\|_\infty\right)\sum_{\alpha\in E_u^{(N)}} \rho_u^{-1}\left(\prod_{i=1}^{|\alpha|} p_{\alpha|i}^q \operatorname{Lip}(S_{\alpha|i})^{\beta(q)}\right)\rho_{\tau(\alpha_{|\alpha|})}+$$

$$2\frac{\varepsilon}{A}\sum_{\alpha\in E_u^{(N)}} \rho_u^{-1}\left(\prod_{i=1}^{|\alpha|} p_{\alpha|i}^q \operatorname{Lip}(S_{\alpha|i})^{\beta(q)}\right)\rho_{\tau(\alpha_{|\alpha|})}X_{u,\alpha,q}(\omega)$$

$$\leq 2\frac{\varepsilon}{AM}\left(\frac{\varepsilon}{A}+\|f\|_\infty\right)M+2\frac{\varepsilon}{A}X_{u,q}(\omega)\leq\varepsilon\,.$$

ii) Obvious.
iii) We have

$$\psi_u(\omega)(1)=\lim_n \sum_{\alpha\in E_u^{(n)}} \rho_u^{-1}\left(\prod_{i=1}^{n} p_{\alpha|i}(\omega)^q \operatorname{Lip}(S_{\alpha|i}(\omega))^{\beta(q)}\right)\rho_{\tau(\alpha_{|\alpha|})}=X_{u,q}(\omega)$$

which proves iii). □

Proof of Theorem 6.1.1

Let $\psi_{u,q}$ and Ω_u'' be as in Lemma 6.1.2. By Riesz representation theorem there exists for each $\omega\in\Omega_u''$ a measure $\mathcal{M}_{u,q}(\omega)$ on X_u such that

$$\psi_{u,q}(\omega)(f)=\int f\,d(\mathcal{M}_{u,q}(\omega))\quad\text{for}\quad f\in C^0(X_u)\,.$$

We will now prove that i), ii) and iii) are satisfied for $\omega\in\Omega_u''$. Fix $\omega\in\Omega_u''$ and write $p_\alpha(\omega)=p_\alpha$ and $S_\alpha(\omega)=S_\alpha$ for $\alpha\in E_u^{(*)}$
i) Obvious since $\mathcal{M}_{u,q}(\omega)(X_u)=\|\psi_{u,q}(\omega)\|=X_{u,q}(\omega)$.
iii) We first prove the following two claims.

Claim 1. If $K\subseteq X_u$ is compact then

$$\mathcal{M}_{u,q}(\omega)(K)\leq \sum_{\alpha\in E_u^{(n)}} \rho_u^{-1}\left(\prod_{i=1}^{|\alpha|} p_{\alpha|i}^q \operatorname{Lip}(S_{\alpha|i})^{\beta(q)}\right)\rho_{\tau(e)}X_{u,\alpha,q}(\omega)$$

Proof of Claim 1. For $n\in\mathbb{N}$ choose $f\in C^0(X_u)$ such that

$$f(x)=\begin{cases}1 & \text{if } x\in K\\ 0 & \text{if } x\in C_\alpha(\omega) \text{ for } \alpha\in E_u^{(n)} \text{ with } K\cap C_\alpha(\omega)=\varnothing\end{cases}$$

We have

$$\mathcal{M}_{u,q}(\omega)(K) \le \int f\, d(\mathcal{M}_{u,q}(\omega)) = \psi_{u,q}(\omega)(f)$$

$$= \lim_m \sum_{\alpha\in E_u^{(m)}} f(x_\alpha(\omega))\rho_u^{-1}\left(\prod_{i=1}^{m} p_{\alpha|i}^q \operatorname{Lip}(S_{\alpha|i})^{\beta(q)}\right)\rho_{\tau(\alpha_{|\alpha|})}$$

$$= \lim_m \sum_{\alpha\in E_u^{(n)}} \sum_{\gamma\in E_{\tau(\alpha_{|\alpha|})}^{(m-n)}} f(x_{\alpha\gamma}(\omega))\rho_u^{-1}\left(\prod_{i=1}^{n} p_{\alpha|i}^q \operatorname{Lip}(S_{\alpha|i})^{\beta(q)}\right)\cdot$$

$$\left(\prod_{i=1}^{m-n} p_{\alpha(\gamma|i)}^q \operatorname{Lip}(S_{\alpha(\gamma|i)})\beta(q)\right)\rho_{\tau(\alpha\gamma_{|\alpha\gamma|})}$$

$$\le \limsup_m \left(\sum_{\substack{\alpha\in E_u^{(n)}\\ C_\alpha(\omega)\cap K\neq\varnothing}} \rho_u^{-1}\left(\prod_{i=1}^{n} p_{\alpha|i}^q \operatorname{Lip}(S_{\alpha|i})^{\beta(q)}\right)\rho_{\tau(\alpha_{|\alpha|})}\cdot\right.$$

$$\left.\sum_{\gamma\in E_{\tau(\alpha_{|\alpha|})}^{m-n}} \rho_{\tau(\alpha_{|\alpha|})}^{-1}\left(\prod_{i=1}^{m-n} p_{\alpha(\gamma|i)}^q \operatorname{Lip}(S_{\alpha(\gamma|i)})\beta(q)\right)\rho_{\tau(\alpha\gamma_{|\alpha\gamma|})}\right)$$

$$= \limsup_m \sum_{\substack{\alpha\in E_u^{(n)}\\ C_\alpha(\omega)\cap K\neq\varnothing}} \rho_u^{-1}\left(\prod_{i=1}^{n} p_{\alpha|i}^q \operatorname{Lip}(S_{\alpha|i})^{\beta(q)}\right)\rho_{\tau(\alpha_{|\alpha|})}X_{u,\alpha,q,m-n}(\omega)$$

$$= \sum_{\substack{\alpha\in E_u^{(n)}\\ C_\alpha(\omega)\cap K\neq\varnothing}} \rho_u^{-1}\left(\prod_{i=1}^{n} p_{\alpha|i}^q \operatorname{Lip}(S_{\alpha|i})^{\beta(q)}\right)\rho_{\tau(\alpha_{|\alpha|})}X_{u,\alpha,q}(\omega).$$

This proves claim 1.

Claim 2. If $K \subseteq X_u$ is compact then

$$\left(\sum_{\substack{\alpha\in E_u^{(n)}\\ C_\alpha(\omega)\cap K\neq\varnothing}} \rho_u^{-1}\left(\prod_{i=1}^{n} p_{\alpha|i}^q \operatorname{Lip}(S_{\alpha|i})^{\beta(q)}\right)\rho_{\tau(\alpha_{|\alpha|})}X_{u,\alpha,q}(\omega)\right)_{n\in\mathbb{N}}$$

is a decreasing sequence.

Proof of Claim 2. Since $\omega \in \Omega_u'' \subseteq \Omega_u'$,

$$X_{u,\alpha,q}(\omega) = \sum_{e\in E_{\tau(\alpha_{|\alpha|})}} \rho_{\tau(\alpha_{|\alpha|})}^{-1} p_{\alpha e}^q \operatorname{Lip}(S_{\alpha e})^{\beta(q)}\rho_{\tau(\alpha e_{|\alpha e|})}X_{u,\alpha e,q}(\omega).$$

The assertion in Claim 2 follows from the equation above. This proves Claim 2.

For compact subsets $K \subseteq X_u$ write

$$\mu(K) := \inf_n \sum_{\substack{\alpha \in E_u^{(n)} \\ C_\alpha(\omega) \cap K \neq \varnothing}} \rho_u^{-1} \left(\prod_{i=1}^{n} p_{\alpha|i}^q \operatorname{Lip}(S_{\alpha|i})^{\beta(q)} \right) \rho_{\tau(\alpha_{|\alpha|})} X_{u,\alpha,q}(\omega) .$$

It follows from Claim 1 and Claim 2 that

$$\sum_{\substack{\alpha \in E_u^{(n)} \\ C_\alpha(\omega) \cap K \neq \varnothing}} \rho_u^{-1} \left(\prod_{i=1}^{n} p_{\alpha|i}^q \operatorname{Lip}(S_{\alpha|i})^{\beta(q)} \right) \rho_{\tau(\alpha_{|\alpha|})} X_{u,\alpha,q}(\omega) \searrow \mu(K) \geq \mathcal{M}_{u,q}(\omega)(K) \tag{6.1.3}$$

for each compact subset $K \subseteq X_u$.

Let $K \subseteq X_u$ be compact. We must now prove that $\mu(K) = \mathcal{M}_{u,q}(\omega)(K)$. Choose compact subsets $K_1, K_2, \ldots$ of $X_u \setminus K$ such that

$$\mathcal{M}_{u,q}(\omega)(K_n) \nearrow \mathcal{M}_{u,q}(\omega)\big(C_u(\omega) \setminus K\big) . \tag{6.1.4}$$

Since $\operatorname{dist}(K_n, K) > 0$,

$$\mu(K_n) + \mu(K) \leq \mu(K_u \cup K) \leq X_{u,q}(\omega) . \tag{6.1.5}$$

Hence

$$\begin{aligned} X_{u,q}(\omega) &= \mathcal{M}_{u,q}(\omega)\big(C_u(\omega)\big) = \mathcal{M}_{u,q}(K) + \mathcal{M}_{u,q}(\omega)\big(C_u(\omega) \setminus K\big) \\ &\leq \mu(K) + \mathcal{M}_{u,q}(\omega)\big(C_u(\omega) \setminus K\big) \leq \mu(K) + \lim_n \mathcal{M}_{u,q}(\omega)(K_n) \\ &\quad [\text{by } (6.1.5)] \\ &= \lim_n \big(\mu(K) + \mathcal{M}_{u,q}(\omega)(K_n)\big) \leq X_{u,q}(\omega) \end{aligned}$$

whence $\mu(K) = \mathcal{M}_{u,q}(\omega)(K)$.

ii) It follows from iii) that

$$\begin{aligned} X_{u,q} &\geq \mathcal{M}_{u,q}(C_u) \\ &= \lim_n \sum_{\substack{\alpha \in E_u^{(n)} \\ C_\alpha(\omega) \cap C_u(\omega) \neq \varnothing}} \rho_u^{-1} \left(\prod_{i=1}^{n} p_{\alpha|i}^q \operatorname{Lip}(S_{\alpha|i})^{\beta(q)} \right) \rho_{\tau(\alpha_{|\alpha|})} X_{u,\alpha,q} \\ &= \lim_n \sum_{\alpha \in E_u^{(n)}} \rho_u^{-1} \left(\prod_{i=1}^{n} p_{\alpha|i}^q \operatorname{Lip}(S_{\alpha|i})^{\beta(q)} \right) \rho_{\tau(\alpha_{|\alpha|})} X_{u,\alpha,q} \\ &= X_{u,q} \quad \Lambda_u\text{-a.s.} \ \square \end{aligned}$$

Corollary 6.1.3. . *Let $q \in \mathbb{R}$ and assume that X_v is compact for all v. Assume that conditions (I) and (IV) are satisfied. Then*

$$\mathcal{M}_{u,q}(\omega)(C_\alpha(\omega)) = \rho_u^{-1}\left(\prod_{i=1}^{|\alpha|} p_{\alpha|i}(\omega)^q \operatorname{Lip}(S_{\alpha|i}(\omega))^{\beta(q)}\right)\rho_{\tau(\alpha_{|\alpha|})}X_{u,\alpha,q}(\omega)$$

for all $u \in V$, $\alpha \in E_u^{(*)}$ and Λ_u-a.a. $\omega \in \Omega_u$.

Proof. By Theorem 6.1.1,

$$\mathcal{M}_{u,q}(\omega)(C_u(\omega)) = \lim_n \sum_{\substack{\gamma\in E_u^{(n+|\alpha|)}\\ C_\gamma(\omega)\cap C_\alpha(\omega)\neq\varnothing}} \rho_u^{-1}\left(\prod_{i=1}^{n+|\alpha|} p_{\gamma|i}(\omega)^q \operatorname{Lip}(S_{\gamma|i}(\omega))^{\beta(q)}\right)\rho_{\tau(\gamma_{|\gamma|})}X_{u,\gamma,q}(\omega) \tag{6.1.6}$$

for Λ_u-a.a. $\omega \in \Omega_u$. However, condition (IV) implies that $\{\gamma \in E_u^{(n+|\alpha|)} \mid C_\gamma(\omega) \cup C_\alpha(\omega) \neq \varnothing\} = \{\alpha\gamma \mid \gamma \in E_{\tau(\alpha_{|\alpha|})}^{(n)}\}$, and (6.1.6) therefore shows that

$$\begin{aligned}
&\mathcal{M}_{u,q}(\omega)(C_u(\omega))\\
&= \lim_n \left(\rho_u^{-1}\left(\prod_{i=1}^{|\alpha|} p_{\alpha|i}(\omega)^q \operatorname{Lip}(S_{\alpha|i}(\omega))^{\beta(q)}\right)\rho_{\tau(\alpha_{|\alpha|})}\cdot\right.\\
&\qquad \left.\sum_{\gamma\in E_{\tau(\alpha_{|\alpha|})}^{(n)}} \rho_{\tau(\alpha_{|\alpha|})}^{-1}\left(\prod_{i=1}^{n+|\alpha|} p_{\gamma|i}(\omega)^q \operatorname{Lip}(S_{\gamma|i}(\omega))^{\beta(q)}\right)\rho_{\tau(\gamma_{|\gamma|})}X_{u,\alpha\gamma,q}(\omega)\right)\\
&=\rho_u^{-1}\left(\prod_{i=1}^{|\alpha|} p_{\alpha|i}(\omega)^q \operatorname{Lip}(S_{\alpha|i}(\omega))^{\beta(q)}\right)\rho_{\tau(\alpha_{|\alpha|})}X_{u,\alpha,q}(\omega)
\end{aligned}$$

for Λ_u-a.a. $\omega \in \Omega_u$. □

Finally observe that if X_v is compact for all v and condition (I) and condition (IV) are satisfied then Theorem 6.1.1 and Theorem 2.3.3 imply that $\mathcal{M}_{u,1}$ and μ_u are equivalent Λ_u-a.s. in the sense that

$$\mathcal{M}_{u,1} \ll \mu_u \quad \Lambda_u\text{-a.s.} \quad \text{and} \quad \mu_u \ll \mathcal{M}_{u,1} \quad \Lambda_u\text{-a.s.}. \tag{6.1.7}$$

6.2 The $Q_{u,q}$ Measure

Recall that if $\omega \in \Omega'_u$ then

$$X_{u,\alpha,q}(\omega) = \sum_{e \in E_{\tau(\alpha_{|\alpha|})}} \rho^{-1}_{\tau(\alpha_{|\alpha|})} p_e(\omega)^q \operatorname{Lip}(S_e(\omega))^{\beta(q)} X_{u,\alpha e,q}(\omega) \tag{6.2.1}$$

for all $\alpha \in E_u^{(*)}$, cf. (5.3.3). For $\omega \in \Omega'_u$ and $\alpha \in E_u^{(*)}$ write

$$\hat{\mathcal{M}}_{u,q}(\omega)([\alpha]) = \rho_u^{-1} \left(\prod_{i=1}^{|\alpha|} p_{\alpha|i}(\omega)^q \operatorname{Lip}(S_{\alpha|i}(\omega))^{\beta(q)} \right) \rho_{\tau(\alpha_{|\alpha|})} X_{u,\alpha,q}(\omega) .$$

If $A \subseteq E_u^{(*)}$ is clopen then there exists $\alpha_1, \ldots, \alpha_n \in E_u^{(*)}$ such that $A = \cup_{i=1}^n [\alpha_i]$. Let $m := \max_i |\alpha_i|$ and $\Gamma := \{\gamma \in E_u^{(m)} \mid \exists i : \alpha_i \prec \gamma\}$. Then

$$A = \cup_{\gamma \in \Gamma} [\gamma]$$
$$[\sigma] \cap [\gamma] = \varnothing \quad \text{for} \quad \sigma, \gamma \in \Gamma \quad \text{with} \quad \sigma \neq \gamma .$$

Now define $\hat{\mathcal{M}}_{u,q}(\omega)(A)$ by

$$\hat{\mathcal{M}}_{u,q}(\omega)(A) = \sum_{\gamma \in \Gamma} \hat{\mathcal{M}}_{u,q}(\omega)([\gamma]) .$$

It follows from (6.2.1) that $\hat{\mathcal{M}}_{u,q}(\omega)$ is a well-defined finitely additive measure on the algebra of clopen subsets of $E_u^{\mathbb{N}}$ for $\omega \in \Omega'_u$. The next lemma shows that $\hat{\mathcal{M}}_{u,q}(\omega)$ is in fact countable additive on the algebra of clopen subsets of $E_u^{\mathbb{N}}$.

Lemma 6.2.1. *Let $\omega \in \Omega'_u$. The set function $\hat{\mathcal{M}}_{u,q}(\omega)$ defined on the algebra $\mathcal{A}$ of clopen subsets of $E_u^{\mathbb{N}}$ is countable additive.*

Proof. Write $\hat{\mathcal{M}} := \hat{\mathcal{M}}_{u,q}(\omega)$. Since $\hat{\mathcal{M}}$ is finitely additive on $\mathcal{A}$ it is sufficient (c.f. [As, Theorem 1.2.8]) to show that $\hat{\mathcal{M}}$ is continuous from above. Let $A_1, A_2, \cdots \in \mathcal{A}$ with $A_n \searrow \varnothing$. Since $E_u^{\mathbb{N}}$ is compact and $A_1, A_2, \ldots$ are closed there exists $N \in \mathbb{N}$ such that $A_n = \varnothing$ for $n \geq N$. This proves the lemma. □

It follows from Lemma (6.2.1) (c.f. [As, 1.3.10]) that $\hat{\mathcal{M}}_{u,q}(\omega)$ can be extended to a measure $\hat{\mathcal{M}}_{u,q}(\omega)$ on the σ-algebra generated by $\mathcal{A}$, i.e. the Borel algebra. Finally put

$$\hat{\mathcal{M}}_{u,q}(\omega) = \mathbf{o}$$

(where $\mathbf{o}$ denotes the zero-measure) for $\omega \in \Omega_u \setminus \Omega'_u$.

The map $\omega \to \hat{\mathcal{M}}_{u,q}(\omega)(A)$, $\omega \in \Omega$, is easily seen to be measurable for $A \in \mathcal{A}$, hence weakly measurable, which implies that there exists a measure $Q_{u,q} \in \mathcal{M}(E_u^{\mathbb{N}} \times \Omega)$ such that

$$Q_{u,q}(A) = \mathbf{E}_{\Lambda_u}\left[\int 1_A \, d\hat{\mathcal{M}}_{u,q}\right] = \int (\hat{\mathcal{M}}_{u,q}(\omega))(A_\omega)\, d\Lambda_u(\omega) \tag{6.2.2}$$

where $A_\omega := \{\sigma \in E_u^{\mathbb{B}} \mid (\sigma, \omega) \in A\}$. Also

$$\int f\, dQ_{u,q} = \int \left(\int f(\sigma,\omega)\, d(\hat{\mathcal{M}}_{u,q}(\omega))(\sigma)\right) d\Lambda_u(\omega) \tag{6.2.3}$$

for $f \in L^1(Q_{u,q})$. Observe that for $q \in]q_{\min}, q_{\max}[$,

$$Q_{u,q}(E_u^{\mathbb{N}} \times \Omega_u) = \int \left(d(\hat{\mathcal{M}}_{u,q}(\omega))(\sigma)\right) d\Lambda_u(\omega) = \int X_{u,q}\, d\Lambda_u = 1$$

i.e. $Q_{u,q}$ is a probability measure.

Lemma 6.2.2. *Let $n \in \mathbb{N}$ and $Z : E_u^{\mathbb{N}} \times \Omega_u \to \mathbb{R}$ be a random variable satisfying*

$$Z(\sigma,\omega) = Z(\pi,\omega)$$

for all $\omega \in \Omega_u$ and all $\sigma, \pi \in E_u^{\mathbb{N}}$ with $\sigma|n = \pi|n$. Then

$$\mathbf{E}_{Q_{u,q}}[Z] = \mathbf{E}_{\Lambda_u}\left[\sum_{\gamma \in E_u^{(n)}} \rho_u^{-1}\left(\prod_{i=1}^{n} p_{\gamma|i}^{q}\, \mathrm{Lip}(S_{\gamma|i})^{\beta(q)}\right) \rho_{\tau(\gamma_{|\gamma|})} X_{u,\gamma,q} Z(\gamma\hat{\gamma}, \cdot)\right]$$

where $\hat{\gamma} \in E_{\tau(\gamma_{|\gamma|})}^{\mathbb{N}}$.

Proof. Follows from (6.2.3). □

Lemma 6.2.3. *If condition (IV) is satisfied then*

$$\hat{\mathcal{M}}_{u,q} \circ \pi_u^{-1} = \mathcal{M}_{u,q} \quad \Lambda_u\text{-a.s.}$$

Proof. Obvious. □

6.3 The Support of $\mathcal{M}_{u,q}$

The purpose of this section is to prove the next theorem which determines the Λ_u almost sure support of $\mathcal{M}_{u,q}$ for $q \in]q_{\min}, q_{\max}[$.

Theorem 6.3.1. *Let $X_v \subseteq \mathbb{R}^d$ be compact for all v and assume that conditions (I), (III), (IV) and (V) are satisfied. Let $u \in V$ and $q \in]q_{\min}, q_{\max}[$. Then the following holds for Λ_u-a.a. $\omega \in \Omega_u$:*
The set

$$\Delta_{\mu_u(\omega)}(\alpha(q))$$

has full $\mathcal{M}_{u,q}(\omega)$ measure, i.e.

$$\alpha_{\mu_u(\omega)}(x) = \alpha(q) \quad \text{for} \quad \mathcal{M}_{u,q}(\omega)\text{-a.a.} \quad x \in C_u(\omega).$$

The proof of Theorem 6.3.1 is based on the next three lemmas.

Lemma 6.3.2. *Let $(X_n)_{n\in\mathbb{N}}$ be an L^2 bounded sequence of random variables on a probability space (Ω, Σ, P). Put $X_0 = 0$ and let*

$$D_k = X_k - \mathbf{E}_P[X_k | X_0, \ldots, X_{k-1}], \quad k \in \mathbb{N}$$

$$Y_n = \sum_{k=1}^{n} \frac{1}{k} D_k$$

Then the following statements hold.

i) *There exists $Y \in L^2$ such that $Y_n \to Y$ P-a.e. and in L^2.*
ii) $\frac{1}{n} \sum_{k=1}^{n} (X_k - \mathbf{E}_P[X_k|X_0, \ldots, X_{k-1}]) \to 0 \quad P\text{-a.e.}.$

Proof. An easy calculation shows that $(Y_n)_n$ is an L^2 martingale. Also

$$\begin{aligned}\sum_n \mathbf{E}_P[(Y_n - Y_{n-1})^2] &= \sum_n \mathbf{E}_P\left[\left(\tfrac{1}{n} D_n\right)^2\right] \\ &\leq \sum_n \tfrac{1}{n^2} 4\mathbf{E}_P\left[X_n^2\right] \leq 4 \sup_m \mathbf{E}_P\left[X_m^2\right] \sum_n \tfrac{1}{n^2} < \infty\end{aligned}$$

and the L^2 martingale convergence theorem (cf. e.g. [Wi, Theorem p. 111]) therefore implies that there exists $Y \in L^2$ such that

$$\sum_{k=1}^{n} \frac{1}{k} D_k = Y_n \to Y \quad P\text{-a.e. and in } L^2.$$

Now Kronecker's Lemma implies that $\frac{1}{n} \sum_{k=1}^{n} D_k \to 0$ P-a.s.. □

We will use the following notation: if $A = (A_{uv})_{u,v\in V}$ is a matrix and $u, v \in V$, A_{u-} denotes the uth row of A and $A_{v|}$ denotes the vth column of A. If A is a matrix, the transpose of A will be denoted by $A^{\mathbf{T}}$.

Lemma 6.3.3. *Let $X_v \subseteq \mathbb{R}^d$ be compact for all v and assume that conditions (I), (III), (IV) and (V) are satisfied. Let $u \in V$ and $q \in]q_{\min}, q_{\max}[$. Then the following statements hold.*

i) *For $Q_{u,q}$-a.a. $(\sigma, \omega) \in E_u^{\mathbb{N}} \times \Omega_u$,*

$$\frac{1}{n}\sum_{k=1}^{n} \log p_{\sigma|k}(\omega) \to \sum_v \sum_u \sum_{e \in E_{uv}} \sigma_u \int p_e^q \operatorname{Lip}(S_e)^{\beta(q)} \log p_e \, d\lambda_u \; \rho_v \, .$$

In particular the following statement holds for Λ_u-a.a. $\omega \in \Omega_u$: the set

$$\left\{ x = \pi_u(\omega)(\sigma) \in C_u(\omega) \,\middle|\, \lim_n \frac{1}{n}\sum_{k=1}^{n} \log p_{\sigma|k}(\omega) = \sum_v \sum_u \sum_{e \in E_{uv}} \sigma_u \int p_e^q \operatorname{Lip}(S_e)^{\beta(q)} \log p_e \, d\lambda_u \; \rho_v \right\}$$

has full $\mathcal{M}_{u,q}(\omega)$ measure.

ii) *For $Q_{u,q}$-a.a. $(\sigma, \omega) \in E_u^{\mathbb{N}} \times \Omega_u$,*

$$\frac{1}{n}\sum_{k=1}^{n} \log(\operatorname{Lip}(S_{\sigma|k}))(\omega) \to \sum_v \sum_u \sum_{e \in E_{uv}} \sigma_u \int p_e^q \operatorname{Lip}(S_e)^{\beta(q)} \log(\operatorname{Lip}(S_e)) \, d\lambda_u \; \rho_v \, .$$

In particular the following statement holds for Λ_u-a.a. $\omega \in \Omega_u$: the set

$$\left\{ x = \pi_u(\omega)(\sigma) \in C_u(\omega) \,\middle|\, \lim_n \frac{1}{n}\sum_{k=1}^{n} \log(\operatorname{Lip}(S_{\sigma|k}))(\omega) = \sum_v \sum_u \sum_{e \in E_{uv}} \sigma_u \int p_e^q \operatorname{Lip}(S_e)^{\beta(q)} \log(\operatorname{Lip}(S_e)) \, d\lambda_u \; \rho_v \right\}$$

has full $\mathcal{M}_{u,q}(\omega)$ measure.

Proof. i) Define for $u \in V$ and $n \in \mathbb{N}$ a random variable $Z_{u,n} : E_u^{\mathbb{N}} \times \Omega_u \to \mathbb{R}$ by

$$Z_{u,n}(\sigma, \omega) = \log p_{\sigma|n}(\omega) \, .$$

Since $|\log T| \leq |Z_{u,n}| \leq |\log \Delta|$, $(Z_{u,n})_n$ is an L^2 bounded sequence of random variables on $(E_u^{\mathbb{N}} \times \Omega_u, \mathcal{B}(E_u^{\mathbb{N}} \times \Omega_u), Q_{u,q})$, and Lemma 6.3.2 therefore implies that

$$\frac{1}{n}\sum_{k=1}^{n}\left(Z_{u,k} - \mathbf{E}_P[Z_{u,k}|Z_{u,0},\ldots,Z_{u,k-1}]\right) \to 0 \quad Q_{u,q}\text{-a.e.}. \tag{6.3.1}$$

A straightforward calculation shows that $(Z_{u,n})_{n\in\mathbb{N}}$ is a $Q_{u,q}$-independent family, whence $\mathbf{E}_P[Z_{u,k}|Z_{u,0},\ldots,Z_{u,k-1}] = \mathbf{E}_P[Z_{u,k}]$. Equation (6.3.1) therefore yields

$$\frac{1}{n}\sum_{k=1}^{n} Z_{u,k} - \frac{1}{n}\sum_{k=1}^{n}\mathbf{E}_P[Z_{u,k}] \to 0 \quad Q_{u,q}\text{-a.e.}. \tag{6.3.2}$$

Now define a vector $\zeta = \zeta(q) = (\zeta_u(q))_{u\in V}$ by

$$\zeta_u(q) = \rho_u^{-1}\sum_{e\in E_u}\int p_e^q \operatorname{Lip}(S_e)^{\beta(q)}\log p_e \, d\lambda_u \;\rho_{\tau(e)}\,.$$

Fix $k \in \mathbb{N}$. By Lemma 6.2.2,

$$\begin{aligned}
&\mathbf{E}_{Q_{u,q}}[Z_{u,k}]\\
&=\mathbf{E}_{\Lambda_u}\left[\sum_{\gamma\in E_u^{(k)}}\rho_u^{-1}\left(\prod_{i=1}^{k} p_{\gamma|i}^q \operatorname{Lip}(S_{\gamma|i})^{\beta(q)}\right)\rho_{\tau(\gamma_k)}X_{u,\gamma,q}\log p_\gamma\right]\\
&\text{[by independence]}\\
&=\sum_{\gamma\in E_u^{(k)}}\rho_u^{-1}\mathbf{E}_{\Lambda_u}\left[\prod_{i=1}^{k} p_{\gamma|i}^q \operatorname{Lip}(S_{\gamma|i})^{\beta(q)}\right]\rho_{\tau(\gamma_{k-1})}\cdot\\
&\qquad\rho_{\tau(\gamma_{k-1})}^{-1}\int p_\gamma^q \operatorname{Lip}(S_\gamma)^{\beta(q)}\log p_\gamma \, d\Lambda_u \rho_{\tau(\gamma_k)}\mathbf{E}_{\Lambda_u}[X_{u,\gamma,q}]\\
&\text{[since } \mathbf{E}_{\Lambda_u}[X_{u,\gamma,q}] = 1 \text{ because } q \in]q_{\min}, q_{\max}[\ ]\\
&=\sum_{v\in V}\sum_{\alpha\in E_{uv}^{(k-1)}}\sum_{e\in E_v}\rho_u^{-1}\mathbf{E}_{\Lambda_u}\left[\prod_{i=1}^{k-1} p_{\alpha|i}^q \operatorname{Lip}(S_{\alpha|i})^{\beta(q)}\right]\rho_v\cdot\\
&\qquad\rho_v^{-1}\int p_e^q \operatorname{Lip}(S_e)^{\beta(q)}\log p_e \, d\lambda_v \;\rho_{\tau(e)}\\
&=\sum_{v\in V}\left(\sum_{\alpha\in E_{uv}^{(k-1)}}\rho_u^{-1}\mathbf{E}_{\Lambda_u}\left[\prod_{i=1}^{k-1} p_{\alpha|i}^q \operatorname{Lip}(S_{\alpha|i})^{\beta(q)}\right]\rho_v\right)\cdot\\
&\qquad\left(\rho_v^{-1}\sum_{e\in E_v}\int p_e^q \operatorname{Lip}(S_e)^{\beta(q)}\log p_e \, d\lambda_u \;\rho_{\tau(e)}\right)
\end{aligned}$$

$$= \sum_v \left(A(q)^{k-1}\right)_{uv} \zeta_v(q)$$
$$= \left(A(q)^{k-1}\right)_{u-} \zeta(q).$$

Hence

$$\frac{1}{n}\sum_{k=1}^{n} \mathbf{E}_{Q_{u,q}}[Z_{u,k}] = \frac{1}{n}\sum_{k=0}^{n-1} \left(A(q)^k\right)_{u-}\zeta(q)$$
$$= \left(\frac{1}{n}\sum_{k=0}^{n-1} A(q)^k\right)_{u-} \zeta(q). \tag{6.3.3}$$

Recall that $\pi(q) = \left(\pi_u(q)\right)_{v\in V} = \left(\sigma_v(q)\rho_u(q)\right)_{v\in V}$. Since

$$\pi(q)A(q) = \pi(q)$$

and the matrix $A(q)$ is irreducible, the ergodic theorem for irreducible Markov chains (cf. e.g. [Wa, Lemma 1.18 and Theorem 1.19]) implies that

$$\frac{1}{n}\sum_{k=0}^{n-1} A(q)^k \to \mathbf{1}\pi(q)^{\mathrm{T}} \tag{6.3.4}$$

where $\mathbf{1} = \begin{pmatrix} 1 \\ \vdots \\ 1 \end{pmatrix} \in \mathbb{R}^{\operatorname{card} V}$. Now (6.3.3) and (6.3.4) give

$$\frac{1}{n}\sum_{k=1}^{n} \mathbf{E}_{Q_{u,q}}[Z_{u,k}] \to \left(\mathbf{1}\pi(q)^{\mathrm{T}}\right)_{u-}\zeta(q) = \pi(q)^{\mathrm{T}}\zeta(q) \quad \text{as} \quad n \to \infty. \tag{6.3.5}$$

Finally putting Lemma 6.3.2 and (6.3.5) together yield

$$\frac{1}{n}\sum_{k=1}^{n} \log p_{\sigma|k}(\omega) = \frac{1}{n}\sum_{k=1}^{n} Z_{u,k}(\sigma,\omega)$$
$$\to \pi(q)^{\mathrm{T}}\zeta(q)$$
$$= \sum_v \sum_u \sum_{e\in E_{uv}} \sigma_u \int p_e^q \operatorname{Lip}(S_e)^{\beta(q)} \log p_e \, d\lambda_u \, \rho_v$$
$$\text{for } Q_{u,q}\text{-a.a. } (\sigma,\omega) \text{ as } n \to \infty.$$

ii) Define for $u \in V$ and $n \in \mathbb{N}$ a random variable $Z_{u,n} : E_u^{\mathbb{N}} \times \Omega_u \to \mathbb{R}$ by

$$Z_{u,n}(\sigma,\omega) = \log(\operatorname{Lip}(S_{\sigma|n}(\omega)))$$

and proceed as in i). □

Lemma 6.3.4. *Let X_v be compact for all v and assume that conditions (IV) and (V) are satisfied. Let $\alpha \geq 0$, $u \in V$, $\omega \in (\Omega_u)_0$ and $\sigma \in E_u^{\mathbb{N}}$. Then the following statements are equivalent.*

i) $\alpha_{\mu_u(\omega)}\big(\pi_u(\omega)(\sigma)\big) = \alpha\,.$

ii) $\frac{\log(p_{\sigma|1}(\omega)\cdot\ldots\cdot p_{\sigma|n}(\omega))}{\log(\mathrm{Lip}(S_{\sigma|1}(\omega))\cdot\ldots\cdot\mathrm{Lip}(S_{\sigma|n}(\omega)))} \to \alpha \quad \text{as} \quad n \to \infty\,.$

Proof. Let $D = \max_v(\operatorname{diam} X_v)$ and write $p_\alpha(\omega) = p_\alpha$ and $S_\alpha(\omega) = S_\alpha$ for $\alpha \in E_u$.
ii)⇒i) Let $0 < r < D$. Choose integers $k, l \in \mathbb{N}$ satisfying

$$\begin{aligned} D\,\mathrm{Lip}(S_{\sigma|1})\cdot\ldots\cdot\mathrm{Lip}(S_{\sigma|k}) < r \leq D\,\mathrm{Lip}(S_{\sigma|1})\cdot\ldots\cdot\mathrm{Lip}(S_{\sigma|k-1})\,, \\ c\,\mathrm{Lip}(S_{\sigma|1})\cdot\ldots\cdot\mathrm{Lip}(S_{\sigma|l}) < r \leq c\,\mathrm{Lip}(S_{\sigma|1})\cdot\ldots\cdot\mathrm{Lip}(S_{\sigma|l-1})\,. \end{aligned}$$

Condition (IV) implies that

$$\begin{gathered} C_{\sigma|k}(\omega) \subseteq B(x,r) \\ C_u(\omega) \cap B(x,r) \subseteq C_{\sigma|l}(\omega) \end{gathered}$$

where $x := \pi_u(\omega)(\sigma)$. Hence

$$\begin{aligned} p_{\sigma|1}\cdot\ldots\cdot p_{\sigma|k} &= \mu_u(\omega)\big(C_{\sigma|k}(\omega)\big) \leq \mu_u(\omega)\big(B(x,r)\big) \\ &\leq \mu_u(\omega)\big(C_{\sigma|l}(\omega)\big) = p_{\sigma|1}\cdot\ldots\cdot p_{\sigma|l} \end{aligned}$$

whence

$$\begin{aligned} \frac{\log \mu_u(\omega)\big(B(x,r)\big)}{\log r} &\leq \frac{\log(p_{\sigma|1}\cdot\ldots\cdot p_{\sigma|k})}{\log(D\,\mathrm{Lip}(S_{\sigma|1})\cdot\ldots\cdot\mathrm{Lip}(S_{\sigma|k-1}))} \\ &\leq \left(\left(1+\frac{\log\Delta}{\log(\prod_{i=1}^k \mathrm{Lip}(S_{\sigma|i}))}\right)^{-1} + \frac{\log D}{\log(\prod_{i=1}^k \mathrm{Lip}(S_{\sigma|i}))}\right)^{-1} \cdot \\ &\qquad \frac{\log(p_{\sigma|1}\cdot\ldots\cdot p_{\sigma|k})}{\log(\mathrm{Lip}(S_{\sigma|1})\cdot\ldots\cdot\mathrm{Lip}(S_{\sigma|k}))} \end{aligned} \tag{6.3.6}$$

and

$$\begin{aligned} \frac{\log \mu_u(\omega)\big(B(x,r)\big)}{\log r} &\geq \frac{\log(p_{\sigma|1}\cdot\ldots\cdot p_{\sigma|l})}{\log(c\,\mathrm{Lip}(S_{\sigma|1})\cdot\ldots\cdot\mathrm{Lip}(S_{\sigma|l}))} \\ &\geq \left(1+\frac{\log c}{\log(\prod_{i=1}^l \mathrm{Lip}(S_{\sigma|i}))}\right)^{-1} \cdot \\ &\qquad \frac{\log(p_{\sigma|1}\cdot\ldots\cdot p_{\sigma|l})}{\log(\mathrm{Lip}(S_{\sigma|1})\cdot\ldots\cdot\mathrm{Lip}(S_{\sigma|l}))} \end{aligned} \tag{6.3.7}$$

It follows from (6.3.6) and (6.3.7) that

$$\frac{\log \mu_u(\omega)(B(x,r))}{\log r} \to \alpha \quad \text{as} \quad r \searrow 0 \, .$$

i) $\Rightarrow$ ii) The proof is similar to the proof of the implication ii)$\Rightarrow$i). □

Proof of Theorem 6.3.1

Since $\hat{\mathcal{M}}_{u,q} \circ \pi_u^{-1} = \mathcal{M}_{u,q}$ Λ_u-a.s., Lemma 6.3.4 shows that it suffices to prove that

$$Q_{u,q}\left(\left\{(\sigma,\omega) \in E_u^{\mathbb{N}} \times \Omega_u \,\middle|\, \frac{\log(p_{\sigma|1}(\omega) \cdot \ldots \cdot p_{\sigma|n}(\omega))}{\log(\mathrm{Lip}(S_{\sigma|1}(\omega)) \cdot \ldots \cdot \mathrm{Lip}(S_{\sigma|n}(\omega)))} \to \alpha(q) \quad \text{as} \quad n \to \infty\right\}\right) = 1 \, . \tag{6.3.8}$$

It follows from Lemma 6.3.3 that

$$\begin{aligned} \frac{\log(p_{\sigma|1}(\omega) \cdot \ldots \cdot p_{\sigma|n}(\omega))}{\log(\mathrm{Lip}(S_{\sigma|1}(\omega)) \cdot \ldots \cdot \mathrm{Lip}(S_{\sigma|n}(\omega)))} &= \frac{\frac{1}{n}\sum_{k=1}^n \log p_{\sigma|k}(\omega)}{\frac{1}{n}\sum_{k=1}^n \log(\mathrm{Lip}(S_{\sigma|k}(\omega)))} \\ &\to \frac{\sum_v \sum_u \sum_{e \in E_{uv}} \sigma_u \int p_e^q \, \mathrm{Lip}(S_e)^{\beta(q)} \log p_e \, d\Lambda_u \; \rho_v}{\sum_v \sum_u \sum_{e \in E_{uv}} \sigma_u \int p_e^q \, \mathrm{Lip}(S_e)^{\beta(q)} \log(\mathrm{Lip}(S_e)) \, d\Lambda_u \; \rho_v} \\ &\quad [\text{by (4.4.8)}] \\ &= \alpha(q) \quad \text{as} \quad n \to \infty \end{aligned}$$

for $Q_{u,q}$-a.a. $(\sigma,\omega) \in E_u^{\mathbb{N}} \times \Omega_u$. This proves (6.3.8). □

Chapter 7

Proofs of Main Results

Throughout this chapter we assume that

1) X_v is a compact metric space for all $v \in V$;
2) conditions (I), (III), (IV) and (V) are satisfied.

Furthermore, c, Δ and T will always denote the constants in conditions (I), (III) and (IV).

7.1 Preliminary Lemmas

This section contains four small technical lemmas which we will need in the proofs of Theorem 2.6.1 through Theorem 2.6.4. We will always write $D = \max_v \operatorname{diam}(X_v)$.

Lemma 7.1.1. *Let $u \in V$, $\omega \in (\Omega_u)_0$ and $x = \pi_u(\omega)(\sigma) \in C_u(\omega)$ for $\sigma \in E_u^{\mathbb{N}}$. Let $r > 0$ and assume that $k, l \in \mathbb{N}$ satisfy*

$$D \operatorname{Lip}(S_{\sigma|1}(\omega)) \cdot \ldots \cdot \operatorname{Lip}(S_{\sigma|k}(\omega)) < r \le D \operatorname{Lip}(S_{\sigma|1}(\omega)) \cdot \ldots \cdot \operatorname{Lip}(S_{\sigma|k-1}(\omega)),$$
$$c \operatorname{Lip}(S_{\sigma|1}(\omega)) \cdot \ldots \cdot \operatorname{Lip}(S_{\sigma|l}(\omega)) < r \le c \operatorname{Lip}(S_{\sigma|1}(\omega)) \cdot \ldots \cdot \operatorname{Lip}(S_{\sigma|l-1}(\omega)).$$

Then

i) $C_{\sigma|k}(\omega) \subseteq B(x, r)$.
ii) $C_u(\omega) \cap B(x, r) \subseteq C_{\sigma|l}(\omega)$.

Proof. Follows easily from conditions (IV) and (V). □

Lemma 7.1.2. *Let $u \in V$, $\alpha \in E_u^{(*)}$ and $\omega \in (\Omega_u)_0$. Then*

i) $\pi_u(\omega)^{-1}(C_\alpha(\omega)) = [\alpha]$.
ii) $C_\alpha(\omega) = \pi_u(\omega)[\alpha]$.

Proof. Follows easily from conditions (IV) and (V). □

Lemma 7.1.3. *Let $u \in V$, $\sigma \in E_u^{(*)}$ and $\omega \in (\Omega_u)_0$. Let $0 < r < c$, $1 \leq a$ and assume that $k, l \in \mathbb{N}$ satisfy*

$$D \operatorname{Lip}(S_{\sigma|1}(\omega)) \cdot \ldots \cdot \operatorname{Lip}(S_{\sigma|k}(\omega)) < r \leq D \operatorname{Lip}(S_{\sigma|1}(\omega)) \cdot \ldots \cdot \operatorname{Lip}(S_{\sigma|k-1}(\omega)), \tag{7.1.1}$$

$$c \operatorname{Lip}(S_{\sigma|1}(\omega)) \cdot \ldots \cdot \operatorname{Lip}(S_{\sigma|l}(\omega)) < ar \leq c \operatorname{Lip}(S_{\sigma|1}(\omega)) \cdot \ldots \cdot \operatorname{Lip}(S_{\sigma|l-1}(\omega)). \tag{7.1.2}$$

Then

$$0 \leq k - l \leq \varphi(a)$$

where

$$\varphi(t) := \frac{\log(tD/c)}{-\log T} + 1 \quad \text{for } 1 \leq t.$$

Proof. Clearly $c \leq \max_v(\operatorname{diam} X_v)$. Hence (since $a \geq 1$), $k \geq l$. Since $k \geq l$, the right hand side of (7.1.1) can be rewritten as

$$r \leq D \operatorname{Lip}(S_{\sigma|1}(\omega)) \cdot \ldots \cdot \operatorname{Lip}(S_{\sigma|l}(\omega)) \operatorname{Lip}(S_{\sigma|l+1}(\omega)) \cdot \ldots \cdot \operatorname{Lip}(S_{\sigma|k}(\omega)).$$

The above inequality together with the left hand side of inequality (7.1.2) imply that

$$\begin{aligned} a^{-1} = \frac{r}{ar} \leq & \frac{D \operatorname{Lip}(S_{\sigma|1}(\omega)) \cdot \ldots \cdot \operatorname{Lip}(S_{\sigma|l}(\omega)) \operatorname{Lip}(S_{\sigma|l+1}(\omega)) \cdot \ldots \cdot \operatorname{Lip}(S_{\sigma|k})}{c \operatorname{Lip}(S_{\sigma|1}(\omega)) \cdot \ldots \cdot \operatorname{Lip}(S_{\sigma|l}(\omega))} \\ \leq & (D/c) T^{k-l-1} \end{aligned}$$

which gives the desired result by taking logarithms. □

Lemma 7.1.4. *Let $u \in V$ and $1 \leq a$. Then*

$$\limsup_{r \searrow 0} \left(\sup_{x \in C_u} \frac{\mu_u(B(x, ar))}{\mu_u(B(x, r))} \right) \leq \Delta^{-\varphi(a)} \quad \Lambda_u\text{-a.s.}$$

Proof. Put

$$\begin{aligned} \Omega_u'' = \{\omega \in \Omega_u' \mid & 1)\ \omega \text{ satisfies (IV)}; \\ & 2)\ \left((S_e(\omega))_{e \in E_u}, (p_e(\omega))_{e \in E_u}\right) \text{ satisfies (I), (III) and (V)}\}. \end{aligned}$$

Then $\Lambda_u(\Omega_u'') = 1$. Now, let $\omega \in \Omega_u''$, $x \in C_u(\omega)$ and $r > 0$. Choose $\sigma \in E_u^{\mathbb{N}}$ such that $\pi_u(\omega)(\sigma) = x$. Next choose integers $k, l \in \mathbb{N}$ satisfying

$$D \operatorname{Lip}(S_{\sigma|1}(\omega)) \cdot \ldots \cdot \operatorname{Lip}(S_{\sigma|k}(\omega)) < r \leq D \operatorname{Lip}(S_{\sigma|1}(\omega)) \cdot \ldots \cdot \operatorname{Lip}(S_{\sigma|k-1}(\omega)), \tag{7.1.3}$$

$$c \operatorname{Lip}(S_{\sigma|1}(\omega)) \cdot \ldots \cdot \operatorname{Lip}(S_{\sigma|l}\omega)) < ar \leq c \operatorname{Lip}(S_{\sigma|1}\omega)) \cdot \ldots \cdot \operatorname{Lip}(S_{\sigma|l-1}(\omega)). \tag{7.1.4}$$

Now Lemma 7.1.1 implies that

$$C_{\sigma|k}(\omega) \subseteq B(x,r), \quad C_u(\omega) \cap B(x,r) \subseteq C_{\sigma|l}(\omega).$$

Hence (by Lemma 7.1.2 and Lemma 7.1.3),

$$\begin{aligned}\frac{\mu_u(\omega)(B(x,ar))}{\mu_u(\omega)(B(x,r))} &\leq \frac{\mu_u(\omega)(C_{\sigma|l}(\omega))}{\mu_u(\omega)(C_{\sigma|k}(\omega))} = \frac{\hat{\mu}_u(\omega)([\sigma|l])}{\hat{\mu}_u(\omega)([\sigma|k])} \\ &\leq \frac{p_{\sigma|1}\cdot\ldots\cdot p_{\sigma|l}}{p_{\sigma|1}\cdot\ldots\cdot p_{\sigma|k}} = \frac{1}{p_{\sigma|l+1}\cdot\ldots\cdot p_{\sigma|k}} \leq \frac{1}{\Delta^{k-l}} \leq \Delta^{-\varphi(a)}\end{aligned}$$

which yields the desired conclusion. □

7.2 Proof of Theorem 2.6.1

The main purpose of this section is to prove Theorem 2.6.1. The proof of Theorem 2.6.1 is based on the next theorem.

Theorem 7.2.1.

i) *Let* $u \in V$ *and consider a real number*

$$q \in \mathbb{R}.$$

Then there exists a number $\overline{K} \in]0, \infty[$ *such that*

$$\mathcal{H}^{q,\beta(q)}_{\mu_u} \llcorner C_u \leq \overline{K} \mathcal{M}_{u,q} \llcorner C_u \quad \Lambda_u\text{-a.a.}$$

ii) *Let* $u \in V$. *For* P_u-a.a. $\mu \in \mathcal{P}(X_u)$,

$$b_\mu \leq \beta$$

We first prove Theorem 2.6.1 and then Theorem 7.2.1.

Proof of Theorem 2.6.1

i) Follows from Theorem 7.2.1.ii).
ii) Follows from Theorem 7.2.1.i) since $X_{u,q} < \infty$ Λ_u-a.s. (because $X_{u,q} \in L^1(\Lambda_u)$) and $\mathcal{M}_{u,q}(C_u) = X_{u,q}$ Λ_u-a.s.
iii) Follows from Theorem 7.2.1.i) since $\Delta_{\mu_u(\omega)}(\alpha(q))$ has full $\mathcal{M}_{u,q}(\omega)$ measure for Λ_u-a.a. $\omega \in \Omega_u$ by Theorem 6.3.1.
iv) Follows from iii). □

We now turn toward the proof of Theorem 7.2.1.

Proof of Theorem 7.2.1

Let $u \in V$ and put

$$\begin{aligned} \Omega''_u = \{\omega \in \Omega'_u \mid & 1)\ \omega \text{ satisfies (IV)}; \\ & 2)\ \big((S_e(\omega))_{e\in E_u}, (p_e(\omega))_{e\in E_u}\big) \text{ satisfies (I), (III) and (V)}; \\ & 3)\ \mathcal{M}_{u,q}(\omega)(C_\alpha(\omega)) = \\ & \qquad \rho_u^{-1} \left(\prod_{i=1}^{|\alpha|} p_{\alpha|i}(\omega)^q \operatorname{Lip}(S_{\alpha|i}(\omega))^{\beta(q)} \right) \rho_{T(\alpha_{|\alpha|})} X_{u,\alpha,q}(\omega) \\ & \quad \text{for all } \alpha \in E_u^{(*)}; \\ & 4)\ \mathcal{M}_{u,q}(\omega)(C_u(\omega)) = X_{u,q}(\omega) < \infty \,\} \end{aligned}$$

It follows from Theorem 6.1.1 and Corollary 6.1.3 that $\Lambda_u(\Omega''_u) = 1$. We will now show that there exists a number $\overline{K}$ such that

$$\mathcal{H}^{q,\beta(q)}_{\mu_u(\omega)} \lfloor C_u(\omega) \le \overline{K} \mathcal{M}_{u,q}(\omega) \lfloor C_u(\omega)$$

for $\omega \in \Omega''_u$. Fix $\omega \in \Omega''_u$. Write $p_\alpha(\omega) = p_\alpha$ and $S_\alpha(\omega) = S_\alpha$ for $\alpha \in E^{(*)}_u$. We divide the proof into three steps.

Step 1. There exists $\overline{K} \in]0, \infty[$ such that

$$\mathcal{H}^{q,\beta(q)}_{\mu_u(\omega)}(C_\alpha(\omega) \cap C_u(\omega)) \le \overline{K} \mathcal{M}_{u,q}(\omega)(C_\alpha(\omega) \cap C_u(\omega))$$

for all $\alpha \in E^{(*)}_u$.

Proof of Step 1. Let $\alpha \in E^{(*)}_u$ and $E \subseteq C_\alpha(\omega)$. Let $\delta > 0$. It follows from Lemma 4.1.1 that there exists an integer $N_\delta \in \mathbb{N}$ such that

$$\sup_{v \in V} \sup_{\gamma \in E^{(n)}_v} \prod_{k=1}^{n} \operatorname{Lip}(S_{\gamma|k}) \le \delta/(D+1) \tag{7.2.1}$$

for $n \ge N_\delta$. Now fix $\Gamma \in \mathbf{Max}(\tau(\alpha_{|\alpha|}))$ with $E^{(N_\delta+1)}_{\tau(\alpha_{|\alpha|})} \prec \Gamma$. Let $\Gamma_E = \{\gamma \in \Gamma \mid \pi_u(\omega)([\alpha\gamma]) \cap E \neq \varnothing\}$ and $\gamma \in \Gamma_E$. Choose $\sigma_\gamma \in [\alpha\gamma]$ such that $x_\gamma := \pi_u(\omega)(\sigma_\gamma) \in E$. Put

$$r_\gamma = \operatorname{Lip}(S_{\alpha|1}) \cdot \ldots \cdot \operatorname{Lip}(S_{\alpha||\alpha|}) \operatorname{Lip}(S_{\alpha(\gamma|1)}) \cdot \ldots \cdot \operatorname{Lip}(S_{\alpha(\gamma||\gamma|-1)}) D\,. \tag{7.2.2}$$

Next choose integers $k_\gamma, l_\gamma \in \mathbb{N}$ such that

$$D \operatorname{Lip}(S_{\sigma_\gamma|1}) \cdot \ldots \cdot \operatorname{Lip}(S_{\sigma_\gamma|k_\gamma}) < r_\gamma \le D \operatorname{Lip}(S_{\sigma_\gamma|1}) \cdot \ldots \cdot \operatorname{Lip}(S_{\sigma|k_\gamma-1}), \tag{7.2.3}$$

$$c \operatorname{Lip}(S_{\sigma_\gamma|1}) \cdot \ldots \cdot \operatorname{Lip}(S_{\sigma_\gamma|l_\gamma}) < r_\gamma \le c \operatorname{Lip}(S_{\sigma_\gamma|1}) \cdot \ldots \cdot \operatorname{Lip}(S_{\sigma_\gamma|l_\gamma-1}). \tag{7.2.4}$$

By Lemma 7.1.1,

$$C_u(\omega) \cap B(x_\gamma, r_\gamma) \subseteq C_{\sigma_\gamma|l_\gamma}(\omega) \tag{7.2.5}$$

$$C_{\sigma_\gamma|k_\gamma}(\omega) \subseteq B(x_\gamma, r_\gamma) \tag{7.2.6}$$

Observe that (7.2.2) implies that

$$|\alpha\gamma| = k_\gamma\,. \tag{7.2.7}$$

The proof of Step 1 is based on the next two claims.

Claim 1. There exists a constant $\overline{K}_0$ only depending on q such that

$$(2r_\gamma)^{\beta(q)} \le \overline{K}_0 \operatorname{Lip}(S_{\sigma_\gamma|1})^{\beta(q)} \cdot \ldots \cdot \operatorname{Lip}(S_{\sigma_\gamma|k_\gamma})^{\beta(q)}.$$

Claim 2. There exists a constant $\overline{K}_1$ only depending on q such that

$$\mu_u(\omega)(B(x_\gamma, r_\gamma))^q \le \overline{K}_1 p^q_{\sigma_\gamma|1} \cdot \ldots \cdot p^q_{\sigma_\gamma|k_\gamma}.$$

Proof of Claim 1. For $\beta(q) < 0$, (7.2.3) implies that

$$(2r_\gamma)^{\beta(q)} \le (2D)^{\beta(q)} \operatorname{Lip}(S_{\sigma_\gamma|1})^{\beta(q)} \cdot \ldots \cdot \operatorname{Lip}(S_{\sigma_\gamma|k_\gamma})^{\beta(q)} \tag{7.2.8}$$

For $0 \le \beta(q)$, (7.2.3) implies that

$$\begin{aligned}(2r_\gamma)^{\beta(q)} &\le (2D)^{\beta(q)} \operatorname{Lip}(S_{\sigma_\gamma|1})^{\beta(q)} \cdot \ldots \cdot \operatorname{Lip}(S_{\sigma_\gamma|k_\gamma-1})^{\beta(q)} \\ &\le (2D/\Delta)^{\beta(q)} \operatorname{Lip}(S_{\sigma_\gamma|1})^{\beta(q)} \cdot \ldots \cdot \operatorname{Lip}(S_{\sigma_\gamma|k_\gamma})^{\beta(q)}\end{aligned} \tag{7.2.9}$$

The assertion in Claim 1 follows from (7.2.8) and (7.2.9). This proves Claim 1.

Proof of Claim 2. For $q < 0$, (7.2.6) and Lemma 7.1.2 imply that

$$\begin{aligned}\mu_u(\omega)(B(x_\gamma, r_\gamma))^q \le& \left((\hat{\mu}_u(\omega) \circ \pi_u(\omega)^{-1}) C_{\sigma_\gamma|k_\gamma}(\omega)\right)^q \\ =& (\hat{\mu}_u(\omega)[\sigma_\gamma|k_\gamma])^q \\ =& p^q_{\sigma_\gamma|1} \cdot \ldots \cdot p^q_{\sigma_\gamma|k_\gamma}\end{aligned} \tag{7.2.10}$$

For $0 \le q$, (7.2.4) through (7.2.5) and Lemma 7.1.2 through Lemma 7.1.3 imply that

$$\begin{aligned}\mu_u(\omega)(B(x_\gamma, r_\gamma))^q \le& \left((\hat{\mu}_u(\omega) \circ \pi_u(\omega)^{-1}) C_{\sigma_\gamma|l_\gamma}(\omega)\right)^q \\ =& (\hat{\mu}_u(\omega)[\sigma_\gamma|l_\gamma])^q \\ =& p^q_{\sigma_\gamma|1} \cdot \ldots \cdot p^q_{\sigma_\gamma|l_\gamma} \le \frac{1}{\Delta^{q(k_\gamma - l_\gamma)}} p^q_{\sigma_\gamma|1} \cdot \ldots \cdot p^q_{\sigma_\gamma|k_\gamma} \\ \le& \frac{1}{\Delta^{q\varphi(1)}} p^q_{\sigma_\gamma|1} \cdot \ldots \cdot p^q_{\sigma_\gamma|k_\gamma}\end{aligned} \tag{7.2.11}$$

The assertion in Claim 2 follows from (7.2.10) and (7.2.11). This proves Claim 2.

Now observe that since $\Gamma \in \mathbf{Max}(\tau(\alpha_{|\alpha|}))$, $C_\alpha(\omega) = \cup_{\gamma \in \Gamma} C_{\alpha\gamma}(\omega)$, whence (using (7.2.7)) $E \subseteq \bigcup_{\gamma \in \Gamma_E} C_{\alpha\gamma}(\omega) = \bigcup_{\gamma \in \Gamma_E} C_{\sigma_\gamma|k_\gamma}(\omega) \subseteq \bigcup_{\gamma \in \Gamma_E} B(x_\gamma, r_\gamma)$. Also (by

(7.2.1)) $r_\gamma \leq \delta$. The family $(B(x_\gamma, r_\gamma))_{\gamma\in\Gamma_E}$ is thus a centered δ-covering of E, and Claim 1 and Claim 2 therefore imply that

$$\begin{aligned}
\overline{\mathcal{H}}^{q,\beta(q)}_{\mu_u(\omega),\delta}(E) &\leq \sum_{\gamma\in\Gamma_E} \mu_u(\omega)(B(x_\gamma, r_\gamma))^q (2r_\gamma)^{\beta(q)} \\
&\leq \overline{K}_0\overline{K}_1 \sum_{\gamma\in\Gamma_E} \left(\prod_{i=1}^{k_\gamma} p^q_{\sigma_\gamma|i} \operatorname{Lip}(S_{\sigma_\gamma|i})^{\beta(q)} \right) \\
&\quad \text{[by (7.2.7)]} \\
&\leq (\overline{\rho}(q)/\underline{\rho}(q))\overline{K}_0\overline{K}_1 \sum_{\gamma\in\Gamma_E} \rho_u(q)^{-1} \left(\prod_{i=1}^{|\alpha\gamma|} p^q_{\sigma_\gamma|i} \operatorname{Lip}(S_{\sigma_\gamma|i})^{\beta(q)} \right) \rho_{\tau(\gamma_{|\gamma|})}(q) \\
&\leq \overline{K} \sum_{\gamma\in\Gamma} \rho_u(q)^{-1} \left(\prod_{i=1}^{|\alpha|} p^q_{\alpha|i} \operatorname{Lip}(S_{\alpha|i})^{\beta(q)} \right) \rho_{\tau(\alpha_{|\alpha|})}(q) \cdot \\
&\qquad \rho_{\tau(\alpha_{|\alpha|})}(q)^{-1} \left(\prod_{i=1}^{|\gamma|} p^q_{\alpha(\gamma|i)} \operatorname{Lip}(S_{\alpha(\gamma|i)})^{\beta(q)} \right) \rho_{\tau(\alpha\gamma_{|\alpha\gamma|})}(q) \\
&\leq \overline{K} \sum_{\gamma\in\Gamma} \rho_u(q)^{-1} \left(\prod_{i=1}^{|\alpha|} p^q_{\alpha|i} \operatorname{Lip}(S_{\alpha|i})^{\beta(q)} \right) \rho_{\tau(\alpha_{|\alpha|})}(q) X_{u,\alpha,q,\Gamma}(\omega)
\end{aligned} \tag{7.2.12}$$

where $\overline{K} = (\overline{\rho}(q)/\underline{\rho}(q))\overline{K}_0\overline{K}_1$. Since $\Gamma \in \mathbf{Max}(\tau(\alpha_{|\alpha|}))$ with $E^{(N_\delta+1)}_{\tau(\alpha_{|\alpha|})} \prec \Gamma$ was arbitrary, the above inequality shows that

$$\overline{\mathcal{H}}^{q,\beta(q)}_{\mu_u(\omega),\delta}(E) \leq \overline{K}\rho_u^{-1} \left(\prod_{i=1}^{|\alpha|} p^q_{\alpha|i} \operatorname{Lip}(S_{\alpha|i})^{\beta(q)} \right) \rho_{\tau(\alpha_{|\alpha|})} \inf_{\substack{\Gamma\in\mathbf{Max}(\tau(\alpha_{|\alpha|})) \\ E^{(N_\delta+1)}_{\tau(\alpha_{|\alpha|})} \prec\Gamma}} X_{u,\alpha,q,\Gamma}(\omega) \tag{7.2.13}$$

whence

$$\begin{aligned}
\overline{\mathcal{H}}^{q,\beta(q)}_{\mu_u(\omega),\delta}(E) \leq \overline{K}\rho_u^{-1} &\left(\prod_{i=1}^{|\alpha|} p^q_{\alpha|i} \operatorname{Lip}(S_{\alpha|i})^{\beta(q)} \right) \rho_{\tau(\alpha_{|\alpha|})} \cdot \\
&\sup_{\Sigma\in\mathbf{Max}(\tau(\alpha_{|\alpha|}))} \inf_{\substack{\Gamma\in\mathbf{Max}(\tau(\alpha_{|\alpha|})) \\ \Sigma\prec\Gamma}} X_{u,\alpha,q,\Gamma}(\omega)
\end{aligned}$$

for $\delta > 0$. Letting $\delta \searrow 0$ now yields

$$\overline{\mathcal{H}}^{q,\beta(q)}_{\mu_u(\omega)}(E) \le \overline{K}\rho_u^{-1}\left(\prod_{i=1}^{|\alpha|} p^q_{\alpha|i}\,\mathrm{Lip}(S_{\alpha|i})^{\beta(q)}\right)\rho_{\tau(\alpha_{|\alpha|})}\cdot \sup_{\Sigma\in\mathbf{Max}(\tau(\alpha_{|\alpha|}))}\ \inf_{\substack{\Gamma\in\mathbf{Max}(\tau(\alpha_{|\alpha|}))\\ \Sigma\prec\Gamma}} X_{u,\alpha,q,\Gamma}(\omega) \tag{7.2.14}$$

for all $E \subseteq C_\alpha(\omega)$. Next observe that

$$\sup_{\Sigma\in\mathbf{Max}(\tau(\alpha_{|\alpha|}))}\ \inf_{\substack{\Gamma\in\mathbf{Max}(\tau(\alpha_{|\alpha|}))\\ \Sigma\prec\Gamma}} X_{u,\alpha,q,\Gamma}(\omega) \le \limsup_n X_{u,\alpha,q,n}(\omega) = X_{u,\alpha,q}(\omega).$$

Hence

$$\overline{\mathcal{H}}^{q,\beta(q)}_{\mu_u(\omega)}(E) \le \overline{K}\rho_u^{-1}\left(\prod_{i=1}^{|\alpha|} p^q_{\alpha|i}\,\mathrm{Lip}(S_{\alpha|i})^{\beta(q)}\right)\rho_{\tau(\alpha_{|\alpha|})}X_{u,\alpha,q}(\omega) = \overline{K}\mathcal{M}_{u,q}(\omega)(C_\alpha(\omega))$$

for all $E \subseteq C_\alpha(\omega)$, which proves step 1.

Step 2. For all relatively open subsets G of $C_u(\omega)$ then

$$\mathcal{H}^{q,\beta(q)}_{\mu_u(\omega)}(G) \le \overline{K}\mathcal{M}_{u,q}(\omega)(G).$$

Proof of Step 2. Let G be a relatively open subset of $C_u(\omega)$ and put $A = \{\alpha \in E_u^{(*)} \mid C_\alpha(\omega) \subseteq G\}$. Since G is open, $G = \cup_{\alpha\in A}(C_\alpha(\omega)\cap C_u(\omega))$. Now we need only cover G once: if $\alpha,\beta \in A$ and $[\alpha]\cap[\beta] \neq \varnothing$, then one of them is contained in the other, so we may discard the smaller one. So there exists a set $A_0 \subseteq A$ such that

$$G = \bigcup_{\alpha\in A_0}(C_\alpha(\omega)\cap C_u(\omega))$$

and $[\alpha]\cap[\beta] = \varnothing$ for $\alpha,\beta\in A_0$. Hence, $\big(C_\alpha(\omega)\big)_{\alpha\in A_0}$ is a pairwise disjoint family. Now (by Step 1),

$$\begin{aligned}
\mathcal{H}^{q,\beta(q)}_{\mu_u(\omega)}(G) &= \mathcal{H}^{q,\beta(q)}_{\mu_u(\omega)}\left(\bigcup_{\alpha\in A_0}(C_\alpha(\omega)\cap C_u(\omega))\right)\\
&\le \sum_{\alpha\in A_0}\mathcal{H}^{q,\beta(q)}_{\mu_u(\omega)}(C_\alpha(\omega)\cap C_u(\omega)) \le \overline{K}\sum_{\alpha\in A_0}\mathcal{M}_{u,q}(\omega)(C_\alpha(\omega))\\
&\le \overline{K}\mathcal{M}_{u,q}(\omega)\left(\bigcup_{\alpha\in A_0}C_\alpha(\omega)\right) \le \overline{K}\mathcal{M}_{u,q}(\omega)(G).
\end{aligned}$$

This proves Step 2.

Step 3. We have

$$\mathcal{H}^{q,\beta(q)}_{\mu_u(\omega)} \lfloor C_u(\omega) \le \overline{K} \mathcal{M}_{u,q}(\omega) \lfloor C_u(\omega) .$$

Proof of Step 3. It follows from Step 1 that

$$\mathcal{H}^{q,\beta(q)}_{\mu_u(\omega)}(C_u(\omega)) \le \overline{K}\mathcal{M}_{u,q}(\omega)(C_u(\omega)) = X_{u,q}(\omega) < \infty .$$

The measures $\mathcal{H}^{q,\beta(q)}_{\mu_u(\omega)}$ and $\mathcal{M}_{u,q}(\omega)$ are thus finite Borel measures when restricted to $C_u(\omega)$, and Step 3 therefore follows from Step 2 by outer regularity. This proves Step 3.
ii) For each $q \in \mathbb{R}$ there exists by i) a subset $\Omega_{u,q}$ of Ω_u satisfying $\Lambda_u(\Omega_{u,q}) = 1$ and

$$\begin{aligned} \mathcal{H}^{q,\beta(q)}_{\mu_u(\omega)} &\le \overline{K}\mathcal{M}_{u,q}(\omega) \quad \text{on} \quad C_u(\omega) \quad \text{for} \quad \omega \in \Omega_{u,q} \\ \mathcal{M}_{u,q}(\omega)(C_u(\omega)) &= X_{u,q}(\omega) < \infty \quad \text{for} \quad \omega \in \Omega_{u,q} \end{aligned}$$

Hence $\mathcal{H}^{q,\beta(q)}_{\mu_u(\omega)}(\operatorname{supp}\mu_u(\omega)) = \mathcal{H}^{q,\beta(q)}_{\mu_u(\omega)}(C_u(\omega)) < \infty$ for $\omega \in \Omega_{u,q}$, i.e.

$$b_{\mu_u(\omega)}(q) \le \beta(q) \quad \text{for} \quad \omega \in \Omega_{u,q} . \tag{7.2.15}$$

Now put $\Omega_u^\circ = \cap_{q\in\mathbb{Q}}\Omega_{u,q}$. Then clearly

$$\Lambda_u(\Omega_u^\circ) = 1 .$$

We claim that

$$b_{\mu_u(\omega)}(p) \le \beta(p) \quad \text{for all} \quad p \in \mathbb{R} \quad \text{and} \quad \omega \in \Omega_u^\circ . \tag{7.2.16}$$

Assume, in order to get a contradiction, that (7.2.16) is not satisfied. We can thus choose $p_0 \in \mathbb{R}$ and $\omega_0 \in \Omega_u^\circ$ with

$$\beta(p_0) < b_{\mu_u(\omega_0)}(p_0) \tag{7.2.17}$$

Next choose a sequence $(q_n)_{n\in\mathbb{N}}$ in $\mathbb{Q}$ such that $q_n \nearrow p_0$. Since β is continuous at p_0 there exists by (7.2.17) an integer N satisfying $\beta(q_N) < b_{\mu_u(\omega_0)}(p_0)$. However, since $q_N \in \mathbb{Q}$ and $\omega_0 \in \Omega_u^\circ \subseteq \Omega_{u,q_N}$, $b_{\mu_u(\omega_0)}(q_N) \le \beta(q_N)$. Hence

$$b_{\mu_u(\omega_0)}(q_N) \le \beta(q_N) < b_{\mu_u(\omega_0)}(p_0) . \tag{7.2.18}$$

Now, since $b_{\mu_u(\omega_0)}$ is decreasing (cf. Proposition 2.1.8), (7.2.18) implies that $p_0 < q_N$, contradicting the fact that $q_N \le p_0$. This proves (7.2.16). It follows from (7.2.16) that

$$b_{\mu_u(\omega)} \le \beta \quad \text{for} \quad \Lambda_u\text{-a.a.} \quad \omega \in \Omega_u .$$

Finally since $P_u = \Lambda_u \circ \mu_u^{-1}$, the above inequality implies that

$$b_\mu \le \beta \quad \text{for} \quad P_u\text{-a.a.} \quad \mu \in \mathcal{P}(X_u) . \ \square$$

7.3 Proof of Theorem 2.6.2

The main purpose of this section is to prove Theorem 2.6.2. The proof of Theorem 2.6.2 is based on the next two theorems.

Theorem 7.3.1.

i) *Let $u \in V$ and consider real numbers*

$$q \in]q_{\min}, q_{\max}[\quad \text{and} \quad \delta > 0 .$$

Then

$$\overline{\mathcal{P}}^{q,\beta(q)+\delta}_{\mu_u} \lfloor C_u = 0 \quad \Lambda_u\text{-a.s.}$$

ii) *Let $u \in V$. For P_u-a.a. $\mu \in \mathcal{P}(X_u)$,*

$$B_\mu \leq \Lambda_\mu \leq \beta \quad \text{on} \quad]q_{\min}, q_{\max}[.$$

Theorem 7.3.2.

i) *Let $u \in V$ and $q \in \mathbb{R}$. Assume that*

$$0 < \min_v \|X_{v,q}\|_{-\infty} .$$

Then there exists a number $\overline{K} \in]0, \infty[$ such that

$$\mathcal{P}^{q,\beta(q)}_{\mu_u} \lfloor C_u \leq \overline{K} \mathcal{M}_{u,q} \lfloor C_u \quad \Lambda_u\text{-a.a.}$$

ii) *Let $u \in V$ and $q \in \mathbb{R}$. Assume that*

$$0 < \min_v \|X_{v,q}\|_{-\infty} .$$

Then there exists a number $\overline{K} \in]0, \infty[$ such that

$$\overline{\mathcal{P}}^{q,\beta(q)}_{\mu_u}(C_u) \leq \overline{K} \mathcal{M}_{u,q}(C_u) \quad \Lambda_u\text{-a.a.}$$

iii) *Assume that*

$$\forall q \in \mathbb{R} : 0 < \min_v \|X_{v,q}\|_{-\infty} .$$

Then for all $u \in V$ and P_u-a.a. $\mu \in \mathcal{P}(X_u)$,

$$B_\mu \leq \Lambda_\mu \leq \beta$$

We first prove Theorem 2.6.2 and then Theorem 7.3.1 and Theorem 7.3.2.

Proof of Theorem 2.6.2

i) Follows from Theorem 7.3.1.ii).
ii) Follows from Theorem 7.3.2.i) since $X_{u,q} < \infty$ Λ_u-a.s. (because $X_{u,q} \in L^1(\Lambda_u)$) and $\mathcal{M}_{u,q}(C_u) = X_{u,q}$ Λ_u-a.s.
iii) Follows from Theorem 7.3.2.i) since $\Delta_{\mu_u(\omega)}(\alpha(q))$ has full $\mathcal{M}_{u,q}(\omega)$ measure for Λ_u-a.a. $\omega \in \Omega_u$ by Theorem 6.3.1.
iv) Follows from iii). □

We now turn toward the proofs of Theorem 7.3.1 and Theorem 7.3.2. We begin with a small lemma.

Lemma 7.3.3. *Let $u \in V$. Let $\varepsilon, \delta > 0$ and $q \in]q_{\min}, q_{\max}[$. Then the following holds for Λ_u-a.a. $\omega \in \Omega_u$:*

There exists an integer $n \in \mathbb{N}$ such that

$$\rho_u^{-1}\left(\prod_{i=1}^{|\gamma|} p_{\gamma|i}(\omega)^q \operatorname{Lip}(S_{\gamma|i}(\omega))^{\beta(q)+\delta}\right)\rho_{\tau(\gamma_{|\gamma|})} \leq$$
$$\varepsilon\rho_u^{-1}\left(\prod_{i=1}^{|\gamma|} p_{\gamma|i}(\omega)^q \operatorname{Lip}(S_{\gamma|i}(\omega))^{\beta(q)}\right)\rho_{\tau(\gamma_{|\gamma|})}X_{u,\gamma,q}(\omega)$$

for all $\gamma \in E_u^{(*)}$ with $|\gamma| \geq n$.

Proof. Choose $p > 0$ such that $a := T^{\delta p}\operatorname{card} E < 1$. Now Chebyshev's inequality and independence give

$$\Lambda_u\left(\left\{\omega \in \Omega_u \,\middle|\, \left(\prod_{i=1}^{|\gamma|} \operatorname{Lip}(S_{\gamma|i}(\omega))^\delta\right) X_{u,\gamma,q}(\omega)^{-1} > \varepsilon\right\}\right)$$
$$\leq \int \left(\prod_{i=1}^{|\gamma|} \operatorname{Lip}(S_{\gamma|i})^{\delta p}\right) X_{u,\gamma,q}^{-p}\, d\Lambda_u \,\frac{1}{\varepsilon^p}$$
$$\leq \int \left(\prod_{i=1}^{|\gamma|} \operatorname{Lip}(S_{\gamma|i})^{\delta p}\right) d\Lambda_u \max_v \left(\int X_{v,q}^{-p}\, d\Lambda_v\right) \frac{1}{\varepsilon^p}$$

for all $\gamma \in E_u^{(*)}$. Hence

$$\sum_n \Lambda_u\left(\left\{\omega \in \Omega_u \,\middle|\, \exists\gamma \in E_u^{(n)} : \left(\prod_{i=1}^{|\gamma|} \operatorname{Lip}(S_{\gamma|i}(\omega))^\delta\right) X_{u,\gamma,q}(\omega)^{-1} > \varepsilon\right\}\right)$$

$$
\begin{aligned}
&= \sum_n \Lambda_u \left(\bigcup_{\gamma \in E_u^{(n)}} \left\{ \omega \in \Omega_u \,\middle|\, \left(\prod_{i=1}^{|\gamma|} \mathrm{Lip}(S_{\gamma|i}(\omega))^\delta \right) X_{u,\gamma,q}(\omega)^{-1} > \varepsilon \right\} \right) \\
&\leq \sum_n \sum_{\gamma \in E_u^{(n)}} \Lambda_u \left(\left\{ \omega \in \Omega_u \,\middle|\, \left(\prod_{i=1}^{|\gamma|} \mathrm{Lip}(S_{\gamma|i}(\omega))^\delta \right) X_{u,\gamma,q}(\omega)^{-1} > \varepsilon \right\} \right) \\
&\leq \sum_n \sum_{\gamma \in E_u^{(n)}} \int \left(\prod_{i=1}^{|\gamma|} \mathrm{Lip}(S_{\gamma|i})^{\delta p} \right) d\Lambda_u \max_v \int X_{v,q}^{-p}\, d\Lambda_v \frac{1}{\varepsilon^p} \\
&\leq \max_v \int X_{v,q}^{-p}\, d\Lambda_v \frac{1}{\varepsilon^p} \sum_n \sum_{\gamma \in E_u^{(n)}} T^{\delta p} \\
&\leq \max_v \int X_{v,q}^{-p}\, d\Lambda_v \frac{1}{\varepsilon^p} \sum_n \left(T^{\delta p} \operatorname{card} E \right)^n < \infty
\end{aligned}
$$

since $\int X_{v,q}^{-p}\, d\Lambda_v < \infty$ for all v (by Proposition 5.3.2) and $T^{\delta p} \operatorname{card} E < 1$. The Borel-Cantelli Lemma therefore implies that

$$
\Lambda_u \left(\bigcup_n \bigcap_{m \geq n} \left\{ \omega \in \Omega_u \,\middle|\, \exists \gamma \in E_u^{(n)} : \left(\prod_{i=1}^{|\gamma|} \mathrm{Lip}(S_{\gamma|i}(\omega))^\delta \right) X_{u,\gamma,q}(\omega)^{-1} > \varepsilon \right\} \right) = 0
$$

Taking complements now yields the desired result. □

We are now ready to prove Theorem 7.3.1 and Theorem 7.3.2.

Proof of Theorem 7.3.1

i) Let $u \in V$ an put

$$\Omega''_u = \{\omega \in \Omega'_u \mid 1)\ \omega \text{ satisfies (IV)};$$
$$2)\ \big((S_e(\omega))_{e\in E_u}, (p_e(\omega))_{e\in E_u}\big) \text{ satisfies (I), (III) and (V)};$$
$$3)\ \mathcal{M}_{u,q}(\omega)(C_\alpha(\omega)) =$$
$$\rho_u^{-1}\left(\prod_{i=1}^{|\alpha|} p_{\alpha|i}(\omega)^q \operatorname{Lip}(S_{\alpha|i}(\omega))^{\beta(q)}\right)\rho_{\tau(\alpha_{|\alpha|})}X_{u,\alpha,q}(\omega)$$
$$\text{for all } \alpha \in E_u^{(*)};$$
$$4)\ \mathcal{M}_{u,q}(\omega)(C_u(\omega)) = X_{u,q}(\omega) < \infty;$$
$$5)\ \text{For each } N \in \mathbb{N} \text{ there exists an integer } n \in \mathbb{N} \text{ such that}$$
$$\rho_u^{-1}\left(\prod_{i=1}^{|\gamma|} p_{\gamma|i}(\omega)^q \operatorname{Lip}(S_{\gamma|i}(\omega))^{\beta(q)+\delta}\right)\rho_{\tau(\gamma_{|\gamma|})} \le$$
$$\frac{1}{N}\rho_u^{-1}\left(\prod_{i=1}^{|\gamma|} p_{\gamma|i}(\omega)^q \operatorname{Lip}(S_{\gamma|i}(\omega))^{\beta(q)}\right)\rho_{\tau(\gamma_{|\gamma|})}X_{u,\gamma,q}(\omega)$$
$$\text{for all } \gamma \in E_u^{(*)} \text{ with } |\gamma| \ge n. \quad \}$$

It follows from Theorem 6.3.1, Corollary 6.1.3 and Lemma 7.3.3 that $\Lambda_u(\Omega''_u) = 1$. We will now prove that

$$\overline{\mathcal{P}}^{q,\beta(q)+\delta}_{\mu_u(\omega)}(C_u(\omega)) = 0$$

for $\omega \in \Omega''_u$. Fix $\omega \in \Omega''_u$ and $N \in \mathbb{N}$. Write $p_\alpha(\omega) = p_\alpha$ and $S_\alpha(\omega) = S_\alpha$. It follows from the definition of Ω''_u that there exists an integer $n \in \mathbb{N}$ satisfying

$$\rho_u^{-1}\left(\prod_{i=1}^{|\gamma|} p^q_{\gamma|i} \operatorname{Lip}(S_{\gamma|i})^{\beta(q)+\delta}\right)\rho_{\tau(\gamma_{|\gamma|})} \le$$
$$\frac{1}{N}\rho_u^{-1}\left(\prod_{i=1}^{|\gamma|} p^q_{\gamma|i} \operatorname{Lip}(S_{\gamma|i})^{\beta(q)}\right)\rho_{\tau(\gamma_{|\gamma|})}X_{u,\gamma,q}(\omega) \tag{7.3.1}$$

for all $\gamma \in E_u^{(*)}$ with $|\gamma| \ge n$. Put

$$\eta_0 = \min\{D \operatorname{Lip}(S_{\gamma|1})\cdot \ldots \operatorname{Lip}(S_{\gamma|n}) \mid \gamma \in E_u^{(*)}\}.$$

Let $0 < \eta < \eta_0$ and $(B(x_i, r_i))_{i \in \mathbb{N}}$ be a centered η-packing of $C_u(\omega)$. For each $i \in \mathbb{N}$ choose $\sigma_i \in E_u^{(*)}$ such that $x_i = \pi_u(\omega)(\sigma_i)$. Next choose integers $k_i, l_i \in \mathbb{N}$ such that

$$D \operatorname{Lip}(S_{\sigma_i|1}) \cdot \ldots \cdot \operatorname{Lip}(S_{\sigma_i|k_i}) < r_i \leq D \operatorname{Lip}(S_{\sigma_i|1}) \cdot \ldots \cdot \operatorname{Lip}(S_{\sigma_i|k_i-1}), \tag{7.3.2}$$

$$c \operatorname{Lip}(S_{\sigma_i|1}) \cdot \ldots \cdot \operatorname{Lip}(S_{\sigma_i|l_i}) < r_i \leq c \operatorname{Lip}(S_{\sigma_i|1}) \cdot \ldots \cdot \operatorname{Lip}(S_{\sigma_i|l_i-1}). \tag{7.3.3}$$

By Lemma 7.1.1,

$$C_u(\omega) \cap B(x_i, r_i) \subseteq C_{\sigma_i|l_i}(\omega) \tag{7.3.4}$$

$$C_{\sigma_i|k_i}(\omega) \subseteq B(x, r) \tag{7.3.5}$$

Also observe that since $r_i < \eta < \eta_0$, .

$$k_i \geq n. \tag{7.3.6}$$

The proof is based on the next two claims.

Claim 1. There exists a constant $\overline{K}_0$ only depending on q and δ such that

$$(2r_i)^{\beta(q)+\delta} \leq \overline{K}_0 \operatorname{Lip}(S_{\sigma_i|1})^{\beta(q)+\delta} \cdot \ldots \cdot \operatorname{Lip}(S_{\sigma_i|k_i})^{\beta(q)+\delta}.$$

Claim 2. There exists a constant $\overline{K}_1$ only depending on q such that

$$\mu_u(\omega)(B(x_i, r_i))^q \leq \overline{K}_1 p_{\sigma_i|1}^q \cdot \ldots \cdot p_{\sigma_i|k_i}^q.$$

Proof of Claim 1. Similar to the proof of Claim 1 in the proof of Theorem 7.2.1. This proves Claim 1.

Proof of Claim 2. Similar to the proof of Claim 2 in the proof of Theorem 7.2.1. This proves Claim 2.

Hence by (7.3.1), (7.3.6), Claim 1 and Claim 2

$$\sum_i \mu_u(\omega)(B(x_i, r_i))^q (2r_i)^{\beta(q)+\delta}$$

$$\leq \overline{K}_0 \overline{K}_1 \sum_i \left(\prod_{j=1}^{k_i} p_{\sigma_i|j}^q \operatorname{Lip}(S_{\sigma_i|j})^{\beta(q)+\delta} \right)$$

$$\le \overline{K}_0 \overline{K}_1 (\overline{\rho}(q)/\underline{\rho}(q)) \sum_i \left(\rho_u^{-1} \prod_{j=1}^{k_i} p_{\sigma_i|j}^q \operatorname{Lip}(S_{\sigma_i|j})^{\beta(q)+\delta} \right) \rho_{\tau((\sigma_i)_{k_i})}$$

[by (7.3.1) and (7.3.6)]

$$\le \overline{K} \frac{1}{N} \sum_i \left(\rho_u^{-1} \prod_{j=1}^{k_i} p_{\sigma_i|j}^q \operatorname{Lip}(S_{\sigma_i|j})^{\beta(q)+\delta} \right) \rho_{\tau((\sigma_i)_{k_i})} X_{u,\sigma_i|k_i,q}(\omega)$$

$$\le \overline{K} \frac{1}{N} \sum_i \mathcal{M}_{u,q}(\omega)(C_{\sigma_i|k_i}(\omega))$$

[by (7.3.5)]

$$\le \overline{K} \frac{1}{N} \sum_i \mathcal{M}_{u,q}(\omega)(B(x_i, r_i))$$

$$\le \overline{K} \frac{1}{N} \mathcal{M}_{u,q}(\bigcup_i B(x_i, r_i))$$

$$\le \overline{K} \frac{1}{N} \mathcal{M}_{u,q}(C_u(\omega)) = \overline{K} \frac{1}{N} X_{u,q}(\omega)$$

whence

$$\overline{\mathcal{P}}^{q,\beta(q)+\delta}_{\mu_u(\omega),\eta}(C_u(\omega)) \le \overline{K} \frac{1}{N} X_{u,q}(\omega)$$

for $0 < \eta < \eta_0$. Letting $\eta \searrow 0$ now yields

$$\overline{\mathcal{P}}^{q,\beta(q)+\delta}_{\mu_u(\omega)}(C_u(\omega)) \le \overline{K} \frac{1}{N} X_{u,q}(\omega) < \infty$$

for $N \in \mathbb{N}$. Letting $N \to \infty$ thus yields $\overline{\mathcal{P}}^{q,\beta(q)+\delta}_{\mu_u(\omega)}(C_u(\omega)) = 0$.
ii) By i) there exists for each $q \in]q_{\min}, q_{\max}[$ and each $n \in \mathbb{N}$ a subset $\Omega_{u,q,n}$ of Ω_u with $\Lambda_u(\Omega_{u,q,n}) = 1$ and

$$\overline{\mathcal{P}}^{q,\beta(q)+\frac{1}{n}}_{\mu_u(\omega)}(C_u(\omega)) = 0 \quad \text{for} \quad \omega \in \Omega_{u,q,n}\,.$$

For fixed $q \in]q_{\min}, q_{\max}[$ write $\Omega_{u,q} = \cap_n \Omega_{u,q,n}$. We have $\Lambda_u(\Omega_{u,q}) = 1$, and $\overline{\mathcal{P}}^{q,\beta(q)+\frac{1}{n}}_{\mu_u(\omega)}(C_u(\omega)) = 0$ for all $\omega \in \Omega_{u,q}$ and $n \in \mathbb{N}$. Hence

$$\Lambda_{\mu_u(\omega)}(q) = \Delta^q_{\mu_u(\omega)}(\operatorname{supp} \mu_u(\omega)) = \Delta^q_{\mu_u(\omega)}(C_u(\omega)) \le \beta(q) + \frac{1}{n}$$

for $\omega \in \Omega_{u,q}$ and $n \in \mathbb{N}$, i.e.

$$\Lambda_{\mu_u(\omega)}(q) \le \beta(q) \quad \text{for} \quad \omega \in \Omega_{u,q}\,. \tag{7.3.7}$$

Now put $\Omega_u^\circ := \bigcap_{q\in]q_{\min},q_{\max}[\cap\mathbb{Q}} \Omega_{u,q}$. Then clearly

$$\Lambda_u(\Omega_u^\circ) = 1\,.$$

We claim that

$$\Lambda_{\mu_u(\omega)}(p) \le \beta(p) \quad \text{for all} \quad p \in]q_{\min}, q_{\max}[\quad \text{and} \quad \omega \in \Omega_u^\circ\,.$$

Fix $p \in]q_{\min}, q_{\max}[$ and $\omega \in \Omega_u^\circ$. Choose a sequence $(q_n)_{n\in\mathbb{N}}$ in $]q_{\min}, q_{\max}[\cap\mathbb{Q}$ such that $q_n \to p$. Since $q_n \in \mathbb{Q}\cap]q_{\min}, q_{\max}[$ and $\omega \in \Omega_u^\circ \subseteq \Omega_{u,q_n}$, (7.3.7) implies that $\Lambda_{\mu_u(\omega)}(q_n) \le \beta(q_n)$ for all $n \in \mathbb{N}$. The continuity of $\Lambda_{\mu_u(\omega)}$ and β show that $\Lambda_{\mu_u(\omega)}(p) = \lim_n \Lambda_{\mu_u(\omega)}(q_n) \le \lim_n \beta(q_n) = \beta(p)$. □

Proof of Theorem 7.3.2

i)-ii) Let $u \in V$ and put

$$\begin{aligned} \Omega_u'' = \{\omega \in \Omega_u' \mid & 1)\ \omega \text{ satisfies (IV)};\\ & 2)\ \big((S_e(\omega))_{e\in E_u}, (p_e(\omega))_{e\in E_u}\big) \text{ satisfies (I), (III), and (V)};\\ & 3)\ \mathcal{M}_{u,q}(\omega)(C_\alpha(\omega)) = \\ & \qquad \rho_u^{-1}\left(\prod_{i=1}^{|\alpha|} p_{\alpha|i}(\omega)^q \operatorname{Lip}(S_{\alpha|i}(\omega))^{\beta(q)}\right)\rho_{\tau(\alpha_{|\alpha|})} X_{u,\alpha,q}(\omega)\\ & \quad \text{for all } \alpha \in E_u^{(*)};\\ & 4)\ \mathcal{M}_{u,q}(\omega)(C_u(\omega)) = X_{u,q}(\omega) < \infty\ \} \end{aligned}$$

Clearly $\Lambda_u(\Omega_u'') = 1$. We will show that there exists a number $\overline{K} \in]0,\infty[$ such that

$$\overline{\mathcal{P}}^{q,\beta(q)}_{\mu_u(\omega)}(C_u(\omega)) \le \overline{K}\mathcal{M}_{u,q}(\omega)(C_u(\omega))$$

and

$$\mathcal{P}^{q,\beta(q)}_{\mu_u(\omega)} \le \overline{K}\mathcal{M}_{u,q}(\omega) \quad \text{on} \quad C_u(\omega)$$

for $\omega \in \Omega_u''$. Fix $\omega \in \Omega_u''$. Write $p_\alpha(\omega) = p_\alpha$ and $S_\alpha(\omega) = S_\alpha$ for $\alpha \in E_u^\alpha$. We divide the proof into three steps.

Step 1. There exists a number $\overline{K} \in]0,\infty[$ such that

$$\mathcal{P}^{q,\beta(q)}_{\mu_u(\omega)}(C_\alpha(\omega)\cap C_u(\omega)) \le \overline{\mathcal{P}}^{q,\beta(q)}_{\mu_u(\omega)}(C_\alpha(\omega)\cap C_u(\omega)) \le \overline{K}\mathcal{M}_{u,q}(\omega)(C_\alpha(\omega)\cap C_u(\omega))$$

for all $\alpha \in E_u^{(*)}$.

Proof of Step 1. Let $\alpha \in E_u^{(*)}$ and $\varepsilon > 0$. The measure $\mathcal{M}_{u,q}(\omega)$ is finite (since $\mathcal{M}_{u,q}(\omega)(X_u) = X_{u,q}(\omega) < \infty$) and thus outer regular. We can therefore choose an open and bounded set G_ε such that $C_\alpha(\omega) \cap C_u(\omega) \subseteq G_\varepsilon$ and $\mathcal{M}_{u,q}(\omega)(G_\varepsilon \setminus (C_\alpha(\omega) \cap C_u(\omega))) \leq \varepsilon$. Clearly $\delta_\varepsilon = \operatorname{dist}(C_\alpha(\omega) \cap C_u(\omega), C_u(\omega) \setminus G_\varepsilon) > 0$. Let $0 < \delta < c \wedge \operatorname{diam}(C_\alpha(\omega)) \wedge \delta_\varepsilon$ and $(B(x_i, r_i))_{i \in \mathbb{N}}$ be a centered δ-packing of $C_\alpha(\omega) \cap C_u(\omega)$. For each $i \in \mathbb{N}$ choose $\sigma_i \in E_u^{(*)}$ such that $x_i = \pi_u(\omega)(\sigma_i)$. Next choose integers $k_i, l_i \in \mathbb{N}$ such that

$$D\,\mathrm{Lip}(S_{\sigma_i|1}) \cdot \ldots \cdot \mathrm{Lip}(S_{\sigma_i|k_i}) < r_i \leq D\,\mathrm{Lip}(S_{\sigma_i|1}) \cdot \ldots \cdot \mathrm{Lip}(S_{\sigma_i|k_i - 1}), \tag{7.3.8}$$

$$c\,\mathrm{Lip}(S_{\sigma_i|1}) \cdot \ldots \cdot \mathrm{Lip}(S_{\sigma_i|l_i}) < r_i \leq c\,\mathrm{Lip}(S_{\sigma_i|1}) \cdot \ldots \cdot \mathrm{Lip}(S_{\sigma_i|l_i - 1}). \tag{7.3.9}$$

By Lemma 7.1.1

$$C_u(\omega) \cap B(x_i, r_i) \subseteq C_{\sigma_i|l_i}(\omega) \tag{7.3.10}$$

$$C_{\sigma_i|k_i}(\omega) \subseteq B(x_i, r_i) \tag{7.3.11}$$

The proof of Step 1 is based on the next two claims.

Claim 1. There exists a constant $\overline{K}_0$ only depending on q such that

$$(2r_i)^{\beta(q)} \leq \overline{K}_0\,\mathrm{Lip}(S_{\sigma_i|1})^{\beta(q)} \cdot \ldots \cdot \mathrm{Lip}(S_{\sigma_i|k_i})^{\beta(q)}.$$

Claim 2. There exists a constant $\overline{K}_1$ only depending on q such that

$$\mu_u(\omega)(B(x_i, r_i))^q \leq \overline{K}_1 p_{\sigma_i|1}^q \cdot \ldots \cdot p_{\sigma_i|k_i}^q.$$

Proof of Claim 1. Similar to the proof of Claim 1 in the proof of Theorem 7.2.1. This proves Claim 1.

Proof of Claim 2. Similar to the proof of Claim 2 in the proof of Theorem 7.2.1. This proves Claim 2.

It follows from Claim 1 and Claim 2 that

$$\sum_i \mu_u(\omega)(B(x_i, r_i))^q (2r_i)^{\beta(q)}$$

$$\leq \overline{K}_0 \overline{K}_1 \sum_i \left(\prod_{j=1}^{k_i} p_{\sigma_i|j}^q\,\mathrm{Lip}(s_{\sigma_i|j})^{\beta(q)} \right)$$

$$
\begin{aligned}
&\leq (\overline{\rho}(q)/\underline{\rho}(q))\overline{K}_0\overline{K}_1(\min_v \|X_{v,q}\|_{-\infty})^{-1} \\
&\qquad \sum_i \rho_u^{-1}\left(\prod_{j=1}^{k_i} p_{\sigma_i|j}^{q}\,\mathrm{Lip}(s_{\sigma_i|j})^{\beta(q)}\right)\rho_{\tau(\sigma_i)_{k_i}}X_{u,\sigma_i|k_i,q} \\
&= \overline{K}\sum_i \mathcal{M}_{u,q}(\omega)(C_{\sigma|k_i}(\omega)) \\
&\text{[by (7.3.11)]} \\
&\leq \overline{K}\sum_i \mathcal{M}_{u,q}(\omega)(B(x_i,r_i)) \\
&\leq \overline{K}\mathcal{M}_{u,q}(\omega)\left(\bigcup_i B(x_i,r_i)\right) \\
&\leq \overline{K}\mathcal{M}_{u,q}(\omega)(G_\varepsilon) \\
&\leq \overline{K}(\mathcal{M}_{u,q}(\omega)(C_u(\omega))+\varepsilon)
\end{aligned}
$$

where $\overline{K}=(\overline{\rho}(q)/\underline{\rho}(q))\overline{K}_0\overline{K}_1(\min_v \|X_{v,q}\|_{-\infty})^{-1}$. Hence

$$\overline{\mathcal{P}}^{q,\beta(q)}_{\mu_u(\omega),\delta}(C_\alpha(\omega)\cap C_u(\omega)) \leq \overline{K}(\mathcal{M}_{u,q}(\omega)(C_\alpha(\omega))+\varepsilon)$$

for $\varepsilon > 0$ and $0 < \delta < c \wedge \mathrm{diam}(C_\alpha(\omega)) \wedge \delta_\varepsilon$. This implies that

$$\overline{\mathcal{P}}^{q,\beta(q)}_{\mu_u(\omega)}(C_\alpha(\omega)\cap C_u(\omega)) \leq \overline{K}\mathcal{M}_{u,q}(\omega)(C_\alpha(\omega))$$

which completes the proof of Step 1.

Step 2. For all relatively open subsets G of $C_u(\omega)$ then

$$\mathcal{P}^{q,\beta(q)}_{\mu_u(\omega)}(G) \leq \overline{K}\mathcal{M}_{u,q}(\omega)(G).$$

Proof of Step 2. Similar to the proof of Step 2 in Theorem 7.2.1.

Step 3. We have

$$\mathcal{P}^{q,\beta(q)}_{\mu_u(\omega)}\lfloor C_u(\omega) \leq \overline{K}\mathcal{M}_{u,q}(\omega)\lfloor C_u(\omega).$$

Proof of Step 3. Similar to the proof of Step 3 in Theorem 7.2.1.
iii) Similar to the proof of ii) in Theorem 7.2.1. □

7.4 Proof of Theorem 2.6.3

The main purpose of this section is to proove Theorem 2.6.3. The proof of Theorem 2.6.3 is based on Theorem 7.4.1 and Theorem 7.4.2. However, we begin with a definition.

If μ, ν are measures on a measurable space (X, Σ) and $\mathcal{A} \subseteq \Sigma$ is a subfamily of Σ we write $\nu \lll \mu$ on $\mathcal{A}$ if and only if

$$\forall A \in \mathcal{A} : \mu(A) < \infty \Rightarrow \nu(A) = 0 .$$

If $\mathcal{A} = \Sigma$ we write $\nu \lll \mu$ instead of $\nu \lll \mu$ on Σ.

We are now ready to state Theorem 7.4.1 and Theorem 7.4.2.

Theorem 7.4.1. *Let $X_v \subseteq \mathbb{R}^d$ for all $v \in V$ with $\lambda^d(X_v) > 0$. Let $u \in V$ and consider real numbers*

$$q \in]q_{\min}, q_{\max}[\quad \textit{and} \quad \delta > 0 .$$

Then

$$\mathcal{M}_{u,q} \lfloor C_u \lll \mathcal{H}^{q,\beta(q)-\delta}_{\mu_u} \lfloor C_u \quad \Lambda_u\text{-a.s.}$$

Theorem 7.4.2. *Let $u \in V$ and $q \in \mathbb{R}$. Assume that*

$$\max_v \|X_{v,q}\|_\infty < \infty .$$

Then there exists $\underline{K} \in]0, \infty[$ such that

$$\underline{K} \mathcal{M}_{u,q} \lfloor C_u \leq \mathcal{H}^{q,\beta(q)}_{\mu_u} \lfloor C_u \quad \Lambda_u\text{-a.a.}$$

We first prove Theorem 2.6.3 and then Theorem 7.4.1 and Theorem 7.4.2.

Proof of Theorem 2.6.3

i) By the definition of P_u it suffices to prove that

$$\beta \leq b_{\mu_u(\omega)} \quad \text{on} \quad]q_{\min}, q_{\max}[\quad \text{for} \quad \Lambda_u\text{-a.a.} \quad \omega \in \Omega_u .$$

For each $q \in]q_{\min}, q_{\max}[$ and $n \in \mathbb{N}$, Proposition 5.2.3 and Theorem 7.4.1 imply that there exists a subset $\Omega_{u,q,n}$ of Ω_u with $\Lambda_u(\Omega_{u,q,n}) = 1$ such that each $\omega \in \Omega_{u,q,n}$ satisfies

$$\mathcal{M}_{u,q}(\omega) \lfloor C_u(\omega) \lll \mathcal{H}^{q,\beta(q)-\frac{1}{n}}_{\mu_u(\omega)} \lfloor C_u(\omega)$$
$$\mathcal{M}_{u,q}(\omega)(C_u(\omega)) = X_{u,q}(\omega) > 0 .$$

Hence $\mathcal{H}^{q,\beta(q)-\frac{1}{n}}_{\mu_u(\omega)}(\operatorname{supp}\mu_u(\omega)) = \mathcal{H}^{q,\beta(q)-\frac{1}{n}}_{\mu_u(\omega)}(C_u(\omega)) = \infty$ for $\omega \in \Omega_{u,q,n}$, whence

$$\beta(q) - \frac{1}{n} \leq b_{\mu_u(\omega)}(q) \quad \text{for} \quad \omega \in \Omega_{u,q,n}\,.$$

For fixed $q \in]q_{\min}, q_{\max}[$ put $\Omega_{u,q} = \cap_n \Omega_{u,q,n}$. Then $\Lambda_u(\Omega_{u,q}) = 1$, and $\beta(q) - \frac{1}{n} \leq b_{\mu_u(\omega)}(q)$ for all $\omega \in \Omega_{u,q}$ and all $n \in \mathbb{N}$. Hence

$$\beta(q) \leq b_{\mu_u(\omega)} \quad \text{for} \quad \omega \in \Omega_{u,q}\,.$$

Now put $\Omega_u^\circ := \cap_{q \in]q_{\min}, q_{\max}[\cap\mathbb{Q}} \Omega_{u,q}$. Then clearly

$$\Lambda_u(\Omega_u^\circ) = 1\,.$$

We claim that

$$\beta(p) \leq b_{\mu_u(\omega)}(p) \quad \text{for all } p \in]q_{\min}, q_{\max}[\text{ and all } \omega \in \Omega_u^\circ.$$

Assume, in order to get a contradiction, that there exists a number $p_0 \in]q_{\min}, q_{\max}[$ and $\omega_0 \in \Omega_u^\circ$ satisfying

$$b_{\mu_u(\omega_0)}(p_0) < \beta(p_0)\,. \tag{7.4.1}$$

Next choose a sequence $(q_n)_n$ in $]q_{\min}, q_{\max}[\cap\mathbb{Q}$ such that $q_n \searrow p_0$. Since β is continuous at p_0, (7.4.1) implies that there exists an integer $N \in \mathbb{N}$ satisfying $b_{\mu_u(\omega_0)}(p_0) < \beta(q_N)$. However, since $q_n \in]q_{\min}, q_{\max}[\cap\mathbb{Q}$ and $\omega_0 \in \Omega_u^\circ \subseteq \Omega_{u,q_N}$, $\beta(q_N) \leq b_{\mu_u(\omega_0)}(q_N)$. Hence

$$b_{\mu_u(\omega_0)}(p_0) < \beta(q_N) \leq b_{\mu_u(\omega_0)}(q_N)\,. \tag{7.4.2}$$

Now, since $b_{\mu_u(\omega_0)}$ is decreasing, (7.4.2) implies that $q_N < p_0$, contradicting the fact that $p_0 \leq q_N$.

ii) Let $q \in]q_{\min}, q_{\max}[$. By Theorem 7.4.2, Proposition 5.2.3 and Theorem 6.3.1 there exists a subset Ω_u'' of Ω_u such that $\Lambda_u(\Omega_u'') = 1$ and each $\omega \in \Omega_u''$ satisfies

$$\underline{K}\mathcal{M}_{u,q}(\omega)\lfloor C_u(\omega) \leq \mathcal{H}^{q,\beta(q)}_{\mu_u(\omega)}\lfloor C_u(\omega)$$
$$\mathcal{M}_{u,q}(\omega)(\Delta_{\mu_u(\omega)}(\alpha(q)) = X_{u,q}(\omega) > 0\,.$$

Hence

$$\mathcal{H}^{q,\beta(q)}_{\mu_u(\omega)}(\Delta_{\mu_u(\omega)}(\alpha(q)) > 0$$

for all $\omega \in \Omega'_u$,. Since $P_u = \Lambda_u \circ \mu_u^{-1}$, the above inequality yields the desired assertion. □

We now turn toward the proofs of Theorem 7.4.1 and Theorem 7.4.2. We begin with two lemmas. Recall that $\mathcal{C}(X) := \{K \subseteq X \mid K \text{ is compact}\}$.

Lemma 7.4.3. *If $\mu, \nu \in \mathcal{M}(\mathbb{R}^d)$ are Borel measures on $\mathbb{R}^d$ with $\nu(\mathbb{R}^d) < \infty$ and $\nu \lll \mu$ on $\mathcal{C}(\mathbb{R}^d)$ then*

$$\nu \lll \mu .$$

Proof. Let $E \in \mathcal{B}(\mathbb{R}^d)$ with $\mu(E) < \infty$. The restriction $\lambda := \mu\lfloor E$ of μ to E is a finite Borel measure on $\mathbb{R}^d$ and thus inner regular. Hence $\infty > \mu(E) = \lambda(E) = \sup_{K\subseteq E, K \text{ compact}} \lambda(K) = \sup_{K\subseteq E, K \text{ compact}} \mu(K)$, whence $\mu(K) < \infty$ for all compact subsets K of E, and consequently $\nu(K) = 0$ for all compact subsets K of E. Finally, since ν finite and thus inner regular, $\nu(E) = \sup_{K\subseteq E, K \text{ compact}} \nu(K) = 0$. □

Lemma 7.4.4. *Let $u \in V$. Let $\varepsilon, \delta > 0$ and $q \in]q_{\min}, q_{\max}[$. Then the following holds for Λ_u-a.a. $\omega \in \Omega_u$:*

The set

$$\begin{aligned}\{x = \pi_u(\omega)(\sigma) \mid \exists n \in \mathbb{N} : \forall m \geq n :& \\ \rho_u^{-1}\left(\prod_{i=1}^{m} p_{\sigma|i}(\omega)^q \operatorname{Lip}(S_{\sigma|i}(\omega))^{\beta(q)}\right) & \rho_{\tau(\sigma_m)} X_{u,\sigma|m,q}(\omega) \leq \\ \varepsilon\rho_u^{-1}\left(\prod_{i=1}^{m} p_{\sigma|i}(\omega)^q \operatorname{Lip}(S_{\sigma|i}(\omega))^{\beta(q)-\delta}\right) & \rho_{\tau(\sigma_m)} \quad \}\end{aligned}$$

has full $\mathcal{M}_{u,q}(\omega)$ measure.

Proof. Since $q \in]q_{\min}, q_{\max}[$ there exists by Theorem 5.2.1 an $\eta \in]0, \varepsilon_q[$ such that

$$\mathbf{E}_{\Lambda_v}\left[X_{v,q}^{1+\eta}\right] < \infty \quad \text{for all } v \in V .$$

Define $Z : E_u^{\mathbb{N}} \times \Omega_u \to [0, \infty]$ by

$$Z(\sigma,\omega) = \sum_{n=1}^{\infty}\left(\left(\prod_{i=1}^{n} \operatorname{Lip}(S_{\sigma|i}(\omega))^{\delta}\right) X_{u,\sigma|n,q}(\omega)\right)^{\eta} .$$

Observe that

$$\begin{aligned}\mathbf{E}_{Q_{u,q}}[Z] &= \sum_{n=1}^{\infty}\int\left(\prod_{i=1}^{n}\operatorname{Lip}(S_{\sigma|i}(\omega))^{\delta}\right)^{\eta} X_{u,\sigma|n,q}(\omega)^{\eta}\, dQ_{u,q}(\sigma,\omega) \\ &\leq \sum_{n=1}^{\infty}\int\left(\prod_{i=1}^{n} T^{\delta}\right)^{\eta} X_{u,\sigma|n,q}(\omega)^{\eta}\, dQ_{u,q}(\sigma,\omega)\end{aligned}$$

$$
\begin{aligned}
&= \sum_{n=1}^{\infty} (T^{\delta\eta})^n \int X_{u,\sigma|n,q}(\omega)^\eta \, dQ_{u,q}(\sigma,\omega) \\
&= \sum_{n=1}^{\infty} (T^{\delta\eta})^n \cdot \\
&\qquad \mathbf{E}_{\Lambda_u}\left[\sum_{\gamma \in E_u^{(n)}} \rho_u^{-1} \left(\prod_{i=1}^{n} p_{\gamma|i} \operatorname{Lip}(S_{\gamma|i})^{\beta(q)}\right) \rho_{\tau(\gamma_{|\gamma|})} X_{u,\gamma,q} X_{u,\gamma,q}^\eta\right] \\
&= \sum_{n=1}^{\infty} (T^{\delta\eta})^n \cdot \\
&\qquad \sum_{\gamma \in E_u^{(n)}} \rho_u^{-1} \mathbf{E}_{\Lambda_u}\left[\prod_{i=1}^{n} p_{\gamma|i} \operatorname{Lip}(S_{\gamma|i})^{\beta(q)}\right] \rho_{\tau(\gamma_{|\gamma|})} \mathbf{E}_{\Lambda_u}\left[X_{u,\gamma,q}^{1+\eta}\right] \\
&\le \max_v \left(\mathbf{E}_{\Lambda_v}\left[X_{v,q}^{1+\eta}\right]\right) \cdot \\
&\qquad \sum_{n=1}^{\infty} (T^{\delta\eta})^n \sum_{\gamma \in E_u^{(n)}} \rho_u^{-1} \mathbf{E}_{\Lambda_u}\left[\prod_{i=1}^{n} p_{\gamma|i} \operatorname{Lip}(S_{\gamma|i})^{\beta(q)}\right] \rho_{\tau(\gamma_{|\gamma|})} \\
&= \max_v \left(\mathbf{E}_{\Lambda_v}\left[X_{v,q}^{1+\eta}\right]\right) \sum_{n=1}^{\infty} (T^{\delta\eta})^n < \infty
\end{aligned}
$$

since $T^{\delta\eta} < 1$ because $\delta, \eta > 0$. Hence $Z(\sigma,\omega) < \infty$ for $Q_{u,q}$-a.a. $(\sigma,\omega) \in E_u^{\mathbb{N}} \times \Omega_u$, whence

$$
\left(\prod_{i=1}^{n} \operatorname{Lip}(S_{\sigma|i}(\omega))^\delta\right) X_{u,\sigma|n,q}(\omega) \to 0 \quad \text{as} \quad n \to \infty \tag{7.4.3}
$$

for $Q_{u,q}$-a.a. $(\sigma,\omega) \in E_u^{\mathbb{N}} \times \Omega_u$. Put

$$
A = \{(\sigma,\omega) \in E_u^{\mathbb{N}} \times \Omega_u \mid \exists n \in \mathbb{N} : \forall m \ge n : \left(\prod_{i=1}^{m} \operatorname{Lip}(S_{\sigma|i}(\omega))^\delta\right) X_{u,\sigma|m,q}(\omega) \le \varepsilon \}
$$

It follows from (7.4.3) that

$$
Q_{u,q}(A) = 1\,.
$$

For $\omega \in \Omega_u$ write $A_\omega = \{\sigma \in E_u^{\mathbb{N}} \mid (\sigma,\omega) \in A\}$ and put $\Omega_u'' = \{\omega \in \Omega_u \mid \hat{\mathcal{M}}_{u,q}(\omega)(A_\omega) = \hat{\mathcal{M}}_{u,q}(\omega)(E_u^{\mathbb{N}})\}$. The fact that $Q_{u,q}(A) = 1$ implies that

$$
\Lambda_u(\Omega_u'') = 1\,.
$$

Now for fixed $\omega \in \Omega''_u$,

$$\begin{aligned} \mathcal{M}_{u,q}(\omega)(\pi_u(\omega)^{-1}(A_\omega)) &= \hat{\mathcal{M}}_{u,q}(\omega)(A_\omega) \\ &= \hat{\mathcal{M}}_{u,q}(\omega)(E_u^{\mathbb{N}}) \\ &= \mathcal{M}_{u,q}(\omega)(\pi_u(\omega)^{-1}(X_u)) \\ &= \mathcal{M}_{u,q}(\omega)(X_u). \end{aligned} \tag{7.4.4}$$

Also observe that all $\sigma \in A_\omega$ satisfy the following: there exists an integer $n \in \mathbb{N}$ such that

$$\left(\prod_{i=1}^{m} \mathrm{Lip}(S_{\sigma|i}(\omega))^{\delta}\right) X_{u,\sigma|m,q}(\omega) \leq \varepsilon$$

for $m \geq n$, i.e.

$$\begin{aligned} \rho_u^{-1} \left(\prod_{i=1}^{m} p_{\sigma|i}(\omega)^q \, \mathrm{Lip}(S_{\sigma|i}(\omega))^{\beta(q)}\right) \rho_{\tau(\sigma_m)} X_{u,\sigma|m,q}(\omega) \leq \\ \varepsilon \rho_u^{-1} \left(\prod_{i=1}^{m} p_{\sigma|i}(\omega)^q \, \mathrm{Lip}(S_{\sigma|i}(\omega))^{\beta(q)-\delta}\right) \rho_{\tau(\sigma_m)} \end{aligned} \tag{7.4.5}$$

for $m \geq n$. The assertion in Lemma 7.4.4 follows from (7.4.4) and (7.4.5). □.

We are now ready to prove Theorem 7.4.1 and Theorem 7.4.2.

Proof of Theorem 7.4.1

Let

$\Omega''_u = \{\omega \in \Omega'_u \mid 1)$ ω satisfies (IV);

2) $\big((S_e(\omega))_{e\in E_u}, (p_e(\omega))_{e\in E_u}\big)$ satisfies (I), (III), and (V);

3) $\mathcal{M}_{u,q}(\omega)\big(C_\alpha(\omega)\big) =$

$$\rho_u^{-1} \left(\prod_{i=1}^{|\alpha|} p_{\alpha|i}(\omega)^q \, \mathrm{Lip}(S_{\alpha|i}(\omega))^{\beta(q)}\right) \rho_{\tau(\alpha_{|\alpha|})} X_{u,\alpha,q}(\omega)$$

for all $\alpha \in E_u^{(*)}$;

4) $\mathcal{M}_{u,q}(\omega)\big(C_u(\omega)\big) = X_{u,q}(\omega) < \infty$;

5) The set

$$\{x = \pi_u(\omega)(\sigma) \mid \exists n \in \mathbb{N} : \forall m \geq n :$$

$$\rho_u^{-1} \left(\prod_{i=1}^{m} p_{\sigma|i}(\omega)^q \operatorname{Lip}(S_{\sigma|i}(\omega))^{\beta(q)} \right) \rho_{\tau(\sigma_m)} X_{u,\sigma|m,q}(\omega) \leq$$

$$\frac{1}{N} \rho_u^{-1} \left(\prod_{i=1}^{m} p_{\sigma|i}(\omega)^q \operatorname{Lip}(S_{\sigma|i}(\omega))^{\beta(q)-\delta} \right) \rho_{\tau(\sigma_m)} \}$$

has full $\mathcal{M}_{u,q}(\omega)$ measure for all $N \in \mathbb{N}$;

6) $$\limsup_{r \searrow 0} \left(\sup_{x \in C_u(\omega)} \frac{\mu_u(\omega)(B(x,2r))}{\mu_u(\omega)(B(x,r))} \right) \leq \Delta^{-\varphi(2)};$$

7) $$X_{u,\alpha,q}(\omega) =$$

$$\sum_{\gamma \in \Gamma} \rho_{\tau(\alpha_{|\alpha|})}^{-1} \left(\prod_{i=1}^{|\gamma|} p_{\alpha(\gamma|i)}(\omega)^q \operatorname{Lip}(S_{\alpha(\gamma|i)}(\omega))^{\beta(q)} \right) \cdot \rho_{\tau(\alpha\gamma_{|\alpha\gamma|})} X_{u,\alpha\gamma,q}(\omega)$$

for $\alpha \in E_u^{(*)}$ and $\Gamma \in \mathbf{Max}(\tau(\alpha_{|\alpha|}))$;

8) If $K \subseteq X_u$ is compact then

$$\sum_{\substack{\alpha \in E_u^{(n)} \\ K \cap C_u(\omega) \neq \varnothing}} \rho_u^{-1} \left(\prod_{i=1}^{|\alpha|} p_{\alpha|i}(\omega)^q \operatorname{Lip}(S_{\alpha|i}(\omega))^{\beta(q)} \right) \rho_{\tau(\alpha_{|\alpha|})} X_{u,\alpha,q}(\omega) \searrow$$

$$\mathcal{M}_{u,q}(\omega)(K) \quad \}$$

It follows from Theorem 6.1.1, Lemma 7.4.4 and Lemma 7.1.4 that $\Lambda_u(\Omega_u'') = 1$. We will now show that

$$\mathcal{M}_{u,q}(\omega) \lfloor C_u(\omega) \lll \mathcal{H}_{\mu_u(\omega)}^{q,\beta(q)-\delta} \lfloor C_u(\omega)$$

for $\omega \in \Omega_u''$. Fix $\omega \in \Omega_u''$. Write $p_\alpha(\omega) = p_\alpha$ and $S_\alpha(\omega) = S_\alpha$ for $\alpha \in E_u^{(*)}$. We divide the proof into two steps.

Step 1. If $K \subseteq C_u(\omega)$ is a compact subset of $C_u(\omega)$ then the following implication holds,

$$\mathcal{H}_{\mu_u(\omega)}^{q,\beta(q)-\delta}(K) < \infty \Rightarrow \mathcal{M}_{u,q}(\omega)(K) = 0\,.$$

Proof of Step 1. Let $K \subseteq C_u(\omega)$ be a compact subset of $C_u(\omega)$ and assume that $\mathcal{H}_{\mu_u(\omega)}^{q,\beta(q)-\delta}(K) < \infty$. By the definition of Ω_u'' there exists a number $r_0 > 0$ satisfying

$$\sup_{x \in C_u(\omega)} \frac{\mu_u(\omega)(B(x,2r))}{\mu_u(\omega)(B(x,r))} \leq \Delta^{-\varphi(2)} := A$$

for $0 < r < r_0$. For $N, n \in \mathbb{N}$ write

$$A_{N,n} := \{x = \pi_u(\omega)(\sigma) \,|\forall m \geq n :$$
$$\rho_u^{-1}\left(\prod_{i=1}^{m} p_{\sigma|i}^q \operatorname{Lip}(S_{\sigma|i})^{\beta(q)}\right)\rho_{\tau(\sigma_m)} X_{u,\sigma|m,q}(\omega) \leq$$
$$\frac{1}{N}\rho_u^{-1}\left(\prod_{i=1}^{m} p_{\sigma|i}^q \operatorname{Lip}(S_{\sigma|i})^{\beta(q)-\delta}\right)\rho_{\tau(\sigma_m)}\,\}. \tag{7.4.6}$$

For $x \in C_u(\omega)$ and $r > 0$ write

$$\Gamma_{x,r} := \{\gamma \in E_u^{(*)} \,|x \in C_\gamma(\omega)\,,$$
$$c\operatorname{Lip}(S_{\gamma|1})\cdot\ldots\cdot\operatorname{Lip}(S_{\gamma||\gamma|}) < r \leq c\operatorname{Lip}(S_{\gamma|1})\cdot\ldots\cdot\operatorname{Lip}(S_{\gamma||\gamma|-1})\}$$

Fix $N, n \in \mathbb{N}$ and let $\varnothing \neq L \subseteq A_{N,n}$ be a compact subset of $A_{N,n}$. Let $\eta = 1 \wedge c \wedge r_0 \wedge \min\{c\operatorname{Lip}(S_{\alpha|1})\cdot\ldots\cdot\operatorname{Lip}(S_{\alpha||\alpha|}) \mid \alpha \in E_u^{(n)}\}$ and $(B(x_i, r_i))_{i\in\mathbb{N}}$ be a centered η-covering of K. For $i \in \mathbb{N}$ write

$$\Gamma_i := \bigcup_{x \in B(x_i,r_i)\cap K\cap L} \Gamma_{x,r_i}\,.$$

Now fix $i \in \mathbb{N}$. Let $\gamma \in \Gamma_i$. There exists $x_\gamma \in B(x_i, r_i) \cap K \cap L$ such that $\gamma \in \Gamma_{x_\gamma, r_i}$. In particular $x_\gamma \in C_\gamma(\omega)$ and we can thus choose $\sigma_\gamma \in [\gamma]$ satisfying $x_\gamma = \pi_u(\omega)(\sigma_\gamma)$. Next, choose integers $k_\gamma, l_\gamma \in \mathbb{N}$ such that

$$D\operatorname{Lip}(S_{\sigma_\gamma|1})\cdot\ldots\cdot\operatorname{Lip}(S_{\sigma_\gamma|k_\gamma}) < r_i \leq D\operatorname{Lip}(S_{\sigma_\gamma|1})\cdot\ldots\cdot\operatorname{Lip}(S_{\sigma_\gamma|k_\gamma-1})\,, \tag{7.4.7}$$

$$c\operatorname{Lip}(S_{\sigma_\gamma|1})\cdot\ldots\cdot\operatorname{Lip}(S_{\sigma_\gamma|l_\gamma}) < r_i \leq c\operatorname{Lip}(S_{\sigma_\gamma|1})\cdot\ldots\cdot\operatorname{Lip}(S_{\sigma_\gamma|l_\gamma-1})\,. \tag{7.4.8}$$

By Lemma 7.1.1,

$$C_u(\omega) \cap B(x_\gamma, r_i) \subseteq C_{\sigma_\gamma|l_\gamma}(\omega) \tag{7.4.9}$$
$$C_{\sigma_\gamma|k_\gamma}(\omega) \subseteq B(x_\gamma, r_i) \tag{7.4.10}$$

Since $\gamma \in \Gamma_i = \cup_{x\in B(x_i,r_i)\cap K\cap L}\Gamma_{x,r_i}$, (7.4.8) implies that

$$|\gamma| = l_\gamma\,. \tag{7.4.11}$$

Hence $c\operatorname{Lip}(S_{\gamma|1})\cdot\ldots\cdot\operatorname{Lip}(S_{\gamma||\gamma|}) < r_i < \eta$, whence (by the definition of η),

$$|\gamma| \geq n\,. \tag{7.4.12}$$

The proof of Step 1 is based on the next four claims.

Claim 1. There exists a constant $\underline{K}_0 \in]0, \infty[$ only depending on q and δ such that

$$\underline{K}_0 \operatorname{Lip}(S_{\sigma_\gamma|1})^{\beta(q)-\delta} \cdot \ldots \cdot \operatorname{Lip}(S_{\sigma_\gamma|l_\gamma})^{\beta(q)-\delta} \leq (2r_i)^{\beta(q)-\delta}.$$

Claim 2. There exists a constant $\underline{K}_1 \in]0, \infty[$ only depending on q such that

$$\underline{K}_1 p^q_{\sigma_\gamma|1} \cdot \ldots \cdot p^q_{\sigma_\gamma|l_\gamma} \leq \mu_u(\omega)(B(x_\gamma, r_i))^q.$$

Claim 3. There exists a constant $\overline{c}_0$ only depending on q such that

$$\mu_u(\omega)(B(x_\gamma, r_i))^q \leq \overline{c}_0 \mu_u(\omega)(B(x_i, r_i))^q.$$

Claim 4. There exists a constant $\overline{c}_1$ only depending on q such that

$$\operatorname{card}(\Gamma_i) \leq \overline{c}_1.$$

Proof of Claim 1. For $\beta(q) - \delta < 0$, (7.4.8) implies that

$$\begin{aligned}
&\operatorname{Lip}(S_{\sigma_\gamma|1})^{\beta(q)-\delta} \cdot \ldots \cdot \operatorname{Lip}(S_{\sigma_\gamma|l_{|\gamma|}})^{\beta(q)-\delta} \\
\leq &\operatorname{Lip}(S_{\sigma_\gamma|1})^{\beta(q)-\delta} \cdot \ldots \cdot \operatorname{Lip}(S_{\sigma_\gamma|l_{|\gamma|}-1})^{\beta(q)-\delta} \Delta^{\beta(q)-\delta} \\
\leq &(\Delta/2c)^{\beta(q)-\delta}(2r_i)^{\beta(q)-\delta}
\end{aligned} \tag{7.4.13}$$

For $0 \leq \beta(q) - \delta$, (7.4.8) implies that

$$\operatorname{Lip}(S_{\sigma_\gamma|1})^{\beta(q)-\delta} \cdot \ldots \cdot \operatorname{Lip}(S_{\sigma_\gamma|l_\gamma})^{\beta(q)-\delta} \leq (2c)^{-(\beta(q)-\delta)}(2r_i)^{\beta(q)-\delta}. \tag{7.4.14}$$

The assertion in Claim 1 follows from (7.4.13) and (7.4.14). This proves Claim 1.

Proof of Claim 2. For $q < 0$, (7.4.9) and Lemma 7.1.2 imply that

$$\begin{aligned}
p^q_{\sigma_\gamma|1} \cdot \ldots \cdot p^q_{\sigma_\gamma|l_\gamma} &= (\hat{\mu}_u(\omega)([\sigma_\gamma|l_\gamma]))^q \\
&= \left((\hat{\mu}_u(\omega) \circ \pi_u(\omega)^{-1})(C_{\sigma_\gamma|l_\gamma}(\omega))\right)^q \\
&\leq \mu_u(\omega)\left(B(x_\gamma, r_i)\right)^q.
\end{aligned} \tag{7.4.15}$$

For $0 \leq q$, (7.4.10), Lemma 7.1.2 and Lemma 7.1.3 imply that

$$\begin{aligned}
p^q_{\sigma_\gamma|1} \cdot \ldots \cdot p^q_{\sigma_\gamma|l_\gamma} &\leq \frac{1}{\Delta^{(k_\gamma - l_\gamma)q}} p^q_{\sigma_\gamma|1} \cdot \ldots \cdot p^q_{\sigma_\gamma|k_\gamma} \\
&\leq \Delta^{-q\varphi(1)} p^q_{\sigma_\gamma|1} \cdot \ldots \cdot p^q_{\sigma_\gamma|k_\gamma} \\
&= \Delta^{-q\varphi(1)} \left(\hat{\mu}_u(\omega)([\sigma_\gamma|k_\gamma])\right)^q
\end{aligned}$$

$$= \Delta^{-q\varphi(1)} \left((\hat{\mu}_u(\omega) \circ \pi_u(\omega)^{-1})(C_{\sigma_\gamma | k_\gamma}(\omega)) \right)^q$$
$$\leq \Delta^{-q\varphi(1)} \mu_u(\omega)(B(x_\gamma, r_i))^q \tag{7.4.16}$$

The assertion in Claim 2 follows from (7.4.15) and (7.4.16). This proves Claim 2.

Proof of Claim 3. Observe

$$B(x_i, r_i) \subseteq B(x_\gamma, 2r_i) \tag{7.4.17}$$
$$B(x_\gamma, r_i) \subseteq B(x_i, 2r_i) \tag{7.4.18}$$

For $q \leq 0$, (7.4.17) implies that

$$\begin{aligned}
\mu_u(\omega)(B(x_\gamma, r_i))^q &= \left(\frac{\mu_u(\omega)(B(x_\gamma, r_i))}{\mu_u(\omega)(B(x_i, r_i))} \right)^q \mu_u(\omega)(B(x_i, r_i))^q \\
&\leq \left(\frac{\mu_u(\omega)(B(x_\gamma, r_i))}{\mu_u(\omega)(B(x_\gamma, 2r_i))} \right)^q \mu_u(\omega)(B(x_i, r_i))^q \\
&\leq \frac{1}{A^q} \mu_u(\omega)(B(x_i, r_i))^q .
\end{aligned} \tag{7.4.19}$$

For $0 < q$, (7.4.18) implies that

$$\begin{aligned}
\mu_u(\omega)(B(x_\gamma, r_i))^q &= \left(\frac{\mu_u(\omega)(B(x_\gamma, r_i))}{\mu_u(\omega)(B(x_i, r_i))} \right)^q \mu_u(\omega)(B(x_i, r_i))^q \\
&\leq \left(\frac{\mu_u(\omega)(B(x_i, 2r_i))}{\mu_u(\omega)(B(x_i, r_i))} \right)^q \mu_u(\omega)(B(x_i, r_i))^q \\
&\leq A^q \mu_u(\omega)(B(x_i, r_i))^q .
\end{aligned} \tag{7.4.20}$$

The assertion in Claim 3 follows from (7.4.19) and (7.4.20). This proves Claim 3.

Proof of Claim 4. Observe that if $y \in B(x_i, r_i)$ then

$$\bigcup_{\gamma \in \Gamma_i} C_\gamma(\omega) \subseteq B(y, (\tfrac{D}{c} + 2)r_i). \tag{7.4.21}$$

Indeed, let $z \in \cup_{\gamma \in \Gamma_i} C_\gamma(\omega)$. There exists $\gamma_0 \in \Gamma_i$ satisfying $z \in C_{\gamma_0}(\omega)$. As $\gamma_0 \in \Gamma_i = \cup_{x \in B(x_i, r_i) \cap K \cap L} \Gamma_{x, r_i}$ there exists $x_0 \in B(x_i, r_i) \cap K \cap L$ such that $\gamma_0 \in \Gamma_{x_0, r_i}$. Hence (since $x_0 \in C_{\gamma_0}(\omega)$ because $\gamma_0 \in \Gamma_{x_0, r_i}$)

$$d(z, y) \leq d(z, x_0) + d(x_0, x_i) + d(x_i, y)$$

$$\le \operatorname{diam}(C_{\gamma_0}(\omega)) + r_i + r_i$$
$$\le D\,\mathrm{Lip}(S_{\gamma_0|1}) \cdot \ldots \cdot \mathrm{Lip}(S_{\gamma_0||\gamma_0|}) + 2r_i$$
$$\le \tfrac{D}{c} r_i + 2r_i$$

where $d(\cdot,\cdot)$ denotes the Euclidean distance in $\mathbb{R}^d$. This proves (7.4.21).

Next, if $\gamma \in \Gamma_{x,r_i}$ then

$$\begin{aligned}
\operatorname{diam} C_\gamma(\omega) &= \mathrm{Lip}(S_{\gamma|1}) \cdot \ldots \cdot \mathrm{Lip}(S_{\gamma||\gamma|}) \operatorname{diam}(X_{\tau(\gamma_{|\gamma|})}) \\
&\ge c\,\mathrm{Lip}(S_{\gamma|1}) \cdot \ldots \cdot \mathrm{Lip}(S_{\gamma||\gamma|-1}) \Delta \min_v (\operatorname{diam} X_v)/c \\
&\ge r_i \Delta \min_v (\operatorname{diam} X_v)/c = c_0 r_i
\end{aligned}$$

where $c_0 = \Delta \min_v(\operatorname{diam} X_v)/c$, whence

$$\forall \gamma \in \Gamma_i : \lambda^d(C_\gamma(\omega)) = \left(\frac{\operatorname{diam}(C_\gamma(\omega))}{\operatorname{diam}(X_{\tau(\gamma_{|\gamma|})})} \right)^d \lambda^d(X_{\tau(\gamma_{|\gamma|})}) \ge \left(\frac{c_0 r_i}{D}\right)^d \min_v \lambda^d(X_v).$$

Also observe that Γ_i is an antichain in $E_u^{\mathbb{N}}$ and $(C_\gamma(\omega))_{\gamma \in \Gamma_i}$ is thus a pairwise disjoint family. Hence

$$\begin{aligned}
\operatorname{card}(\Gamma_i) \left(\frac{c_0 r_i}{D}\right)^d \min_v \lambda^d(X_v) &\le \sum_{\gamma \in \Gamma_i} \lambda^d(C_\gamma(\omega)) = \lambda^d \left(\bigcup_{\gamma \in \Gamma_i} C_\gamma(\omega) \right) \\
&\quad [\text{by } (7.4.21)] \\
&\le \lambda^d(B(y, (\tfrac{D}{c} + 2) r_i)) = \lambda^d(B(0,1))((\tfrac{D}{c} + 2) r_i))^d
\end{aligned}$$

and so

$$\operatorname{card}(\Gamma_i) \le \frac{D \lambda^d(B(0,1))(\frac{D}{c} + 2)^d}{c_0 \min_v \lambda^d(X_v)} := \bar{c}_1 .$$

This proves claim 4.

It follows from (7.4.6), (7.4.11), (7.4.12) and Claim 1 through Claim 4 that

$$\begin{aligned}
&\sum_{\gamma \in \Gamma_i} \rho_u^{-1} \left(\prod_{i=1}^{|\gamma|} p_{\gamma|i}^q \,\mathrm{Lip}(S_{\gamma|i})^{\beta(q)} \right) \rho_{\tau(\gamma_{|\gamma|})} X_{u,\gamma,q}(\omega) \\
&[\text{by } (7.4.11)] \\
&\le \sum_{\gamma \in \Gamma_i} \rho_u^{-1} \left(\prod_{i=1}^{l_\gamma} p_{\sigma_\gamma|i}^q \,\mathrm{Lip}(S_{\sigma_\gamma|i})^{\beta(q)} \right) \rho_{\tau((\sigma_\gamma)_{l_\gamma})} X_{u,\sigma_\gamma|l_\gamma,q}(\omega)
\end{aligned}$$

$$\left[\text{by (7.4.6) and (7.4.12) since } \pi_u(\omega)(\sigma_\gamma) = x_\gamma \in L \subseteq A_{N,n}\right]$$

$$\leq \sum_{\gamma\in\Gamma_i} \frac{1}{N}\rho_u^{-1}\left(\prod_{i=1}^{l_\gamma} p^q_{\sigma_\gamma|i}\operatorname{Lip}(S_{\sigma_\gamma|i})^{\beta(q)-\delta}\right)\rho_{\tau((\sigma_\gamma)_{l_\gamma})}$$

$$\left[\text{by Claim 1 and Claim 2}\right]$$

$$\leq \sum_{\gamma\in\Gamma_i} \frac{1}{N}(\overline{\rho}(q)/\underline{\rho}(q))\underline{K_0}^{-1}\underline{K_1}^{-1}\mu_u(\omega)\big(B(x_\gamma,r_i)\big)^q(2r_i)^{\beta(q)-\delta}$$

$$\left[\text{by Claim 3}\right]$$

$$\leq \sum_{\gamma\in\Gamma_i} \frac{1}{N}(\overline{\rho}(q)/\underline{\rho}(q))\underline{K_0}^{-1}\underline{K_1}^{-1}\overline{c}_0\mu_u(\omega)\big(B(x_i,r_i)\big)^q(2r_i)^{\beta(q)-\delta}$$

$$= \operatorname{card}(\Gamma_i)\frac{1}{N}(\overline{\rho}(q)/\underline{\rho}(q))\underline{K_0}^{-1}\underline{K_1}^{-1}\overline{c}_0\mu_u(\omega)\big(B(x_i,r_i)\big)^q(2r_i)^{\beta(q)-\delta}$$

$$\left[\text{by Claim 4}\right]$$

$$\leq \frac{1}{N}(\overline{\rho}(q)/\underline{\rho}(q))\underline{K_0}^{-1}\underline{K_1}^{-1}\overline{c}_0\overline{c}_1\mu_u(\omega)\big(B(x_i,r_i)\big)^q(2r_i)^{\beta(q)-\delta}$$

$$= \frac{1}{N}C\mu_u(\omega)\big(B(x_i,r_i)\big)^q(2r_i)^{\beta(q)-\delta} \tag{7.4.22}$$

for $i \in \mathbb{N}$ (here $C := (\overline{\rho}(q)/\underline{\rho}(q))\underline{K_0}^{-1}\underline{K_1}^{-1}\overline{c}_0\overline{c}_1$).

Since $L \cap K \cap B(x_i,r_i)$ is compact,

$$\mathcal{M}_{u,q}(\omega)(L\cap K\cap B(x_i,r_i)) \leq \sum_{\substack{\gamma\in E_u^{(|\Gamma_i|)}\\ L\cap K\cap B(x_i,r_i)\cap C_\gamma(\omega)\neq\varnothing}} \rho_u^{-1}\left(\prod_{i=1}^{|\gamma|} p^q_{\gamma|i}\operatorname{Lip}(S_{\gamma|i})^{\beta(q)}\right)\rho_{\tau(\gamma_{|\gamma|})}X_{u,\gamma,q}(\omega)\,. \tag{7.4.23}$$

Now note that

$$\{\gamma \in E_u^{(|\Gamma_i|)} \mid L\cap K\cap B(x_i,r_i)\cap C_\gamma(\omega) \neq \varnothing\} \subseteq \{\gamma\alpha \mid \gamma\in\Gamma_i,\ \alpha \in E^{(|\Gamma_i|-|\gamma|)}_{\tau(\gamma_{|\gamma|})}\}\,. \tag{7.4.24}$$

Indeed, let $\gamma \in E_u^{(|\Gamma_i|)}$ with $L\cap K\cap B(x_i,r_i)\cap C_\gamma(\omega) \neq \varnothing$. Choose $x \in L\cap K\cap B(x_i,r_i)\cap C_\gamma(\omega)$. As $x \in C_\gamma$ there exists $\sigma \in [\gamma]$ such that $x = \pi_u(\omega)(\sigma)$. Observe that

$$c\operatorname{Lip}(S_{\gamma|1})\cdot\ldots\cdot\operatorname{Lip}(S_{\gamma||\gamma|}) < r_i\,. \tag{7.4.25}$$

Otherwise there exists an integer $j > |\gamma|$ satisfying

$$c\operatorname{Lip}(S_{\sigma|1})\cdot\ldots\cdot\operatorname{Lip}(S_{\sigma|j}) < r_i \leq c\operatorname{Lip}(S_{\sigma|1})\cdot\ldots\cdot\operatorname{Lip}(S_{\sigma|j-1})\,. \tag{7.4.26}$$

Since $x = \pi_u(\omega)(\sigma) \in C_{\sigma|j}(\omega)$, (7.4.26) shows that $\sigma|j \in \Gamma_{x,r_i} \subseteq \Gamma_i$. Hence $|\Gamma_i| = |\gamma| < j = |\sigma|j| \le |\Gamma_i|$ which is a contradiction. This proves (7.4.25). By (7.4.25) there exists an integer $m \le |\gamma|$ satisfying

$$c\,\mathrm{Lip}(S_{\gamma|1}) \cdot \ldots \cdot \mathrm{Lip}(S_{\gamma|m}) < r_i \le c\,\mathrm{Lip}(S_{\gamma|1}) \cdot \ldots \cdot \mathrm{Lip}(S_{\gamma|m-1}) .$$

As $x \in C_\gamma(\omega) \subseteq C_{\gamma|m}(\omega)$, the above inequality implies that $\gamma|m \in \Gamma_{x,r_i} \subseteq \Gamma_i$, whence $\gamma = \pi\alpha$ where $\pi := \gamma|m \in \Gamma_i$ and $\alpha := \gamma_m \ldots \gamma_{|\gamma|}$. This proves (7.4.24).

Also

$$X_{u,\alpha,q}(\omega) = \sum_{\gamma\in\Gamma} \rho^{-1}_{\tau(\alpha_{|\alpha|})} \left(\prod_{i=1}^{|\gamma|} p^q_{\alpha(\gamma|i)} \mathrm{Lip}(S_{\alpha(\gamma|i)})^{\beta(q)} \right) \rho_{\tau(\alpha\gamma_{|\alpha\gamma|})} X_{u,\alpha\gamma,q}(\omega) . \tag{7.4.27}$$

for all $\alpha \in E_u^{(*)}$ and $\Gamma \in \mathbf{Max}(\tau(\alpha_{|\alpha|}))$.

Now (7.4.23), (7.4.24) and (7.4.27) give

$$\begin{aligned} &\mathcal{M}_{u,q}(\omega)(L \cap K \cap B(x_i, r_i)) \\ &\quad \le \sum_{\substack{\pi=\gamma\alpha \\ \gamma\in\Gamma_i,\, \alpha\in E^{(*)}_{\tau(\gamma_{|\gamma|})}}} \rho_u^{-1} \left(\prod_{i=1}^{|\pi|} p^q_{\pi|i} \mathrm{Lip}(S_{\pi|i})^{\beta(q)} \right) \rho_{\tau(\pi_{|\pi|})} X_{u,\pi,q}(\omega) \\ &\quad \le \sum_{\gamma\in\Gamma_i} \rho_u^{-1} \left(\prod_{i=1}^{|\gamma|} p^q_{\gamma|i} \mathrm{Lip}(S_{\gamma|i})^{\beta(q)} \right) \rho_{\tau(\gamma_{|\gamma|})} X_{u,\gamma,q}(\omega) \end{aligned} \tag{7.4.28}$$

for all $i \in \mathbb{N}$.

By putting equation (7.4.22) and (7.4.28) together,

$$\begin{aligned} \mathcal{M}_{u,q}(\omega)(L \cap K) &\le \sum_i \mathcal{M}_{u,q}(\omega)(L \cap K \cap B(x_i, r_i)) \\ &\le \sum_i \sum_{\gamma\in\Gamma_i} \rho_u^{-1} \left(\prod_{i=1}^{|\gamma|} p^q_{\gamma|i} \mathrm{Lip}(S_{\gamma|i})^{\beta(q)} \right) \rho_{\tau(\gamma_{|\gamma|})} X_{u,\gamma,q}(\omega) \\ &\le \frac{1}{N} C \mu_u(\omega)(B(x_i, r_i))^q (2r_i)^{\beta(q)-\delta} . \end{aligned}$$

Hence

$$\mathcal{M}_{u,q}(\omega)(L \cap K) \le \frac{1}{N} C \overline{\mathcal{H}}^{q,\beta(q)-\delta}_{\mu_u(\omega),\eta}(K) \le \frac{1}{N} C \mathcal{H}^{q,\beta(q)-\delta}_{\mu_u(\omega),\eta}(K) \tag{7.4.29}$$

for all $n, N \in \mathbb{N}$ and compact subsets L of $A_{N,n}$. Since the restriction $\mathcal{M}_{u,q}(\omega)\lfloor K$ of $\mathcal{M}_{u,q}(\omega)$ to K is a finite Borel measure on $\mathbb{R}^d$ and thus inner regular, (7.4.29) implies that

$$\begin{aligned}\mathcal{M}_{u,q}(\omega)(K \cap A_{N,n}) &= \sup_{\substack{L \subseteq A_{N,n}\\ L \text{ compact}}} \mathcal{M}_{u,q}(\omega)(K \cap L)\\ &\leq \sup_{\substack{L \subseteq A_{N,n}\\ L \text{ compact}}} \frac{1}{N} C\mathcal{H}^{q,\beta(q)-\delta}_{\mu_u(\omega)}(K) = \frac{1}{N} C\mathcal{H}^{q,\beta(q)-\delta}_{\mu_u(\omega)}(K)\end{aligned}$$

for all $n, N \in \mathbb{N}$. Finally since $A_{N,n} \nearrow \cup_m A_{N,m}$ and $\cup_m A_{N,m}$ has full $\mathcal{M}_{u,q}(\omega)$ measure, the previous inequality implies that

$$\begin{aligned}\mathcal{M}_{u,q}(\omega)(K) = \mathcal{M}_{u,q}\left(K \bigcap \bigcup_n A_{N,n}\right) &= \sup_n \mathcal{M}_{u,q}\left(K \bigcap A_{N,n}\right)\\ &\leq \frac{1}{N} C\mathcal{H}^{q,\beta(q)-\delta}_{\mu_u(\omega)}(K)\end{aligned}$$

for all $N \in \mathbb{N}$. Letting $N \to \infty$ thus yields $\mathcal{M}_{u,q}(\omega)(K) = 0$ which proves Step 1.

Step 2. We have

$$\mathcal{M}_{u,q}(\omega)\lfloor C_u(\omega) \lll \mathcal{H}^{q,\beta(q)-\delta}_{\mu_u(\omega)}\lfloor C_u(\omega).$$

Proof of Step 2. Follows immediately from Step 1 and Lemma 7.4.3.
This completes the proof of Theorem 7.4.1. □

Proof of Theorem 7.4.2

Let $u \in V$ and put

$$\Omega''_u = \{\omega \in \Omega'_u \mid 1) \ \omega \text{ satisfies (IV)};$$

$$2) \ \big((S_e(\omega))_{e\in E_u}, (p_e(\omega))_{e\in E_u}\big) \text{ satisfies (I), (III) and (V)};$$

$$3) \ \mathcal{M}_{u,q}(\omega)(C_\alpha(\omega)) =$$

$$\rho_u^{-1}\left(\prod_{i=1}^{|\alpha|} p_{\alpha|i}(\omega)^q \operatorname{Lip}(S_{\alpha|i}(\omega))^{\beta(q)}\right)\rho_{\tau(\alpha_{|\alpha|})} X_{u,\alpha,q}(\omega)$$

$$\text{for all } \alpha \in E_u^{(*)};$$

$$4) \ \mathcal{M}_{u,q}(\omega)(C_u(\omega)) = X_{u,q}(\omega) < \infty \ \}$$

Clearly $\Lambda_u(\Omega''_u) = 1$. We will now prove that there exists a number $\underline{K} \in]0,\infty[$ such that

$$\underline{K}\mathcal{M}_{u,q}(\omega)\lfloor C_u(\omega) \le \mathcal{H}^{q,\beta(q)}_{\mu_u(\omega)}\lfloor C_u(\omega)$$

for $\omega \in \Omega''_u$. Now fix $\omega \in \Omega''_u$. Write $p_\alpha(\omega) = p_\alpha$ and $S_\alpha(\omega) = S_\alpha$ for $\alpha \in E_u^{(*)}$. We divide the proof into two steps.

Step 1. There exists $\underline{K} \in]0,\infty[$ such that

$$\underline{K}\mathcal{M}_{u,q}(\omega)(C_\alpha(\omega) \cap C_u(\omega)) \le \mathcal{H}^{q,\beta(q)}_{\mu_u(\omega)}(C_\alpha(\omega) \cap C_u(\omega))$$

for all $\alpha \in E_u^{(*)}$.

Proof of Step 1. Let $\delta \in]0,c[$ and $(B(x_i,r_i))_{i\in\mathbb{N}}$ be a centered δ-covering of $\big(C_\alpha(\omega) \cap C_u(\omega)\big)$. For each $i \in \mathbb{N}$ choose $\sigma_i \in [\alpha]$ such that $x_i = \pi_u(\omega)(\sigma_i)$. Next choose integers $k_i, l_i \in \mathbb{N}$ satisfying

$$D\operatorname{Lip}(S_{\sigma_i|1})\cdot\ldots\cdot\operatorname{Lip}(S_{\sigma_i|k_i}) < r_i \le D\operatorname{Lip}(S_{\sigma_i|1})\cdot\ldots\cdot\operatorname{Lip}(S_{\sigma_i|k_i-1}), \tag{7.4.30}$$

$$c\operatorname{Lip}(S_{\sigma_i|1})\cdot\ldots\cdot\operatorname{Lip}(S_{\sigma_i|l_i}) < r_i \le c\operatorname{Lip}(S_{\sigma_i|1})\cdot\ldots\cdot\operatorname{Lip}(S_{\sigma_i|l_i-1}). \tag{7.4.31}$$

By Lemma 7.1.1,

$$C_{\sigma_i|k_i}(\omega) \subseteq B(x_i,r_i) \tag{7.4.32}$$

$$C_u(\omega) \cap B(x_i,r_i) \subseteq C_{\sigma_i|l_i}(\omega) \tag{7.4.33}$$

The proof of Step 1 is based on the next two claims.

Claim 1. There exists a constant $\underline{K}_0 \in]0,\infty[$ only depending on q such that

$$\underline{K}_0 \operatorname{Lip}(S_{\sigma_i|1})^{\beta(q)} \cdot \ldots \cdot \operatorname{Lip}(S_{\sigma_i|l_i})^{\beta(q)} \leq (2r_i)^{\beta(q)}.$$

Claim 2. There exists a constant $\underline{K}_1 \in]0,\infty[$ only depending on q such that

$$\underline{K}_1 p^q_{\sigma_i|1} \cdot \ldots \cdot p^q_{\sigma_i|l_i} \leq \mu_u(\omega)(B(x_i,r_i))^q.$$

Proof of Claim 1. Similar to the proof of Claim 1 in the proof of Theorem 7.4.1. This proves Claim 1.

Proof of Claim 2. Similar to the proof of Claim 2 in the proof of Theorem 7.4.1. This proves Claim 2.

It follows from Claim 1 and Claim 2 that

$$\begin{aligned}
&\sum_i \mu_u(\omega)(B(x_i,r_i))^q(2r_i)^{\beta(q)} \\
&\geq \underline{K}_0\underline{K}_1 \sum_i \left(\prod_{j=1}^{l_i} p^q_{\sigma_i|j} \operatorname{Lip}(S_{\sigma_i|j})^{\beta(q)} \right) \\
&\geq \underline{K}_0\underline{K}_1(\underline{\rho}(q)/\overline{\rho}(q))(\max_v \|X_{v,q}\|_\infty)^{-1} \cdot \\
&\qquad \sum_i \rho_u^{-1} \left(\prod_{j=1}^{l_i} p^q_{\sigma_i|j} \operatorname{Lip}(S_{\sigma_i|j})^{\beta(q)} \right) \rho_{\tau((\sigma_i)_{l_i})} X_{u,\sigma_i|l_i,q}(\omega) \\
&= \underline{K} \sum_i \mathcal{M}_{u,q}(\omega)(C_{\sigma_i|l_i}(\omega)) \\
&\quad [\text{by } (7.4.33)] \\
&\geq \underline{K} \sum_i \mathcal{M}_{u,q}(\omega)(B(x_i,r_i) \cap C_u(\omega)) \\
&\geq \underline{K}\mathcal{M}_{u,q}(\omega)((\cup_i B(x_i,r_i)) \cap C_u(\omega)) \\
&\geq \underline{K}\mathcal{M}_{u,q}(\omega)(C_\alpha(\omega) \cap C_u(\omega))
\end{aligned}$$

where $\underline{K} = \underline{K}_0\underline{K}_1(\underline{\rho}(q)/\overline{\rho}(q))(\max_v \|X_{v,q}\|_\infty)^{-1}$. Hence

$$\begin{aligned}
\underline{K}\mathcal{M}_{u,q}(\omega)(C_\alpha(\omega) \cap C_u(\omega)) &\leq \overline{\mathcal{H}}^{q,\beta(q)}_{\mu_u(\omega),\delta}(C_\alpha(\omega) \cap C_u(\omega)) \\
&\leq \overline{\mathcal{H}}^{q,\beta(q)}_{\mu_u(\omega)}(C_\alpha(\omega) \cap C_u(\omega)) \\
&\leq \mathcal{H}^{q,\beta(q)}_{\mu_u(\omega)}(C_\alpha(\omega) \cap C_u(\omega))
\end{aligned}$$

which proves Step 1.

Step 2. For all relatively open subsets G of $C_u(\omega)$ then

$$\underline{K}\mathcal{M}_{u,q}(\omega)(G) \leq \mathcal{H}^{q,\beta(q)}_{\mu_u(\omega)}(G).$$

Proof of Step 2. The proof is similar to the proof of Step 2 in Theorem 7.2.1. This proves Step 2.

Step 3. We have

$$\underline{K}\mathcal{M}_{u,q}(\omega)\lfloor C_u(\omega) \leq \mathcal{H}^{q,\beta(q)}_{\mu_u(\omega)}\lfloor C_u(\omega).$$

Proof of Step 3. Since (by Theorem 7.2.1) $\mathcal{H}^{q,\beta(q)}_{\mu_u(\omega)}(C_u(\omega)) \leq \overline{K}\mathcal{M}_{u,q}(\omega)(C_u(\omega)) = \overline{K}X_{u,q}(\omega) < \infty$, $\mathcal{H}^{q,\beta(q)}_{\mu_u(\omega)}$ and $\mathcal{M}_{u,q}(\omega)$ are finite Borel measures when restricted to $C_u(\omega)$ and thus outer regular. The assertion in Step 3 now follows from Step 2 and outer regularity. This proves Step 3.

This completes the proof of Theorem 7.4.2. □

7.5 Proof of Theorem 2.6.4

The main purpose of this section is to prove Theorem 2.6.4. The proof of Theorem 2.6.4 is based on the next theorem.

Theorem 7.5.1. *Let $u \in V$ and consider the real number*

$$q \in \mathbb{R}.$$

Then there exists $\underline{K} \in]0, \infty[$ such that

$$\underline{K} X_{u,q} \leq \mathcal{P}_{\mu_u}^{q,\beta(q)}(C_u) \quad \Lambda_u\text{-a.a.}$$

We first prove Theorem 2.6.4 and then Theorem 7.5.1.

Proof of Theorem 2.6.4

i) Follows from Theorem 2.6.3.i) since $b_\mu \leq B_\mu$ for all $\mu \in \mathcal{P}(X_u)$.
ii) Follows from Theorem 7.5.1 since Proposition 5.2.3 shows that $X_{u,q}(\omega) > 0$ for Λ_u-a.a. $\omega \in \Omega_u$ for $q \in]q_{\min}, q_{\max}[$. □

We now turn toward the proof of Theorem 7.5.1.

Proof of Theorem 7.5.1

Let $u \in V$ and write

$$\begin{aligned} \Omega_u'' = \{\omega \in \Omega_u' \mid & 1)\ \omega \text{ satisfies (IV)}; \\ & 2)\ \big((S_e(\omega))_{e \in E_u}, (p_e(\omega))_{e \in E_u}\big) \text{ satisfies (I), (III), and (V)}; \\ & 3)\ X_{u,q}(\omega) = \lim_n X_{u,q,n}(\omega) \text{ exists}\} \end{aligned}$$

Clearly $\Lambda_u(\Omega_u'') = 1$. We will now show that there exists a number $\underline{K} \in]0, \infty[$ such that

$$\underline{K} X_{u,q}(\omega) \leq \mathcal{P}_{\mu_u(\omega)}^{q,\beta(q)}(C_u(\omega))$$

for all $\omega \in \Omega_u''$. Fix $\omega \in \Omega_u''$. Write $p_\alpha(\omega) = p_\alpha$ and $S_\alpha(\omega) = S_\alpha$ for $\alpha \in E_u^{(*)}$. Let $\delta > 0$. It follows from Lemma 4.1.1 that there exists an integer $N_\delta \in \mathbb{N}$ such that

$$\sup_{\alpha \in E_u^{(n)}} \prod_{k=1}^{n} \mathrm{Lip}(S_{\alpha|k}) \leq \delta / c$$

for $n \geq N$.

Let $\varnothing \neq E \subseteq C_u(\omega)$ and $\Gamma \in \mathbf{Max}(u)$ with $E_u^{(N_\delta)} \prec \Gamma$. Write $\Gamma_E = \{\gamma \in \Gamma \mid C_\gamma(\omega) \cap E \neq \varnothing$. Fix $\gamma \in \Gamma_E$ and choose $\sigma_\gamma \in [\gamma]$ such that

$$x_\gamma := \pi_u(\omega)(\sigma_\gamma) \in E\,.$$

Put

$$r_\gamma := c\,\mathrm{Lip}(S_{\gamma|1}) \cdot \ldots \cdot \mathrm{Lip}(S_{\gamma||\gamma|-1}) \tag{7.5.1}$$

Next choose integers $k_\gamma, l_\gamma \in \mathbb{N}$ such that

$$D\,\mathrm{Lip}(S_{\sigma_\gamma|1}) \cdot \ldots \cdot \mathrm{Lip}(S_{\sigma_\gamma|k_\gamma}) < r_\gamma \leq D\,\mathrm{Lip}(S_{\sigma_\gamma|1}) \cdot \ldots \cdot \mathrm{Lip}(S_{\sigma_\gamma|k_\gamma-1}) \tag{7.5.2}$$

$$c\,\mathrm{Lip}(S_{\sigma_\gamma|1}) \cdot \ldots \cdot \mathrm{Lip}(S_{\sigma_\gamma|l_\gamma}) < r_\gamma \leq c\,\mathrm{Lip}(S_{\sigma_\gamma|1}) \cdot \ldots \cdot \mathrm{Lip}(S_{\sigma_\gamma|l_\gamma-1})\,. \tag{7.5.3}$$

By Lemma 7.1.1,

$$C_{\sigma_\gamma|k_\gamma}(\omega) \subseteq B(x_\gamma, r_\gamma) \tag{7.5.4}$$

$$C_u(\omega) \cap B(x_\gamma, r_\gamma) \subseteq C_{\sigma_\gamma|l_\gamma}(\omega) \tag{7.5.5}$$

Observe that (7.5.1) implies that

$$|\gamma| = l_\gamma\,. \tag{7.5.6}$$

The proof is based on the next two claims.

Claim 1. There exists a constant $\underline{K}_0 \in]0, \infty[$ only depending on q such that

$$\underline{K}_0\,\mathrm{Lip}(S_{\sigma_\gamma|1}^{\beta(q)} \cdot \ldots \cdot \mathrm{Lip}(S_{\sigma_\gamma|l_\gamma})^{\beta(q)} \leq (2r_i)^{\beta(q)}\,.$$

Claim 2. There exists a constant $\underline{K}_1 \in]0, \infty[$ only depending on q such that

$$\underline{K}_1 p_{\sigma_\gamma|1}^q \cdot \ldots \cdot p_{\sigma_\gamma|l_\gamma}^q \leq \mu_u(\omega)(B(x_\gamma, r_\gamma))^q\,.$$

Proof of Claim 1. Similar to the proof of Claim 1 in the proof of Theorem 7.4.1. This proves Claim 1.

Proof of Claim 2. Similar to the proof of Claim 2 in the proof of Theorem 7.4.1. This proves Claim 2.

It follows from (7.5.5) and (7.5.6) that

$$B(x_\gamma, r_\gamma) \subseteq C_{\sigma_\gamma|l_\gamma}(\omega) = C_\gamma(\omega)\,. \tag{7.5.7}$$

Since $\Gamma \in \mathbf{Max}(u)$ and $\Gamma_E \subseteq \Gamma$, $(C_\gamma(\omega))_{\gamma\in\Gamma_E}$ is a pairwise disjoint family, and (7.5.7) therefore implies that $(B(x_\gamma, r_\gamma))_{\gamma\in\Gamma_E}$ is a centered packing of E. Also (since $|\gamma| \geq N_\delta$), $r_\gamma = c\operatorname{Lip}(S_{\gamma|1})\cdot\ldots\cdot\operatorname{Lip}(S_{\gamma||\gamma|-1}) < \delta$. The family $(B(x_\gamma, r_\gamma))_{\gamma\in\Gamma_E}$ is thus a centered δ-packing of E, and Claim 1 and Claim 2 therefore show that,

$$\begin{aligned}
\overline{\mathcal{P}}^{q,\beta(q)}_{\mu_u(\omega),\delta}(E) &\geq \sum_{\gamma\in\Gamma_E} \mu_u(\omega)\big(B(x_\gamma, r_\gamma)\big)^q (2r_\gamma)^{\beta(q)} \\
&\geq \underline{K}_0\underline{K}_1 \sum_{\gamma\in\Gamma_E}\left(\prod_{j=1}^{l_\gamma} p^q_{\sigma_\gamma|j}\operatorname{Lip}(S_{\sigma_\gamma|j})^{\beta(q)}\right) \\
&\geq \underline{K}_0\underline{K}_1(\underline{\rho}(q)/\overline{\rho}(q)) \sum_{\gamma\in\Gamma_E} \rho_u^{-1}\left(\prod_{j=1}^{l_\gamma} p^q_{\sigma_\gamma|j}\operatorname{Lip}(S_{\sigma_\gamma|j})^{\beta(q)}\right)\rho_{\tau(\gamma_{|\gamma|})} \\
&\quad \big[\text{by } (7.5.6)\big] \\
&\geq \underline{K}\sum_{\gamma\in\Gamma_E}\rho_u^{-1}\left(\prod_{j=1}^{|\gamma|} p^q_{\gamma|j}\operatorname{Lip}(S_{\gamma|j})^{\beta(q)}\right)\rho_{\tau(\gamma_{|\gamma|})}
\end{aligned}$$

where $\underline{K} = \underline{K}_0\underline{K}_1(\underline{\rho}(q)/\overline{\rho}(q))$. Since $\Gamma \in \mathbf{Max}(u)$ with $E_u^{(N_\delta)} \prec \Gamma$ was arbitrary, the above inequality implies that

$$\begin{aligned}
&\overline{\mathcal{P}}^{q,\beta(q)}_{\mu_u(\omega),\delta}(E) \\
&\geq \underline{K} \sup_{\substack{\Gamma\in\mathbf{Max}(u)\\ E_u^{(N_\delta)}\prec\Gamma}} \sum_{\substack{\gamma\in\Gamma\\ C_\gamma(\omega)\cap E\neq\varnothing}} \rho_u^{-1}\left(\prod_{j=1}^{|\gamma|} p^q_{\gamma|j}\operatorname{Lip}(S_{\gamma|j})^{\beta(q)}\right)\rho_{\tau(\gamma_{|\gamma|})}.
\end{aligned}$$

Hence

$$\begin{aligned}
&\overline{\mathcal{P}}^{q,\beta(q)}_{\mu_u(\omega),\delta}(E) \\
&\geq \underline{K} \inf_{\Sigma\in\mathbf{Max}(u)} \sup_{\substack{\Gamma\in\mathbf{Max}(u)\\ \Sigma\prec\Gamma}} \sum_{\substack{\gamma\in\Gamma\\ C_\gamma(\omega)\cap E\neq\varnothing}} \rho_u^{-1}\left(\prod_{j=1}^{|\gamma|} p^q_{\gamma|j}\operatorname{Lip}(S_{\gamma|j})^{\beta(q)}\right)\rho_{\tau(\gamma_{|\gamma|})}
\end{aligned}$$

for $\delta > 0$. Letting $\delta \searrow 0$ now yields

$$\begin{aligned}
&\overline{\mathcal{P}}^{q,\beta(q)}_{\mu_u(\omega)}(E) \\
&\geq \underline{K} \inf_{\Sigma\in\mathbf{Max}(u)} \sup_{\substack{\Gamma\in\mathbf{Max}(u)\\ \Sigma\prec\Gamma}} \sum_{\substack{\gamma\in\Gamma\\ C_\gamma(\omega)\cap E\neq\varnothing}} \rho_u^{-1}\left(\prod_{j=1}^{|\gamma|} p^q_{\gamma|j}\operatorname{Lip}(S_{\gamma|j})^{\beta(q)}\right)\rho_{\tau(\gamma_{|\gamma|})} \qquad (7.5.8)
\end{aligned}$$

for $E \subseteq C_u(\omega)$.

Let $(E_i)_{i\in\mathbb{N}}$ be a cover of $C_u(\omega)$, i.e. $C_u(\omega) \subseteq \cup_i E_i$. Then (7.5.8) implies that

$$\begin{aligned}
&\sum_i \overline{\mathcal{P}}^{q,\beta(q)}_{\mu_u(\omega)}(E_i) \\
\geq &\sum_i \overline{\mathcal{P}}^{q,\beta(q)}_{\mu_u(\omega)}(E_i \cap C_u(\omega)) \\
\geq &\underline{K} \sum_i \left(\inf_{\Sigma\in\mathbf{Max}(u)} \sup_{\substack{\Gamma\in\mathbf{Max}(u)\\ \Sigma\prec\Gamma}} \sum_{\substack{\gamma\in\Gamma\\ C_\gamma(\omega)\cap C_u(\omega)\cap E_i\neq\varnothing}} \rho_u^{-1} \left(\prod_{j=1}^{|\gamma|} p^q_{\gamma|j} \operatorname{Lip}(S_{\gamma|j})^{\beta(q)} \right) \rho_{\tau(\gamma_{|\gamma|})} \right) \\
\geq &\underline{K} \inf_{\Sigma\in\mathbf{Max}(u)} \sup_{\substack{\Gamma\in\mathbf{Max}(u)\\ \Sigma\prec\Gamma}} \left(\sum_i \sum_{\substack{\gamma\in\Gamma\\ C_\gamma(\omega)\cap C_u(\omega)\cap E_i\neq\varnothing}} \rho_u^{-1} \left(\prod_{j=1}^{|\gamma|} p^q_{\gamma|j} \operatorname{Lip}(S_{\gamma|j})^{\beta(q)} \right) \rho_{\tau(\gamma_{|\gamma|})} \right) \\
\geq &\underline{K} \inf_{\Sigma\in\mathbf{Max}(u)} \sup_{\substack{\Gamma\in\mathbf{Max}(u)\\ \Sigma\prec\Gamma}} \left(\sum_{\gamma\in\Gamma} \rho_u^{-1} \left(\prod_{j=1}^{|\gamma|} p^q_{\gamma|j} \operatorname{Lip}(S_{\gamma|j})^{\beta(q)} \right) \rho_{\tau(\gamma_{|\gamma|})} \right)
\end{aligned}$$

whence

$$\begin{aligned}
\mathcal{P}^{q,\beta(q)}_{\mu_u(\omega)}(C_u(\omega)) \geq &\underline{K} \inf_{\Sigma\in\mathbf{Max}(u)} \sup_{\substack{\Gamma\in\mathbf{Max}(u)\\ \Sigma\prec\Gamma}} \left(\sum_{\gamma\in\Gamma} \rho_u^{-1} \left(\prod_{j=1}^{|\gamma|} p^q_{\gamma|j} \operatorname{Lip}(S_{\gamma|j})^{\beta(q)} \right) \rho_{\tau(\gamma_{|\gamma|})} \right) \\
\geq &\underline{K} \limsup_n \sum_{\gamma\in E_u^{(n)}} \rho_u^{-1} \left(\prod_{j=1}^{n} p^q_{\gamma|j} \operatorname{Lip}(S_{\gamma|j})^{\beta(q)} \right) \rho_{\tau(\gamma_{|\gamma|})} \\
= &\underline{K} \limsup_n X_{u,q,n}(\omega) \\
= &\underline{K} X_{u,q}(\omega). \quad \square
\end{aligned}$$

7.6 Proofs of Theorem 2.6.5, Theorem 2.6.6 and Theorem 2.6.7

First recall the following result.

Lemma 7.6.1.

i) *If* $-\infty < q_{\min}$ *then graph of* β *has the line that passes through the origin with slope* $\frac{\beta(q_{\min})}{q_{\min}}$ *as tangent at the point* $(q_{\min}, \beta(q_{\min}))$.

ii) *If* $q_{\max} < \infty$ *then graph of* β *has the line that passes through the origin with slope* $\frac{\beta(q_{\max})}{q_{\max}}$ *as tangent at the point* $(q_{\max}, \beta(q_{\max}))$.

Proof. Follows immediately from Proposition 2.5.6. □

We divide the proof of Theorem 2.6.5 into two parts. First we prove Theorem 2.6.5, Case 1 and Case 2, i)-ii), then we prove Lemma 7.6.2 and finally we prove Theorem 2.6.5, Case 2, iii)-vii).

Proof of Theorem 2.6.5, Case 1 and Case 2, i)-ii)

Case 1. i) Follows from Theorem 2.1.11, Theorem 2.1.12, Lemma 7.1.4, Theorem 2.6.2 and Theorem 2.6.3 since $q_{\min} = -\infty$ and $q_{\max} = \infty$ in Case 1.

ii) Let

$$\Omega''_u = \left(\bigcap_{n \in \mathbb{N}} \bigcap_{\gamma \in E^{(n)}} \{\underline{a}_n \leq \ell_\gamma \leq \overline{a}_n\} \right) \bigcap \Omega'_u \,.$$

Clearly $\Lambda_u(\Omega'_u) = 1$. We claim that

$$\Delta_{\mu_u(\omega)}(\alpha) = \begin{cases} \varnothing & \text{for } \alpha \neq a \\ \operatorname{supp} \mu_u(\omega) & \text{for } \alpha = a \end{cases} \tag{7.6.1}$$

for $\omega \in \Omega''_u$. Now fix $\omega \in \Omega''_u$ and let $x \in \operatorname{supp} \mu_u(\omega) = C_u(\omega)$. Choose $\sigma \in E_u^{\mathbb{N}}$ with $x = \pi_u(\omega)(\sigma)$. We have $\underline{a}_n \leq \ell_{\sigma|n}(\omega) \leq \overline{a}_n$ for all $n \in \mathbb{N}$, whence (by Lemma 4.4.2)

$$\frac{\log(p_{\sigma|n}(\omega) \cdot \ldots \cdot p_{\sigma|n}(\omega))}{\log(\operatorname{Lip}(S_{\sigma|1}(\omega)) \cdot \ldots \cdot \operatorname{Lip}(S_{\sigma|n}(\omega)))} = \ell_\gamma(\omega) \to a \quad \text{as} \quad n \to \infty \,. \tag{7.6.2}$$

Finally Lemma 6.3.4 and (7.6.2) imply that

$$\alpha_{\mu_u(\omega)}(x) = \alpha_{\mu_u(\omega)}(\pi_u(\sigma)(\omega)) = a$$

which proves (7.6.1).

iii) Let

$$\begin{aligned} \Omega''_u = \{\omega \in \Omega'_u \mid & 1)\ \Delta_{\mu_u(\omega)}(\alpha(0)) \text{ has full } \mathcal{M}_{u,0}(\omega) \text{ measure;} \\ & 2)\ \mathcal{M}_{u,q}(\omega)(X_u) = X_{u,q}(\omega); \\ & 3)\ X_{u,0}(\omega) > 0 \,\} \,. \end{aligned}$$

Theorem 6.3.1 implies that $\Lambda_u(\Omega'_u) = 1$. We claim that $f_{\mu_u(\omega)}(a) = F_{\mu_u(\omega)}(a) = \beta(0)$ for all $\omega \in \Omega''_u$. Let $\omega \in \Omega''_u$. Since $a = \beta(0)$, Proposition 2.1.4 implies that

$$\begin{aligned}\mathcal{P}^{a+\delta}(\Delta_{\mu_u(\omega)}(a)) &= \mathcal{P}^{0\cdot a+(\beta(0)+\frac{\delta}{2})+\frac{\delta}{2}}(\Delta_{\mu_u(\omega)}(a)) \\ &\le 2^{0\cdot a+\frac{\delta}{2}}\mathcal{P}^{0,\beta(0)+\frac{\delta}{2}}_{\mu_u(\omega)}(\operatorname{supp}\mu_u(\omega)) \\ &= 0\end{aligned}$$

for all $\delta > 0$. Hence $\operatorname{Dim}(\Delta_{\mu_u(\omega)}(a)) \le a+\delta$ for all $\delta > 0$. Letting $\delta \searrow 0$ yields

$$F_{\mu_u(\omega)}(a) := \operatorname{Dim}(\Delta_{\mu_u(\omega)}(a)) \le a\,. \tag{7.6.3}$$

Since $\omega \in \Omega''_u$, $\mathcal{M}_{u,0}(\omega)(\Delta_{\mu_u(\omega)}(\alpha(0))) = \mathcal{M}_{u,0}(\omega)(X_u) = X_{u,q}(\omega) > 0$ whence (by Theorem 7.4.1)

$$\mathcal{H}^{0,\beta(0)}_{\mu_u(\omega)}(\Delta_{\mu_u(\omega)}(\alpha(0))) = \infty\,. \tag{7.6.4}$$

Now, since $a = \beta(0) > 0$, Proposition 2.1.5 and (7.6.4) imply that

$$\begin{aligned}\infty = \mathcal{H}^{0,\beta(0)}_{\mu_u(\omega)}(\Delta_{\mu_u(\omega)}(\alpha(0))) &\le 2^{\beta(0)}\mathcal{H}^{\alpha(0)\cdot 0+\beta(0)-\delta}(\Delta_{\mu_u(\omega)}(\alpha(0))) \\ = 2^{\beta(0)}\mathcal{H}^{\beta(0)-\delta}(\Delta_{\mu_u(\omega)}(\alpha(0)))&\end{aligned}$$

for $0 < \delta < \beta(0)$. Hence $\beta(0) - \delta \le \dim(\Delta_{\mu_u(\omega)}(a))$ for $0 < \delta < \beta(0)$. Letting $\delta \searrow 0$ now yields

$$a = \beta(0) \le \dim(\Delta_{\mu_u(\omega)}(a)) := f_{\mu_u(\omega)}(a)\,. \tag{7.6.5}$$

The desired conclusion now follows from (7.6.3) and (7.6.5).

Case 2. i) It follows immediately from Theorem 2.6.2 and Theorem 2.6.3 that

$$b_\mu = B_\mu = \beta \quad \text{on }]q_{\min}, q_{\max}[$$

for P_u-a.a. $\mu \in \mathcal{P}(X_u)$, and it is thus sufficient to prove the following four claims.

Claim 1. If $-\infty < q_{\min}$ then P_u-a.a. $\mu \in \mathcal{P}(X_u)$ satisfy

$$B_\mu = \hat{\beta} \quad \text{on} \quad]-\infty, q_{\min}]\,.$$

Claim 2. If $q_{\max} < \infty$ then P_u-a.a. $\mu \in \mathcal{P}(X_u)$ satisfy

$$B_\mu = \hat{\beta} \quad \text{on} \quad [q_{\max}, \infty[\,.$$

Claim 3. If $-\infty < q_{\min}$ then P_u-a.a. $\mu \in \mathcal{P}(X_u)$ satisfy

$$b_\mu = \hat{\beta} \quad \text{on} \quad]-\infty, q_{\min}] .$$

Claim 4. If $q_{\max} < \infty$ then P_u-a.a. $\mu \in \mathcal{P}(X_u)$ satisfy

$$b_\mu = \hat{\beta} \quad \text{on} \quad [q_{\max}, \infty[.$$

Proof of Claim 1. Let

$$M := \{\mu \in \mathcal{P}(X_u) \mid B_\mu = \beta \quad \text{on} \quad]q_{\min}, q_{\max}[\} .$$

Clearly $P_u(M) = 1$. We will now prove that $b_\mu = \hat{\beta}$ on $]-\infty, q_{\min}]$ for each $\mu \in M$. Fix $\mu \in M$. Since $B_\mu = \beta$ on $]q_{\min}, q_{\max}[$, the continuity of β and B_μ (recall that B_μ is convex and thus continuous) imply that $B_\mu = \beta$ on $[q_{\min}, q_{\max}[$ and Lemma 7.6.1 therefore implies that:

$$\left.\begin{array}{l} \text{the graph of } B_\mu \text{ has the line that passes through} \\ \text{the origin with slope } \dfrac{B_\mu(q_{\min})}{q_{\min}} \text{ as a tangent from the} \\ \text{right at the point } (q_{\min}, B_\mu(q_{\min})). \end{array}\right\} \tag{7.6.6}$$

Moreover, it follows from Proposition 2.1.8 and Proposition 2.1.10 that B_μ has the following properties:

$$\left.\begin{array}{c} B_\mu \text{ is convex and has an affine asymptote} \\ q \to -\overline{A}q + \overline{E} \\ \text{as } q \to -\infty \text{ with } \overline{A}, \overline{E} \geq 0; \text{ in fact} \\ B_\mu(q) + \overline{A}q \searrow \overline{E} \quad \text{as} \quad q \to -\infty \end{array}\right\} \tag{7.6.7}$$

Now (7.6.6) and (7.6.7) yield

$$B_\mu(q) = \frac{B_\mu(q_{\min})}{q_{\min}} q = \hat{\beta}(q) \quad \text{for} \quad q \leq q_{\min} .$$

This proves Claim 1.

Proof of Claim 2. Similar to the proof of Claim 1.

Proof of Claim 3. Let

$$\begin{aligned} M := \{\mu \in \mathcal{P}(X_u) \mid & 1)\ B_\mu = \hat{\beta}; \\ & 2)\ b_\mu = \beta \text{ on }]q_{\min}, q_{\max}[; \\ & 3)\ \limsup_{r \searrow 0} \left(\sup_{x \in \operatorname{supp} \mu} \frac{\mu(B(x,2r))}{\mu(B(x,r))} \right) \leq \Delta^{-\varphi(2)} \} \end{aligned}$$

It follows from Lemma 7.1.4, Claim 1 and Claim 2 that $P_u(M) = 1$. Let $\mu \in M$. We will now prove that

$$b_\mu = \hat{\beta} \quad \text{on} \quad]-\infty, q_{\min}] .$$

Assume, in order to get a contradiction, that there exists $q_0 \leq q_{\min}$ such that $b_\mu(q_0) \neq \hat{\beta}(q_0) = B_\mu(q_0)$. Hence (since $b_\mu \leq B_\mu$ by Proposition 2.1.2),

$$b_\mu(q_0) < B_\mu(q_0) . \tag{7.6.8}$$

We first prove that

$$q_0 < q_{\min} .$$

Indeed, if $q_0 = q_{\min}$ then $b_\mu(q_0) = b_\mu(q_{\min}) \geq b_\mu(q) = B_\mu(q)$ for $q \in]q_{\min}, q_{\max}[$ (since b_μ is decreasing), and the continuity of B_μ therefore implies that $b_\mu(q_0) \geq B_\mu(q_0)$ contradicting (7.6.8). This proves that $q_0 < q_{\min}$.

Since $q_0 < q_{\min}$ and the functions β and B_μ coincide on $[q_{\min}, q_{\max}[$, Lemma 7.6.1 and the strict convexity of $\beta = B_\mu$ on $]q_{\min}, q_{\max}[$ show that there exists a $q_1 \in]q_{\min}, q_{\max}[$ such that

$$f(q_{\min}) < B_\mu(q_{\min}) \tag{7.6.9}$$

where $f : \mathbb{R} \to \mathbb{R}$ is the unique affine function whose graph contains the points:

$$(q_0, b_\mu(q_0)) \quad \text{and} \quad (q_1, B_\mu(q_1)) .$$

Since $q_{\min} \in]q_0, q_1[$ and $\mu \in \mathcal{P}_0$ (by Lemma 7.1.4), Proposition 2.1.9 implies that

$$\begin{aligned} B_\mu(q_{\min}) &= b_\mu(q_{\min}) & \left[\text{since } q_0 \neq q\right] \\ &= b_\mu\left(\tfrac{q_1 - q_{\min}}{q_1 - q_0} q_0 + \left(1 - \tfrac{q_1 - q_{\min}}{q_1 - q_0}\right) q_1\right) \\ &\leq \tfrac{q_1 - q_{\min}}{q_1 - q_0} b_\mu(q_0) + \left(1 - \tfrac{q_1 - q_{\min}}{q_1 - q_0}\right) B_\mu(q_1) = f(q_{\min}) \end{aligned}$$

contradicting (7.6.9).

Proof of Claim 4. Similar to the proof of Claim 3.
ii) It follows from Theorem 2.6.2 that

$$\Lambda_u \leq \beta \quad \text{on }]q_{\min}, q_{\max}[\text{ for } P_u\text{-a.a. } \mu \in \mathcal{P}(X_u) \tag{7.6.10}$$

Let $q \in]q_{\min}, q_{\max}[$. Theorem 2.6.4 implies that there exists a subset $M_{u,q}$ of $\mathcal{P}(X_u)$ such that $P_u(M_{u,q}) = 1$ and $0 < \mathcal{P}_\mu^{q,\beta(q)}(\operatorname{supp}\mu) \leq \overline{\mathcal{P}}_\mu^{q,\beta(q)}(\operatorname{supp}\mu)$ for all $\mu \in M_{u,q}$. Hence

$$\beta(q) \leq \Delta_\mu^q(\operatorname{supp}\mu) := \Lambda_\mu(q) \quad \text{for } \mu \in M_{u,q} .$$

Now put $M_u := \bigcap_{q\in\mathbb{Q}\cap]q_{\min},q_{\max}[} M_{u,q}$. Clearly $P_u(M_u) = 1$. We claim that

$$\beta(p) \le \Lambda_\mu(p) \quad \text{for all } p \in]q_{\min}, q_{\max}[\text{ and } \mu \in M_u \tag{7.6.11}$$

Indeed, let $p \in]q_{\min}, q_{\max}[$ and $\mu \in M_u$. Next choose a sequence $(q_n)_{n\in\mathbb{N}}$ in $\mathbb{Q}\cap]q_{\min}, q_{\max}[$ such that $q_n \to p$. Since $q_n \in \mathbb{Q}\cap]q_{\min}, q_{\max}[$ and $\mu \in M_u \subseteq M_{u,q_n}$, $\beta(q_n) \le \Lambda_\mu(q_n)$ for all $n \in \mathbb{N}$ and the continuity of β and Λ_μ therefore imply that $\beta(p) = \lim_n \beta(q_n) \le \lim_n \Lambda_\mu(q_n) = \Lambda_\mu(p)$.

Finally, equations (7.6.10) and (7.6.11) and Theorem 2.1.11, Theorem 2.1.12 and Lemma 7.1.4 yield the desired result.

This completes the proof of Theorem 2.6.5, Case 1 and Case 2, i)-ii). □

Before continuing the proof of Theorem 2.6.5 we prove a small lemma. Define random variables $\underline{a}_u$, $\overline{a}_u$, $\underline{A}_u$, $\overline{A}_u : \mathcal{P}(X_u) \to \mathbb{R}_+$ by

$$\underline{a}_u(\mu) = \sup_{0<q}\left(-\frac{b_\mu(q)}{q}\right), \quad \overline{a}_u(\mu) = \inf_{q<0}\left(-\frac{b_\mu(q)}{q}\right) \tag{7.6.12}$$

$$\underline{A}_u(\mu) = \sup_{0<q}\left(-\frac{B_\mu(q)}{q}\right), \quad \overline{A}_u(\mu) = \inf_{q<0}\left(-\frac{B_\mu(q)}{q}\right) \tag{7.6.13}$$

Lemma 7.6.2. *For P_u-a.a. $\mu \in \mathcal{P}(X_u)$,*

$$\underline{a}_u(\mu) = \underline{A}_u(\mu) = a_{\min}, \quad \overline{a}_u(\mu) = \overline{A}_u(\mu) = a_{\max}.$$

Proof. It follows from Theorem 2.6.5.ii) that for P_u-a.a. $\mu \in \mathcal{P}(X_u)$ then

$$\overline{A}_u(\mu) = \inf_{q<0}\left(-\frac{B_\mu(q)}{q}\right) = \inf_{q<0}\left(-\frac{\hat{\beta}(q)}{q}\right)$$
$$= \begin{cases} \inf_{q<0}\left(-\frac{\beta(q)}{q}\right) & \text{for } -\infty = q_{\min} \\ \inf_{q_{\min}<q<0}\left(-\frac{\beta(q)}{q}\right) \wedge \inf_{q\le q_{\min}}\left(-\frac{\beta(q_{\min})}{q_{\min}}\right) & \text{for } -\infty < q_{\min} \end{cases} \tag{7.6.14}$$

However, if $q < q_{\min}$, Lemma 7.6.1 and the strict convexity of β imply that $\beta(q) > \frac{\beta(q_{\min})}{q_{\min}} q$, whence

$$-\frac{\beta(q)}{q} > -\frac{\beta(q_{\min})}{q_{\min}} \quad \text{for } q < q_{\min}. \tag{7.6.15}$$

It follows from (7.6.14) and (7.6.15) that

$$\overline{A}_u(\mu) = \inf_{q<0}\left(-\frac{\beta(q)}{q}\right) = a_{\max}.$$

The proofs of the other assertions in Lemma 7.6.2 are similar. □

We now continue the proof of Theorem 2.6.5.

Proof of Theorem 2.6.5, Case 2, iii)-vii)

Put

$$M := \{\mu \in \mathcal{P}(X_u) \mid b_\mu = B_\mu = \hat{\beta}\,,\ \underline{a}_u(\mu) = \underline{A}_u(\mu) = a_{\min}\,,\ \overline{a}_u(\mu) = \overline{A}_u(\mu) = a_{\max}\,\}$$

Then $P_u(M) = 1$.
iii) Let $\mu \in M$. Theorem 2.1.7 implies that

$$\Delta_\mu(\alpha) = \varnothing \quad \text{for} \quad \alpha \notin [\underline{a}_u(\mu), \overline{a}_u(\mu)] = [a_{\min}, a_{\max}]\,.$$

iv) Let $\alpha \in]a_{\min}, a_{\max}[$. It follows from Proposition 2.5.6 that we can choose $q \in\,]q_{\min}, q_{\max}[$ such that

$$\alpha = \alpha(q)\,.$$

Now put

$$\begin{aligned}\Omega_{u,q} = \mu_u^{-1}(M)\bigcap\{\omega \in \Omega_u' \mid & 1)\ \Delta_{\mu_u(\omega)}(\alpha(q)) \text{ has full } \mathcal{M}_{u,q}(\omega) \text{ measure;}\\ & 2)\ X_{u,q}(\omega) > 0;\\ & 3)\ \mathcal{M}_{u,q}(\omega)(X_u) = X_{u,q}(\omega)\ \}\end{aligned}$$

Clearly $\Lambda_u(\Omega_{u,q}) = 1$. We will now prove that

$$f_{\mu_u(\omega)}(\alpha) = F_{\mu_u(\omega)}(\alpha) = \beta^*(\alpha)$$

for all $\omega \in \Omega_{u,q}$. Let $\omega \in \Omega_{u,q}$. Then $\mathcal{M}_{u,q}(\omega)(\Delta_{\mu_u(\omega)}(\alpha(q))) = \mathcal{M}_{u,q}(\omega)(X_u) = X_{u,q}(\omega) > 0$ and since $0 < \beta^*(\alpha) = \beta^*(\alpha(q)) = q\alpha(q) + \beta(q)$, Theorem 7.4.1 and Proposition 2.1.5 imply that

$$\infty = \mathcal{H}^{q,\beta(q)}_{\mu_u(\omega)}(\Delta_{\mu_u(\omega)}(\alpha(q))) \le 2^{\beta(q)}\mathcal{H}^{q\alpha(q)+\beta(q)-\delta}(\Delta_{\mu_u(\omega)}(\alpha(q)))$$

for $0 < \delta < q\alpha(q) + \beta(q) = \beta^*(\alpha)$. Hence $\beta^*(\alpha) - \delta \le \dim \Delta_{\mu_u(\omega)}(\alpha(q))$ for $0 < \delta < q\alpha(q) + \beta(q) = \beta^*(\alpha)$. Letting $\delta \searrow 0$ now yields

$$\begin{aligned}\beta^*(\alpha) &\le \dim \Delta_{\mu_u(\omega)}(\alpha(q)) = f_{\mu_u(\omega)}(\alpha) \le F_{\mu_u(\omega)}(\alpha)\\ &\quad [\text{by Theorem 2.1.7}]\\ &\le B^*_{\mu_u(\omega)}(\alpha) = \beta^*(\alpha)\,.\end{aligned}$$

v) Let $\mu \in M$. For $\delta > 0$, Proposition 2.1.4 and the equation $\beta^*(q_{\max}) = q_{\max}\alpha(q_{\max}) + \beta(q_{\max})$ imply that

$$\begin{aligned}\mathcal{P}^{\delta}(\Delta_\mu(a_{\min})) &= \mathcal{P}^{\beta^*(q_{\max})+\delta}(\Delta_\mu(\alpha(q_{\max})))\\ &= \mathcal{P}^{q_{\max}\alpha(q_{\max})+(\beta(q_{\max})+\frac{\delta}{2})+\frac{\delta}{2}}(\Delta_\mu(\alpha(q_{\max})))\\ &\leq 2^{q_{\max}\alpha(q_{\max})+\frac{\delta}{2}}\mathcal{P}_\mu^{q_{\max},\beta(q_{\max})+\frac{\delta}{2}}(\operatorname{supp}\mu) = 0\,.\end{aligned}$$

Hence

$$f_\mu(a_{\min}) \leq F_\mu(a_{\min}) = \operatorname{Dim}(\Delta_\mu(a_{\min})) \leq \delta$$

for $\delta > 0$. Letting $\delta \searrow 0$ now yields the desired result since $P_u(M) = 1$.
vi) Similar to the proof of v).
vii) Let $\mu \in M$. Theorem 2.1.7 implies that

$$f_\mu \leq F_\mu \leq B_\mu^* = \beta^* \quad \text{on} \quad]\underline{a}_u(\mu), \overline{a}_u(\mu)[=]a_{\min}, a_{\max}[$$

This proves the assertion since $P_u(M) = 1$.
viii) Clearly $\mathcal{P}^t(C_u(\omega)) = \mathcal{P}^{0,t}_{\mu_u(\omega)}(C_u(\omega))$ for all $t \geq 0$ and $\omega \in \Omega_u$, and Theorem 7.3.1 therefore implies that for each fixed integer $n \in \mathbb{N}$, $\mathcal{P}^{\beta(q)+\frac{1}{n}}(C_u(\omega)) = \mathcal{P}^{0,\beta(0)+\frac{1}{n}}_{\mu_u(\omega)}(C_u(\omega)) = 0$ for Λ_u-a.a. $\omega \in \Omega_u$. Hence for each fixed integer $n \in \mathbb{N}$,

$$\operatorname{Dim}(C_u(\omega)) \leq \beta(0) + \frac{1}{n} \quad \text{for } \Lambda_u\text{-a.a. } \omega \in \Omega_u\,.$$

This clearly implies that

$$\operatorname{Dim}(C_u(\omega)) \leq \beta(0) \quad \text{for } \Lambda_u\text{-a.a. } \omega \in \Omega_u\,.$$

It follows from [Tsu2, Theorem 3.11] (cf. also Graf [Gra1, Theorem 7.6]) that

$$\beta(0) = \dim(C_u(\omega)) \quad \text{for } \Lambda_u\text{-a.a. } \omega \in \Omega_u\,.$$

Hence, $\beta(0) = \dim(C_u(\omega)) \leq \operatorname{Dim}(C_u(\omega)) \leq \beta(0)$ for Λ_u-a.a. $\omega \in \Omega_u$. □

Proof of Theorem 2.6.6

Let

$$\begin{aligned}M := \{\mu \in \mathcal{P}(X_u)\,|\,&1)\ \Delta_\mu(\alpha(q)) \text{ has full } \mathcal{H}_\mu^{q,\beta(q)}\lfloor \operatorname{supp}\mu \text{ measure;}\\ &2)\ f_\mu(\alpha(q)) = \beta^*(\alpha(q)) = F_\mu(\alpha(q))\ \}\,.\end{aligned}$$

Theorem 2.6.1 and Theorem 2.6.5 imply that $P_u(M) = 1$. Now fix $\mu \in M$ and $E \subseteq \operatorname{supp}\mu$ with $\mathcal{H}_\mu^{q,\beta(q)}(E) > 0$. Proposition 2.1.5 shows that for each $\delta > 0$,

$$\begin{aligned} 0 < \mathcal{H}_\mu^{q,\beta(q)}(E) &= \mathcal{H}_\mu^{q,\beta(q)}(E \cap \Delta_\mu(\alpha(q))) \\ &\leq 2^{\beta(q)} \mathcal{H}^{q\alpha(q)+\beta(q)-\delta}(E \cap \Delta_\mu(\alpha(q))) \end{aligned}$$

whence (since $\beta^*(\alpha(q)) = q\alpha(q) + \beta(q)$)

$$\begin{aligned} f_\mu(\alpha(q)) - \delta = \beta^*(\alpha(q)) - \delta &= q\alpha(q) + \beta(q) - \delta \\ &\leq \dim(E \cap \Delta_\mu(\alpha(q))) \leq \dim E \end{aligned}$$

for $\delta > 0$. Letting $\delta \searrow 0$ thus yields $f_\mu(\alpha(q)) \leq \dim E$. Finally, since $E \subseteq \operatorname{supp}\mu$ with $\mathcal{H}_\mu^{q,\beta(q)}(E) > 0$ was arbitrary this inequality implies that

$$f_\mu(\alpha(q)) \leq \dim \mathcal{H}_\mu^{q,\beta(q)} \lfloor \operatorname{supp}\mu \tag{7.6.16}$$

However, Theorem 2.6.3 implies that $\mathcal{H}_\mu^{q,\beta(q)}(\Delta_\mu(\alpha(q))) > 0$ whence

$$\operatorname{Dim} \mathcal{H}_\mu^{q,\beta(q)} \lfloor \operatorname{supp}\mu \leq \operatorname{Dim} \Delta_\mu(\alpha(q)) = F_\mu(\alpha(q)) . \tag{7.6.17}$$

It follows from (7.6.16) and (7.6.17) that

$$\begin{aligned} f_\mu(\alpha(q)) &\leq \dim \mathcal{H}_\mu^{q,\beta(q)} \lfloor \operatorname{supp}\mu \\ &\leq \operatorname{Dim} \mathcal{H}_\mu^{q,\beta(q)} \lfloor \operatorname{supp}\mu \\ &\leq F_\mu(\alpha(q)) = f_\mu(\alpha(q)) \end{aligned}$$

for all $\mu \in M$. $\square$

Proof of Theorem 2.6.7

i) Recall that $1 \in]q_{\min}, q_{\max}[$. Now Theorem 6.3.1 implies that

$$\Lambda_u\left(\left\{\omega \in \Omega_u \middle| \alpha_{\mu_u(\omega)} = \alpha(1) \quad \mathcal{M}_{u,1}(\omega)\text{-a.e.}\right\}\right) = 1 \tag{7.6.18}$$

It is easily seen that $\mu \leq \mathcal{H}_\mu^{1,0} \lfloor \operatorname{supp}\mu$ for all $\mu \in \mathcal{P}(X_u)$. Hence (by Theorem 7.2.1)

$$\mu_u(\omega) \leq \mathcal{H}_{\mu_u(\omega)}^{1,0} \lfloor \operatorname{supp}\mu_u(\omega) \leq \overline{K}\mathcal{M}_{u,1}(\omega) \lfloor \operatorname{supp}\mu_u(\omega) \tag{7.6.19}$$

for Λ_u-a.a. $\omega \in \Omega_u$. It follows from (7.6.18) and (7.6.19) that

$$\begin{aligned} 1 &\leq \Lambda_u\left(\left\{\omega \in \Omega_u \middle| \alpha_{\mu_u(\omega)} = \alpha(1) \quad \mu_u(\omega)\text{-a.e.}\right\}\right) \\ &= P_u\left(\left\{\mu \in \mathcal{P}(X_u) \middle| \alpha_\mu = \alpha(1) \quad \mu\text{-a.e.}\right\}\right) . \end{aligned}$$

ii) Follows from i) by [Haa3, Proposition 1] (cf. also [Yo, Theorem 4.4].) $\square$

7.7 Proofs of the Theorems in section 2.7

The purpose of this section is to prove Theorem 2.7.1 and Theorem 2.7.2. The proofs are based on the next three lemmas.

Define for $u \in V$, $q \in \mathbb{R}$ and $n \in \mathbb{N}$ a random variable $Z_{u,q,n} : \Omega_u \to \mathbb{R}$ by

$$Z_{u,q,n}(\omega) = \inf_{\substack{\Gamma \in \mathbf{Max}(u)\setminus\{\{\varnothing\}\} \\ |\Gamma| \leq n}} X_{u,q,\Gamma}(\omega) .$$

Lemma 7.7.1. *Let $u \in V$ and $q \in \mathbb{R}$.*

i) *$(Z_{u,q,n})_n$ is a decreasing sequence of random variables.*

ii) *For all $n \in \mathbb{N}$,*

$$Z_{u,q,n+1} = \sum_{e \in E_u} \rho_u^{-1} p_e^q \operatorname{Lip}(S_e)^{\beta(q)} \rho_{\tau(q)} \left(1 \wedge Z_{\tau(u),q,n}(\mathcal{S}_e)\right) .$$

iii) $Z_{u,q} := \inf_n Z_{u,q,n} = \inf_{\Gamma \in \mathbf{Max}(u)\setminus\{\{\varnothing\}\}} X_{u,q,\Gamma}$.

Proof. i) Obvious.
ii) For each $\omega \in \Omega_u$,

$$X_{u,q,\Gamma}(\omega) = \sum_{e \in E_u} \rho_u^{-1} p_e(\omega)^q \operatorname{Lip}(S_e(\omega))^{\beta(q)} \rho_{\tau(e)} X_{\tau(e),q,\Gamma(e)}(\mathcal{S}_e(\omega)) .$$

Hence

$$\begin{aligned}
& Z_{u,q,n+1}(\omega) \\
&= \inf_{\substack{\Gamma \in \mathbf{Max}(u)\setminus\{\{\varnothing\}\} \\ |\Gamma| \leq n+1}} X_{u,q,\Gamma}(\omega) \\
&= \inf_{\substack{\Gamma \in \mathbf{Max}(u)\setminus\{\{\varnothing\}\} \\ |\Gamma| \leq n+1}} \left(\sum_{e \in E_u} \rho_u^{-1} p_e(\omega)^q \operatorname{Lip}(S_e(\omega))^{\beta(q)} \rho_{\tau(e)} X_{\tau(e),q,\Gamma(e)}(\mathcal{S}_e(\omega)) \right) \\
&= \sum_{e \in E_u} \rho_u^{-1} p_e(\omega)^q \operatorname{Lip}(S_e(\omega))^{\beta(q)} \rho_{\tau(e)} \inf_{\substack{\Gamma \in \mathbf{Max}(u)\setminus\{\{\varnothing\}\} \\ |\Gamma| \leq n+1}} X_{\tau(e),q,\Gamma(e)}(\mathcal{S}_e(\omega)) \\
&= \sum_{e \in E_u} \rho_u^{-1} p_e(\omega)^q \operatorname{Lip}(S_e(\omega))^{\beta(q)} \rho_{\tau(e)} \cdot \\
&\qquad \left(1 \wedge \inf_{\substack{\Sigma \in \mathbf{Max}(\tau(e))\setminus\{\{\varnothing\}\} \\ |\Sigma| \leq n}} X_{\tau(e),q,\Sigma(e)}(\mathcal{S}_e(\omega)) \right) \\
&= \sum_{e \in E_u} \rho_u^{-1} p_e(\omega)^q \operatorname{Lip}(S_e(\omega))^{\beta(q)} \rho_{\tau(e)} \left(1 \wedge Z_{\tau(e),q,n}(\mathcal{S}_e(\omega))\right) .
\end{aligned}$$

iii) Obvious. □

Lemma 7.7.2. *Let $u \in V$ and $q \in \mathbb{R}$. Then the following statements are equivalent.*

i) *There exists a $u \in V$ such that*

$$\Lambda_u(Z_{u,q} > 0) > 0\,.$$

ii) *There exists a $u \in V$ such that*

$$\Lambda_u(Z_{u,q} > 0) = 1\,.$$

iii) *For all $v \in V$ then*

$$\rho_v^{-1}\left(\sum_{e\in E_v} p_e^q \operatorname{Lip}(S_e)^{\beta(q)}\right)\rho_{\tau(e)} = 1 \quad \lambda_v\text{-a.s.}$$

Proof. i) $\Rightarrow$ iii) It follows from Lemma 7.7.1 that

$$Z_{v,q}(\omega) = \sum_{e\in E_v} \rho_v^{-1} p_e(\omega)^q \operatorname{Lip}(S_e(\omega))^{\beta(q)} \rho_{\tau(q)}\left(1 \wedge Z_{\tau(e),q}(S_e(\omega))\right) \tag{7.7.1}$$

for $v \in V$ and $\omega \in \Omega_v$. Since $\rho_v^{-1} p_e^q \operatorname{Lip}(S_e)^{\beta(q)} \rho_{\tau(q)} \leq (\overline{\rho}(q)/\underline{\rho}(q))(1 \vee \Delta^q)(1 \vee \Delta^{\beta(q)})$, (7.7.1) shows that $Z_{v,q}$ is bounded and thus Λ_v-integrable.

Claim 1. We have

$$\forall v \in V : Z_{v,q} \leq 1 \quad \Lambda_v\text{-a.s.}.$$

Proof of Claim 1. We have for $v \in V$ (by Lemma 4.3.1),

$$\begin{aligned}
\rho_v \int (1 \wedge Z_{v,q})\, d\Lambda_v &\leq \rho_v \int Z_{v,q}\, d\Lambda_v \\
&= \sum_{e\in E_v} \int p_e^q \operatorname{Lip}(S_e)^{\beta(q)} \rho_{\tau(e)} \left(1 \wedge Z_{\tau(e),q}(S_e)\right) d\Lambda_v \\
&= \sum_{e\in E_v} \iint p_e^q \operatorname{Lip}(S_e)^{\beta(q)} \rho_{\tau(e)} \cdot \\
&\qquad \left(1 \wedge Z_{\tau(e),q}(\omega_{\tau(e)})\right) d\lambda_v((S_\varepsilon)_{\varepsilon\in E_v}, (p_\varepsilon)_{\varepsilon\in E_v}) \cdot \\
&\qquad\qquad d\left(\prod_{\varepsilon\in E_v} \Lambda_{\tau(\varepsilon)}\right)\left(\left(\omega_{\tau(\varepsilon)}\right)_{\varepsilon\in E_v}\right) \\
&= \sum_{e\in E_v} \int p_e^q \operatorname{Lip}(S_e)^{\beta(q)} \rho_{\tau(e)}\, d\lambda_v \cdot \\
&\qquad \int \left(1 \wedge Z_{\tau(e),q}(\omega_{\tau(e)})\right) d\left(\prod_{\varepsilon\in E_v} \Lambda_{\tau(\varepsilon)}\right)\left(\left(\omega_{\tau(\varepsilon)}\right)_{\varepsilon\in E_v}\right)
\end{aligned}$$

$$= \sum_{e\in E_v} \int p_e^q \operatorname{Lip}(S_e)^{\beta(q)} \rho_{\tau(e)}\, d\lambda_v \int \left(1 \wedge Z_{\tau(e),q}\right) d\Lambda_{\tau(e)}$$

$$= \sum_w \left(\sum_{e\in E_{vw}} \int p_e^q \operatorname{Lip}(S_e)^{\beta(q)}\, d\lambda_v \right) \rho_w \int \left(1 \wedge Z_{w,q}\right) d\Lambda_w . \tag{7.7.2}$$

Hence

$$A(q) \left(\rho_v \int (1 \wedge Z_{v,q})\, d\Lambda_v \right)_{v\in V} \geq \left(\rho_v \int (1 \wedge Z_{v,q})\, d\Lambda_v \right)_{v\in V} . \tag{7.7.3}$$

Since $A(q)$ is irreducible with spectral radius $\Phi(q, \beta(q)) = 1$ and the following assertions hold, $\left(\rho_v \int (1 \wedge Z_{v,q})\, d\Lambda_v\right)_{v\in V} \geq \mathbf{0}$ (where $\mathbf{0} = \begin{pmatrix} 0 \\ \vdots \\ 0 \end{pmatrix} \in \mathbb{R}^{\operatorname{card} V}$) with $\rho_v \int (1 \wedge Z_{v,q})\, d\Lambda_v < \infty$ and $\left(\rho_v \int (1 \wedge Z_{v,q})\, d\Lambda_v\right)_{v\in V} \neq \mathbf{0}$ (since $\rho_u \int (1 \wedge Z_{u,q})\, d\Lambda_u > 0$ because $\Lambda_u(Z_{u,q} > 0) > 0$ by i)), (7.7.3) and the superinvariance part of Perron-Frobinus Theorem (cf. e.g. [Se, Exercise 1.17, p. 29]) imply that

$$A(q) \left(\rho_v \int (1 \wedge Z_{v,q})\, d\Lambda_v \right)_{v\in V} = \left(\rho_v \int (1 \wedge Z_{v,q})\, d\Lambda_v \right)_{v\in V}$$

i.e. equality holds in (7.7.2). Hence

$$\rho_v \int (1 \wedge Z_{v,q})\, d\Lambda_v = \rho_v \int Z_{v,q}\, d\Lambda_v \quad \text{for all } v \in V ,$$

and so

$$\int (1 \wedge Z_{v,q})\, d\Lambda_v = \int Z_{v,q}\, d\Lambda_v \quad \text{for all } v \in V$$

which clearly implies that $Z_{v,q} \leq 1$ Λ_u-a.s. for all $v \in V$. This proves Claim 1.

It follows from Claim 1 that

$$\|Z_{v,q}\|_\infty \leq 1 \quad \text{for all } v \in V ,$$

and (7.7.1) and Lemma 4.3.1 therefore imply that

$$\left.\begin{array}{l} \displaystyle\sum_{e\in E_v} \rho_v^{-1} p_e^q \operatorname{Lip}(S_e)^{\beta(q)} \rho_{\tau(e)} Z_{\tau(e),q}(\omega_{\tau(e)}) \leq \|Z_{v,q}\|_\infty \\ \text{for all } v \in V \\ \text{and } \lambda_v\text{-a.a. } \left((S_e)_{e\in E_v}, (p_e)_{e\in E_v}\right) \in \Xi_v \\ \text{and } \displaystyle\prod_{e\in E_v} \Lambda_{\tau(e)}\text{-a.a. } \left(\omega_{\tau(e)}\right)_{e\in E_v} \in \prod_{e\in E_v} \Omega_{\tau(e)} \end{array}\right\} \tag{7.7.4}$$

It follows from (7.7.4) that

$$\left.\begin{array}{l}\displaystyle\sum_{e\in E_v} \rho_v^{-1} p_e^q \operatorname{Lip}(S_e)^{\beta(q)} \rho_{\tau(e)} \|Z_{\tau(e),q}\|_\infty \le \|Z_{v,q}\|_\infty \\ \text{for all } v \in V \text{ and } \lambda_v\text{-a.a. } \left((S_e)_{e\in E_v}, (p_e)_{e\in E_v}\right) \in \Xi_v\end{array}\right\} \tag{7.7.5}$$

whence

$$\begin{aligned}&\sum_w \left(\sum_{e\in E_{vw}} \int p_e^q \operatorname{Lip}(S_e)^{\beta(q)}\, d\lambda_w \right) \rho_w \|Z_{w,q}\|_\infty \\ &\qquad = \sum_{e\in E_v} \int p_e^q \operatorname{Lip}(S_e)^{\beta(q)}\, d\lambda_v \rho_{\tau(e)} \|Z_{\tau(e),q}\|_\infty \\ &\qquad \le \rho_v \|Z_{v,q}\|_\infty\end{aligned}$$

for all $v \in V$. Hence

$$A(q) \left(\rho_v \|Z_{v,q}\|_\infty\right)_{v\in V} \le \left(\rho_v \|Z_{v,q}\|_\infty\right)_{v\in V} . \tag{7.7.6}$$

Since $A(q)$ is irreducible with spectral radius $\Phi(q, \beta(q)) = 1$ and $\left(\rho_v \|Z_{v,q}\|_\infty\right)_{v\in V} \ge \mathbf{0}$ with $\rho_v \|Z_{v,q}\|_\infty < \infty$ for all $v \in V$ and $\left(\rho_v \|Z_{v,q}\|_\infty\right)_{v\in V} \neq \mathbf{0}$ (since $\rho_u \|Z_{u,q}\|_\infty > 0$ because $\Lambda_u(Z_{u,q} > 0) > 0$ by i)), (7.7.6) and the Perron-Frobenius Theorem (cf. [Se]) imply that there exists a constant $c \in]0, \infty[$ satisfying

$$\left(\rho_v \|Z_{v,q}\|_\infty\right)_{v\in V} = c(\rho_v)_{v\in V}$$

i.e.

$$\|Z_{v,q}\|_\infty = c \in]0, \infty[\quad \text{for all} \quad v . \tag{7.7.7}$$

It follows immediately from (7.7.5) and (7.7.7) that

$$\sum_{e\in E_v} \rho_v^{-1} p_e^q \operatorname{Lip}(S_e)^{\beta(q)} \rho_{\tau(e)} \le 1 \quad \lambda_v\text{-a.s. for all } v \in V. \tag{7.7.8}$$

Also, by the definition of $\beta(q)$,

$$\sum_{e\in E_v} \int p_e^q \operatorname{Lip}(S_e)^{\beta(q)}\, d\lambda_v \rho_{\tau(e)} = 1 \quad \text{for all } v \in V.$$

Finally, (7.7.8) and the above inequality give

$$\sum_{e\in E_v} \rho_v^{-1} p_e^q \operatorname{Lip}(S_e)^{\beta(q)} \rho_{\tau(e)} = 1 \quad \lambda_v\text{-a.s.}$$

for all $v \in V$.

iii) $\Rightarrow$ ii) Condition iii) and the definition of $Z_{v,1}$ clearly imply that

$$Z_{v,q,1} = 1 \quad \lambda_v\text{-a.s. for all } v \in V .$$

Hence (by Lemma 7.7.1.ii))

$$Z_{v,q,n} = 1 \quad \lambda_v\text{-a.s. for all } n \in \mathbb{N} \text{ and all } v \in V .$$

This clearly implies that $Z_{v,q} = 1 > 0$ λ_v-a.s. for all v.

ii) $\Rightarrow$ i) Obvious. □

Lemma 7.7.3. *Let $q \in \mathbb{R}$. Then the following statements are equivalent.*

i) *For all $v \in V$ then*

$$\rho_v^{-1}\left(\sum_{e\in E_v} p_e^q \operatorname{Lip}(S_e)^{\beta(q)}\right)\rho_{\tau(e)} = 1 \quad \lambda_v\text{-a.s.}$$

ii) *There exists a $u \in V$ such that*

$$\Lambda_u\left(\sup_{\Sigma\in\mathbf{Max}(u)} \inf_{\substack{\Gamma\in\mathbf{Max}(u)\\ \Sigma\prec\Gamma}} X_{u,q,\Gamma} > 0\right) = 1\,.$$

iii) *There exists a $u \in V$ such that*

$$\Lambda_u\left(\sup_{\Sigma\in\mathbf{Max}(u)} \inf_{\substack{\Gamma\in\mathbf{Max}(u)\\ \Sigma\prec\Gamma}} X_{u,q,\Gamma} > 0\right) > 0\,.$$

Proof. i) $\Rightarrow$ ii) It follows immediately from i) that for each $\Gamma \in \mathbf{Max}(v)$,

$$X_{v,q,\Gamma} = \sum_{\gamma\in\Gamma} \rho_u^{-1}\left(\prod_{i=1}^{|\gamma|} p_{\gamma|i}^q \operatorname{Lip}(S_{\gamma|i})^{\beta(q)}\right)\rho_{\tau(\gamma_{|\gamma|})} = 1 \quad \Lambda_v\text{-a.s.}$$

for all $v \in V$. Hence

$$\sup_{\Sigma\in\mathbf{Max}(v)} \inf_{\substack{\Gamma\in\mathbf{Max}(v)\\ \Sigma\prec\Gamma}} X_{v,q,\Gamma} = 1 \quad \Lambda_v\text{-a.s.}$$

for all v. This proves ii).
ii) $\Rightarrow$ iii) Obvious.
iii) $\Rightarrow$ i) If $\Sigma \in \mathbf{Max}(u)$ and $\omega \in \Omega_u$ then

$$\inf_{\substack{\Gamma\in\mathbf{Max}(u)\\ \Sigma\prec\Gamma}} X_{u,q,\Gamma}(\omega)$$

$$= \inf_{\substack{\Gamma\in\mathbf{Max}(u)\\ \Sigma\prec\Gamma}} \sum_{\gamma\in\Gamma} \rho_u^{-1}\left(\prod_{i=1}^{|\gamma|} p_{\gamma|i}(\omega)^q \operatorname{Lip}(S_{\gamma|i}(\omega))^{\beta(q)}\right)\rho_{\tau(\gamma_{|\gamma|})}$$

$$= \inf_{\substack{\Gamma\in\mathbf{Max}(u)\\ \Sigma\prec\Gamma}} \sum_{\sigma\in\Sigma}\left(\rho_u^{-1}\left(\prod_{i=1}^{|\sigma|} p_{\sigma|i}(\omega)^q \operatorname{Lip}(S_{\sigma|i}(\omega))^{\beta(q)}\right)\rho_{\tau(\sigma_{|\sigma|})}\cdot\right.$$

$$\left.\sum_{\alpha\in\Gamma(\sigma)} \rho_{\tau(\sigma_{|\sigma|})}^{-1}\left(\prod_{i=1}^{|\alpha|} p_{\sigma(\alpha|i)}(\omega)^q \operatorname{Lip}(S_{\sigma(\alpha|i)}(\omega))^{\beta(q)}\right)\rho_{\tau(\sigma\alpha_{|\sigma\alpha|})}\right)$$

$$= \sum_{\sigma\in\Sigma} \rho_u^{-1} \left(\prod_{i=1}^{|\sigma|} p_{\sigma|i}(\omega)^q \operatorname{Lip}(S_{\sigma|i}(\omega))^{\beta(q)} \right) \rho_{\tau(\sigma_{|\sigma|})} \cdot$$

$$\inf_{\substack{\Gamma\in\mathbf{Max}(u)\\ \Sigma\prec\Gamma}} \left(\sum_{\alpha\in\Gamma(\sigma)} \rho_{\tau(\sigma_{|\sigma|})}^{-1} \left(\prod_{i=1}^{|\alpha|} p_{\sigma(\alpha|i)}(\omega)^q \operatorname{Lip}(S_{\sigma(\alpha|i)}(\omega))^{\beta(q)} \right) \rho_{\tau(\sigma\alpha_{|\sigma\alpha|})} \right)$$

$$= \sum_{\sigma\in\Sigma} \rho_u^{-1} \left(\prod_{i=1}^{|\sigma|} p_{\sigma|i}(\omega)^q \operatorname{Lip}(S_{\sigma|i}(\omega))^{\beta(q)} \right) \rho_{\tau(\sigma_{|\sigma|})} \cdot$$

$$\left(1 \wedge \inf_{\Pi\in\mathbf{Max}(\tau(\sigma_{|\sigma|}))\setminus\{\{\varnothing\}\}} \sum_{\alpha\in\Pi(\sigma)} \rho_{\tau(\sigma_{|\sigma|})}^{-1} \left(\prod_{i=1}^{|\alpha|} p_{\sigma(\alpha|i)}(\omega)^q \operatorname{Lip}(S_{\sigma(\alpha|i)}(\omega))^{\beta(q)} \right) \cdot \rho_{\tau(\sigma\alpha_{|\sigma\alpha|})} \right)$$

$$= \sum_{\sigma\in\Sigma} \rho_u^{-1} \left(\prod_{i=1}^{|\sigma|} p_{\sigma|i}(\omega)^q \operatorname{Lip}(S_{\sigma|i}(\omega))^{\beta(q)} \right) \rho_{\tau(\sigma_{|\sigma|})} \left(1 \wedge Z_{\tau(\sigma_{|\sigma|}),q}(S_\sigma(\omega)) \right) . \tag{7.7.9}$$

By iii) there exists $\Omega_u'' \subseteq \Omega_u$ with $\Lambda_u(\Omega_u'') > 0$ satisfying the following condition:

$$\begin{gathered}\text{for each } \omega \in \Omega_u'' \text{ then there exists}\\ \text{an antichain } \Sigma_\omega \in \mathbf{Max}(u) \text{ with}\\ \inf_{\substack{\Gamma\in\mathbf{Max}(u)\\ \Sigma_\omega\prec\Gamma}} X_{u,q,\Gamma}(\omega) > 0 .\end{gathered}$$

Equation (7.7.9) therefore implies that for each $\omega \in \Omega_u''$ there exists a $\Sigma_\omega \in \mathbf{Max}(u)$ and a $\sigma_\omega \in \Sigma_\omega$ such that

$$Z_{\tau((\sigma_\omega)_{|\sigma_\omega|}),q}(S_{\sigma_\omega}(\omega)) > 0 . \tag{7.7.10}$$

For $\gamma \in E_u^{(*)}$ write

$$\Omega_u(\gamma) = \{\omega \in \Omega_u \mid Z_{\tau(\gamma_{|\gamma|}),q}(S_\gamma(\omega)) > 0\} .$$

By (7.7.10), $\bigcup_{\gamma\in E_u^{(*)}} \Omega_u(\gamma) \supseteq \Omega_u''$, whence

$$\sum_{\gamma\in E_u^{(*)}} \Lambda_u(\Omega_u(\gamma)) \geq \Lambda_u \left(\bigcup_{\gamma\in E_u^{(*)}} \Omega_u(\gamma) \right) \geq \Lambda_u(\Omega_u'') > 0 .$$

We can thus choose $\gamma_0 \in E_u^{(*)}$ with

$$0 < \Lambda_u(\Omega_u(\gamma_0)) = \Lambda_u(Z_{\tau((\gamma_0)_{|\gamma_0|}),q}(\mathcal{S}_{\gamma_0}) > 0)$$
$$= (\Lambda_u \circ \mathcal{S}_{\gamma_0}^{-1})(Z_{\tau((\gamma_0)_{|\gamma_0|}),q} > 0) = \Lambda_{\tau((\gamma_0)_{|\gamma_0|})}(Z_{\tau((\gamma_0)_{|\gamma_0|}),q} > 0)$$

and Lemma 7.7.2 therefore implies that condition i) is satisfied. □

We are now ready to prove Theorem 2.7.1 and Theorem 2.7.2.

Proof of Theorem 2.7.1

We must prove that

$$\mathcal{H}_{\mu_v(\omega)}^{q,\beta(q)}(C_v(\omega)) = 0$$

for all $v \in V$ and Λ_v-a.a. $\omega \in \Omega_v$. Equation (7.2.14) in the proof of Theorem 7.2.1 shows that there exists a constant $\overline{K} \in]0,\infty[$ such that

$$\mathcal{H}_{\mu_v(\omega)}^{q,\beta(q)}(C_v(\omega)) \le \overline{K} \sup_{\Sigma \in \mathbf{Max}(v)} \inf_{\substack{\Gamma \in \mathbf{Max}(v) \\ \Sigma \prec \Sigma}} X_{v,q,\Gamma}(\omega) \tag{7.7.11}$$

for all $v \in V$ and Λ_v-a.a. $\omega \in \Omega_v$. Also, by Lemma 7.7.3,

$$\Lambda_v \left(\sup_{\Sigma \in \mathbf{Max}(v)} \inf_{\substack{\Gamma \in \mathbf{Max}(v) \\ \Sigma \prec \Sigma}} X_{v,q,\Gamma}(\omega) > 0 \right) = 0 \tag{7.7.12}$$

for all $v \in V$. Now equations (7.7.11) and (7.7.12) imply that

$$\mathcal{H}_{\mu_v(\omega)}^{q,\beta(q)}(C_u(\omega)) = 0$$

for all $v \in V$ and Λ_v-a.a. $\omega \in \Omega_v$. □

Proof of Theorem 2.7.2

We will prove the following implications

$$\begin{array}{ccccc} \text{i)} & \Rightarrow & \text{ii)} & \Rightarrow & \text{iii)} \\ & & \Downarrow & & \Downarrow \\ & & \text{iv)} & \Rightarrow & \text{v)} & \Rightarrow & \text{i)} \end{array}$$

i) ⇒ ii) It follows from Theorem 7.4.2 that there exists a constant $\underline{K} \in]0,\infty[$ such that

$$\underline{K}\mathcal{M}_{v,q}(\omega) \le \mathcal{H}_{\mu_v(\omega)}^{q,\beta(q)} \quad \text{on } C_u(\omega) \text{ for all } v \text{ and } \Lambda_v\text{-a.a. } \omega \in \Omega_v. \tag{7.7.13}$$

Theorem 6.3.1 and Proposition 5.2.3 imply that

$$\mathcal{M}_{v,q}(\omega)(\Delta_{\mu_v(\omega)}(\alpha(q))) = X_{v,q}(\omega) > 0 \quad \text{for all } v \text{ and } \Lambda_v\text{-a.a. } \omega \in \Omega_v. \qquad (7.7.14)$$

Equations (7.7.13) and (7.7.14) imply that

$$\mathcal{H}^{q,\beta(q)}_{\mu_v(\omega)}(\Delta_{\mu_v(\omega)}(\alpha(q))) \geq \underline{K} X_{v,q}(\omega) > 0$$

for all v and Λ_v-a.a. $\omega \in \Omega_v$. This proves ii).
ii) $\Rightarrow$ iii) Obvious.
ii) $\Rightarrow$ iv) Obvious.
iii) $\Rightarrow$ v) Obvious.
iv) $\Rightarrow$ v) Obvious.
v) $\Rightarrow$ i) Follows immediately from Theorem 2.7.1. □

7.8 Proofs of the Theorems in section 2.8

The purpose of this section is to prove Theorem 2.8.1. The proof is based on the next two lemmas.

Lemma 7.8.1. *Let $X_v \subseteq \mathbb{R}^d$ for all $v \in V$. Let $q \in \mathbb{R}$, $u, v \in V$ and $e \in E_{uv}$.*

i) $\liminf_{r \searrow 0} \inf_{x \in \operatorname{supp} \mu_v(\mathcal{S}_e(\omega))} \left(\frac{\mu_u(\omega)\big(S_e(\omega)(B(x,r))\big)}{\mu_v(\mathcal{S}_e(\omega))\big(B(x,r)\big)} \right)^q$
$= p_e^q(\omega)$ for Λ_u-a.a. $\omega \in \Omega_u$.

ii) $\limsup_{r \searrow 0} \sup_{x \in \operatorname{supp} \mu_v(\mathcal{S}_e(\omega))} \left(\frac{\mu_u(\omega)\big(S_e(\omega)(B(x,r))\big)}{\mu_v(\mathcal{S}_e(\omega))\big(B(x,r)\big)} \right)^q$
$= p_e^q(\omega)$ for Λ_u-a.a. $\omega \in \Omega_u$.

Proof. First observe that

$$C_u(\omega) = \bigcup_{e \in E_u} S_e(\omega)\Big(C_{\tau(e)}(\mathcal{S}_e(\omega))\Big) \tag{7.8.1}$$

for $\omega \in \Omega_u$. Let

$$\Omega_u'' = \{\omega \in \Omega_u' \mid 1) \ \omega \text{ satisfies (IV)};$$
$$2) \ \big((S_e(\omega))_{e \in E_u}, (p_e(\omega))_{e \in E_u}\big) \text{ satisfies (I), (III) and (V)}\}$$

Then $\Lambda_u(\Omega_u'') = 1$. Fix $\omega \in \Omega_u''$. Write $S_\varepsilon(\omega) = S_\varepsilon$ for $\varepsilon \in E_u$. Let $0 < r < \frac{c}{2T}$ and $x \in C_v(\mathcal{S}_e(\omega))$. We claim that

$$\frac{\mu_u(\omega)(S_e(U(x,r))}{\mu_v(\mathcal{S}_e(\omega))(U(x,r))} = p_e(\omega) \tag{7.8.2}$$

where $U(x,r)$ denotes the open ball with center x and radius r. The conclusion in the lemma follows immediately from (7.8.2).

Let $A := \{\alpha \in E_u^{(*)} \mid C_\alpha(\mathcal{S}_e(\omega)) \subseteq U(x,r)\}$. Since $U(x,r)$ is open, $U(x,r) \cap C_v(\mathcal{S}_e(\omega)) = \bigcup_{\alpha \in A}\big(C_\alpha(\mathcal{S}_e(\omega)) \cap C_v(\mathcal{S}_e(\omega))\big)$. Now we need only cover $U(x,r)$ once: if $\alpha, \beta \in A$ and $[\alpha] \cap [\beta] \neq \varnothing$ then one of them is contained in the other so we may discard the smaller one. So there is a subset $A_0 \subseteq A$ such that

$$U(x,r) \bigcap C_v(\mathcal{S}_e(\omega)) = \bigcup_{\alpha \in A} \Big(C_\alpha(\mathcal{S}_e(\omega)) \cap C_v(\mathcal{S}_e(\omega))\Big)$$

and $[\alpha] \cap [\beta] \neq \varnothing$ for $\alpha, \beta \in A_0$, i.e. $\big(C_\alpha(\mathcal{S}_e(\omega))\big)_{\alpha \in A_0}$ is a pairwise disjoint family by (IV).

For $a > 0$ write $B(S_e(X_v), a) := \{z \in \mathbb{R}^d \mid \mathrm{dist}(z, S_e(X_v)) \leq a\}$. Observe that

$$\left.\begin{array}{l} B(S_e(X_v), \frac{c}{2}) \bigcap S_\varepsilon(C_w(S_\varepsilon(\omega))) \subseteq B(S_e(X_v), \frac{c}{2}) \bigcap S_\varepsilon(X_w) = \varnothing \\ \text{for } w \in V \text{ and } \varepsilon \in E_{uw} \text{ with } (e, v) \neq (\varepsilon, w) \end{array}\right\} \tag{7.8.3}$$

Since in particular $\bigcup_{\alpha \in A_0} C_{e\alpha}(\omega) = \bigcup_{\alpha \in A_0} S_e\big(C_\alpha(S_e(\omega))\big) = S_e\big(\bigcup_{\alpha \in A_0} C_\alpha(S_e(\omega))\big) \subseteq S_e(U(x, r)) = U\big(S_e(x), \mathrm{Lip}(S_e) r\big) \subseteq B(S_e(X_v), \frac{c}{2})$, (7.8.3) implies that

$$\left.\begin{array}{l} \left(\bigcap_{\alpha \in A_0} C_{e\alpha}(\omega)\right) \bigcap S_\varepsilon(C_w(S_\varepsilon(\omega))) = \varnothing \\ \text{for } w \in V \text{ and } \varepsilon \in E_{uw} \text{ with } (e, v) \neq (\varepsilon, w) \end{array}\right\} \tag{7.8.4}$$

We will now prove that

$$S_e(U(x, r)) \bigcap C_u(\omega) = S_e\left(U(x, r) \bigcap C_v(S_e(\omega))\right). \tag{7.8.5}$$

Indeed, it is clear that $S_e\big(U(x, r)\big) \cap C_u(\omega) \supseteq S_e\big(U(x, r) \cap C_v(S_e(\omega))\big)$ by (7.8.1). Now let $y \in S_e\big(U(x, r)\big) \cap C_u(\omega)$. We must prove that $y \in S_e\big(C_v(S_e(\omega))\big)$. We have (using (7.8.1))

$$\begin{aligned} y &\in S_e\big(U(x, r)\big) \bigcap C_u(\omega) \\ &\subseteq U\big(S_e(x), \mathrm{Lip}(S_e) r\big) \bigcap \left(\bigcup_{w \in V} \bigcup_{\varepsilon \in E_{vw}} S_\varepsilon\big(C_w(S_\varepsilon(\omega))\big)\right) \\ &\subseteq \bigcup_{w \in V} \bigcup_{\varepsilon \in E_{vw}} \left(B(S_e(X_v), \tfrac{c}{2}) \bigcap S_\varepsilon\big(C_w(S_\varepsilon(\omega))\big)\right) \end{aligned}$$

and (7.8.3) therefore implies that $y \in S_e\big(C_v(S_e(\omega))\big)$.

It follows from (7.8.4) and (7.8.5) that

$$\begin{aligned} \frac{\mu_u(\omega)(S_e(U(x, r))}{\mu_v(S_e(\omega))(U(x, r))} &= \frac{\mu_u(\omega)\Big(S_e\big(U(x, r) \cap C_v(S_e(\omega))\big)\Big)}{\mu_v(S_e(\omega))\Big(U(x, r) \cap C_v(S_e(\omega))\Big)} \qquad \text{[by (7.8.5)]} \\ &= \frac{\mu_u(\omega)\Big(S_e\Big(\bigcup_{\alpha \in A_0}\big(C_\alpha(S_e(\omega)) \cap C_v(S_e(\omega))\big)\Big)\Big)}{\mu_v(S_e(\omega))\Big(\bigcup_{\alpha \in A_0}\big(C_\alpha(S_e(\omega)) \cap C_v(S_e(\omega))\big)\Big)} \\ &= \frac{\mu_u(\omega)\Big(\Big(\bigcup_{\alpha \in A_0} S_e\big(C_\alpha(S_e(\omega))\big)\Big) \cap S_e\big(C_v(S_e(\omega))\big)\Big)}{\mu_v(S_e(\omega))\Big(\Big(\bigcup_{\alpha \in A_0} C_\alpha(S_e(\omega))\Big) \cap C_v(S_e(\omega))\Big)} \end{aligned}$$

$$= \frac{\mu_u(\omega)\Big(\Big(\bigcup_{\alpha\in A_0} C_{e\alpha}(\omega)\Big) \cap S_e(C_v(\mathcal{S}_e(\omega)))\Big)}{\mu_v(\mathcal{S}_e(\omega))\Big(\Big(\bigcup_{\alpha\in A_0} C_\alpha(\mathcal{S}_e(\omega))\Big) \cap C_v(\mathcal{S}_e(\omega))\Big)}$$

$$\left[\begin{array}{l}\text{by (7.8.4) since } \operatorname{supp}\mu_u(\omega) = C_u(\omega) = \\ \bigcup_w \bigcup_{\varepsilon\in E_{uw}} S_\varepsilon(C_w(\mathcal{S}_\varepsilon(\omega)))\end{array}\right]$$

$$= \frac{\mu_u(\omega)\Big(\bigcup_{\alpha\in A_0} C_{e\alpha}(\omega)\Big)}{\mu_v(\mathcal{S}_e(\omega))\Big(\bigcup_{\alpha\in A_0} C_\alpha(\mathcal{S}_e(\omega))\Big)}$$

$$= \frac{\sum_{\alpha\in A_0} p_e(\omega) p_{e(\alpha|1)}(\omega)\cdot\ldots\cdot p_{e(\alpha||\alpha|)}(\omega)}{\sum_{\alpha\in A_0} p_{e(\alpha|1)}(\omega)\cdot\ldots\cdot p_{e(\alpha||\alpha|)}(\omega)} = p_e(\omega) \quad \square$$

For $q \in \mathbb{R}$ define $U_q : \prod_{u\in V} L^1(\Omega_u, \Lambda_u) \to \prod_{u\in V} L^1(\Omega_u, \Lambda_u)$ by

$$\big(U_q((f_v)_v)\big)_u(\omega) = \sum_w \sum_{e\in E_{vw}} p_e(\omega)^q \operatorname{Lip}(S_e(\omega))^{\beta(q)} f_w(\mathcal{S}_e(\omega)).$$

Lemma 7.8.2. *Let $X_v \subseteq \mathbb{R}^d$ for all $v \in V$ Let $q \in \mathbb{R}$ and assume that*

$$\rho_v^{-1}\left(\sum_{e\in E_v} p_e^q \operatorname{Lip}(S_e)^{\beta(q)}\right)\rho_{\tau(e)} = 1 \quad \lambda_v\text{-a.s.} \tag{7.8.6}$$

for all $v \in V$. Let $(f_u)_u \in \prod_{u\in V} L^1(\Omega_u, \Lambda_u)_+$ with $(f_u)_u \neq 0$ be U_q invariant, i.e.

$$U_q((f_u)_u) = (f_u)_u .$$

Then there exists a vector $(c_{u,q})_u \in \mathbb{R}^{\operatorname{card} V}$ such that

$$f_u = c_{u,q} \quad \Lambda_u\text{-a.e.}.$$

Proof. We have (by independence)

$$\begin{aligned}
\mathbf{E}_{\Lambda_u}[f_u] &= \mathbf{E}_{\Lambda_u}[U_q((f_v)_v)_u] \\
&= \mathbf{E}_{\Lambda_u}\left[\sum_v \sum_{e\in E_{uv}} p_e^q \operatorname{Lip}(S_e)^{\beta(q)} f_v(\mathcal{S}_e)\right] \\
&= \sum_v \sum_{e\in E_{uv}} \int p_e^q \operatorname{Lip}(S_e)^{\beta(q)}\, d\lambda_u \mathbf{E}_{\Lambda_u}[f_v(\mathcal{S}_e)] \\
&= \sum_v \sum_{e\in E_{uv}} A_{uv}(q) \mathbf{E}_{\Lambda_v}[f_v].
\end{aligned}$$

for all $u \in V$. Hence

$$A(q)\left(\mathbf{E}_{\Lambda_u}[f_u]\right)_u = \left(\mathbf{E}_{\Lambda_u}[f_u]\right)_u . \tag{7.8.7}$$

Since $A(q)$ is irreducible with spectral radius $\Phi(q,\beta(q)) = 1$ and $\left(\mathbf{E}_{\Lambda_u}[f_u]\right)_u \geq \mathbf{0}$ (where $\mathbf{0} = \begin{pmatrix} 0 \\ \vdots \\ 0 \end{pmatrix} \in \mathbb{R}^{\operatorname{card} V}$) with $\left(\mathbf{E}_{\Lambda_u}[f_u]\right)_u \neq \mathbf{0}$ (since $(f_u)_u \neq 0$ and $f_u \geq 0$), (7.8.7) and Perron-Frobenius Theorem imply that there exists $c_q > 0$ such that

$$\left(\mathbf{E}_{\Lambda_u}[f_u]\right)_u = c_q(\rho_u(q))_u . \tag{7.8.8}$$

For each $n \in \mathbb{N}$ then clearly

$$\left(U_q((f_v)_v)\right)_u(\omega) = \sum_{\gamma \in E_u^{(n)}} \left(\prod_{i=1}^{n} p_{\gamma|i}(\omega)^q \operatorname{Lip}(S_{\gamma|i}(\omega))^{\beta(q)} \right) f_{\tau(\gamma_n)}(S_\gamma(\omega)) \tag{7.8.9}$$

for $u \in V$ and $\omega \in \Omega_u$. Also observe that (7.8.6) implies that

$$\rho_u^{-1} \sum_{\gamma \in E_u^{(n)}} \left(\prod_{i=1}^{n} p_{\gamma|i}^q \operatorname{Lip}(S_{\gamma|i})^{\beta(q)} \right) \rho_{\tau(\gamma_n)} = 1 \quad \Lambda_u\text{-a.s.} \tag{7.8.10}$$

for all $u \in V$ and $n \in \mathbb{N}$. It follows from (7.8.8) through (7.8.10) and independence that

$$\begin{aligned}
\mathbf{E}_{\Lambda_u}\left[f_u \,|\mathcal{A}_{u,n}\right] =& \mathbf{E}_{\Lambda_u}\left[\left(U_q((f_v)_v)\right) |\mathcal{A}_{u,n}\right] \\
=& \sum_{\gamma \in E_u^{(n)}} \left(\prod_{i=1}^{n} p_{\gamma|i}^q \operatorname{Lip}(S_{\gamma|i})^{\beta(q)} \right) \mathbf{E}_{\Lambda_u}\left[f_{\tau(\gamma_n)}(S_\gamma) \,|\mathcal{A}_{u,n}\right] \\
=& \sum_{\gamma \in E_u^{(n)}} \left(\prod_{i=1}^{n} p_{\gamma|i}^q \operatorname{Lip}(S_{\gamma|i})^{\beta(q)} \right) \mathbf{E}_{\Lambda_u}\left[f_{\tau(\gamma_n)}(S_\gamma)\right] \\
=& \sum_{\gamma \in E_u^{(n)}} \left(\prod_{i=1}^{n} p_{\gamma|i}^q \operatorname{Lip}(S_{\gamma|i})^{\beta(q)} \right) \mathbf{E}_{\Lambda_{\tau(\gamma_n)}}\left[f_{\tau(\gamma_n)}\right] \\
=& c_q \sum_{\gamma \in E_u^{(n)}} \left(\prod_{i=1}^{n} p_{\gamma|i}^q \operatorname{Lip}(S_{\gamma|i})^{\beta(q)} \right) \rho_{\tau(\gamma_n)} \\
=& c_q \rho_u \\
=& \mathbf{E}_{\Lambda_u}[f_u] .
\end{aligned} \tag{7.8.11}$$

Finally note that

$$\mathbf{E}_{\Lambda_u}[f_u \,|\mathcal{A}_{u,n}] \to \mathbf{E}_{\Lambda_u}[f_u \,|\sigma(\cup_m \mathcal{A}_{u,m})]$$
$$= f_u \quad \text{as} \quad n \to \infty\,. \tag{7.8.12}$$

Now (7.8.11) and (7.8.12) show that

$$f_u = \mathbf{E}_{\Lambda_u}[f_u] \quad \Lambda_u\text{-a.e.}. \ \square$$

We are now ready to prove Theorem 2.8.1.

Proof of Theorem 2.8.1

We divide the proof into two cases

Case A. There exists a $u \in V$ such that

$$\lambda_u\left(\rho_u^{-1}\left(\sum_{e\in E_u} p_e^q \operatorname{Lip}(S_e)^{\beta(q)}\right)\rho_{\tau(e)} \neq 1\right) > 0\,.$$

Proof of Theorem 2.8.1 in Case A. Theorem 2.7.1 implies that

$$\mathcal{H}_\mu^{q,\beta(q)}(\operatorname{supp}\mu) = 0$$

for all v and P_v-a.a. $\mu \in \mathcal{P}(X_v)$.

Case B. For all $v \in V$,

$$\lambda_v\left(\rho_v^{-1}\left(\sum_{e\in E_v} p_e^q \operatorname{Lip}(S_e)^{\beta(q)}\right)\rho_{\tau(e)} = 1\right) = 1\,.$$

Proof of Theorem 2.8.1 in Case B. For each $u \in V$ define $f_u : \Omega_u \to \mathbb{R}_+$ by

$$f_u(\omega) = \mathcal{H}_{\mu_u(\omega)}^{q,\beta(q)}(\operatorname{supp}\mu_u(\omega)) = \mathcal{H}_{\mu_u(\omega)}^{q,\beta(q)}(C_u(\omega))\,.$$

It suffices by Lemma 7.8.2 to prove the following two claims

Claim 1. $f_u \in L^1(\Omega_u, \Lambda_u)_+$ and $(f_u)_u \neq 0$.

Claim 2. $(f_u)_u$ is U_q invariant.

Proof of Claim 1. Theorem 7.2.1 implies that,

$$f_u(\omega) = \mathcal{H}^{q\beta(q)}_{\mu_u(\omega)}(C_u(\omega)) \le \overline{K}\mathcal{M}_{u,q}(\omega)(X_u) = \overline{K}X_{u,q}(\omega) \quad \text{for } \Lambda_u\text{-a.a. } \omega \in \Omega_u$$

for all $u \in V$. Since $X_{u,q} \in L^1(\Omega_u, \Lambda_u)$, the above inequality shows that $f_u \in L^1(\Omega_u, \Lambda_u)$. Proposition 5.2.3 implies that $f_u > 0$ λ_u-a.e. for all u, whence $(f_u)_u \neq 0$. This proves Claim 1.

Proof of Claim 2. Fix $u \in V$. It is easily seen that

$$C_u(\omega) = \bigcup_{e \in E_u} S_e(\omega)\Big(C_{\tau(e)}(\mathcal{S}_e(\omega))\Big)\,. \tag{7.8.13}$$

for $\omega \in \Omega_u$. Now (7.8.13), Lemma 7.8.1 and [Ol, Lemma 4.3] imply that

$$\begin{aligned} f_u(\omega) &= \mathcal{H}^{q,\beta(q)}_{\mu_u(\omega)}\left(\bigcup_{e \in E_u} S_e(\omega)\Big(C_{\tau(e)}(\mathcal{S}_e(\omega))\Big)\right) \\ &\le \sum_{e \in E_u} \mathcal{H}^{q,\beta(q)}_{\mu_u(\omega)}\Big(S_e(\omega)\Big(C_{\tau(e)}(\mathcal{S}_e(\omega))\Big)\Big) \\ &= \sum_{e \in E_u} p_e(\omega)^q \operatorname{Lip}(S_e(\omega))^{\beta(q)} \mathcal{H}^{q,\beta(q)}_{\mu_u(\mathcal{S}_e(\omega))}(C_{\tau(e)}(\mathcal{S}_e(\omega))) \\ &= \sum_v \sum_{e \in E_{uv}} p_e(\omega)^q \operatorname{Lip}(S_e(\omega))^{\beta(q)} f_v(\mathcal{S}_e(\omega)) \end{aligned} \tag{7.8.14}$$

for Λ_u-a.a. $\omega \in \Omega_u$, whence (by independence)

$$\begin{aligned} \mathbf{E}_{\Lambda_u}[f_u] &\le \sum_v \sum_{e \in E_{uv}} \mathbf{E}_{\Lambda_u}\left[p_e^q \operatorname{Lip}(S_e)^{\beta(q)} f_v(\mathcal{S}_e)\right] \\ &= \sum_v \sum_{e \in E_{uv}} \int p_e^q \operatorname{Lip}(S_e)^{\beta(q)}\, d\lambda_u \mathbf{E}_{\Lambda_u}[f_v(\mathcal{S}_e)] \\ &= \sum_v \sum_{e \in E_{uv}} A_{uv}(q)\mathbf{E}_{\Lambda_v}[f_v]\,. \end{aligned} \tag{7.8.15}$$

for all $u \in V$. Hence

$$A(q)\left(\mathbf{E}_{\Lambda_u}[f_u]\right)_u \le \left(\mathbf{E}_{\Lambda_u}[f_u]\right)_u\,. \tag{7.8.16}$$

Since $A(q)$ is irreducible with spectral radius $\Phi(q, \beta(q)) = 1$ and $\left(\mathbf{E}_{\Lambda_u}[f_u]\right)_u \ge \mathbf{0}$ with $\left(\mathbf{E}_{\Lambda_u}[f_u]\right)_u \neq \mathbf{0}$ (since $(f_u)_u \neq 0$ by Claim A), (7.8.16) and Perron-Frobenius Theorem imply that

$$A(q)\left(\mathbf{E}_{\Lambda_u}[f_u]\right)_u = \left(\mathbf{E}_{\Lambda_u}[f_u]\right)_u\,.$$

i.e. equality-sign holds in (7.8.15). Hence

$$\mathbf{E}_{\Lambda_u}[f_u] = \mathbf{E}_{\Lambda_u}\left[\sum_v \sum_{e\in E_{uv}} p_e^q \operatorname{Lip}(S_e)^{\beta(q)} f_v(\mathcal{S}_e)\right] \tag{7.8.17}$$

for all $u \in V$. By combining (7.8.14) and (7.8.17),

$$f_u = \sum_v \sum_{e\in E_{uv}} p_e^q \operatorname{Lip}(S_e)^{\beta(q)} f_v(\mathcal{S}_e) \quad \Lambda_u\text{-a.s.}$$

for all $u \in V$. This proves Claim 2.

This completes the proof of Theorem 2.8.1. □

7.9 Proofs of the Theorems in section 2.9

The purpose of this section is to prove Theorem 2.9.1.

Proof of Theorem 2.9.1

Case 1. i) Follows immediately from Theorem 2.6.5.
Case 2. i) Follows immediately from Theorem 2.6.5.
ii) Let

$$M := \{\mu \in \mathcal{P}(X_u) \mid b_\mu = B_\mu = \hat{\beta}\}. \tag{7.9.1}$$

Then $P_u(M) = 1$ by Theorem 2.6.5. It is obvious from the construction of $\hat{\beta}$ and $q_{\min}$ and $q_{\max}$ that $\hat{\beta}$ is differentiable. The functions b_μ and B_μ are thus differentiable for $\mu \in M$, and

$$b'_\mu(q) = B'_\mu(q) = \hat{\beta}'(q) = \begin{cases} \frac{\beta(q_{\min})}{q_{\min}} & -\infty < q \le q_{\min} \\ \beta'(q) = -\alpha(q) & q_{\min} < q < q_{\max} \\ \frac{\beta(q_{\max})}{q_{\max}} & q_{\max} \le q < \infty \end{cases} \tag{7.9.2}$$

However, if $\mu \in M$ then $b_\mu = B_\mu = \hat{\beta}$ cannot be infinitely differentiable at $q_{\min}$. Otherwise there exists a measure $\mu \in M$ such that the functions $b_\mu = B_\mu = \hat{\beta}$ are infinitely differentiable at $q_{\min}$, and (7.9.2) therefore implies that

$$-\beta^{(n+1)}(q_{\min}) = \alpha^{(n)}(q_{\min}) = 0 \quad \text{for } n \in \mathbb{N},$$

whence $\beta(q) = \beta(0) + \beta'(0)q$ for all q (since β is real analytic), contradicting Proposition 2.5.5. Hence, all $\mu \in M$ have a Hausdorff phase transition of order larger than 2 at $q_{\min}$ and a packing phase transition of order larger than 2 at $q_{\min}$. For $\mu \in M$ let n_μ denote the order of the Hausdorff phase transition of μ at $q_{\min}$, and let N_μ denote the order of the packing phase transition of μ at $q_{\min}$. Since $b_\mu = B_\mu = \hat{\beta}$ for $\mu \in M$, the numbers n_μ and N_μ coincide for all $\mu \in M$, i.e. there exists an integer n such that $n_\mu = N_\mu = n$ for all $\mu \in M$.
iii) Define M as in (7.9.1). Then $P_u(M) = 1$ by Theorem 2.6.5. We conclude as above that the functions b_μ and B_μ are differentiable for $\mu \in M$, and

$$b'_\mu(q) = B'_\mu(q) = \hat{\beta}'(q) = \begin{cases} \frac{\beta(q_{\min})}{q_{\min}} & -\infty < q \le q_{\min} \\ \beta'(q) = -\alpha(q) & q_{\min} < q < q_{\max} \\ \frac{\beta(q_{\max})}{q_{\max}} & q_{\max} \le q < \infty \end{cases} \tag{7.9.3}$$

However, if $\mu \in M$ then $b'_\mu = B'_\mu = \hat{\beta}'$ cannot be differentiable at $q_{\min}$. Otherwise there exists a measure $\mu \in M$ such that the functions $b'_\mu = B'_\mu = \hat{\beta}'$ are differentiable at $q_{\min}$, and (7.9.3) therefore implies that

$$\alpha'(q_{\min}) = 0,$$

contradicting the assumption that $\alpha'(q_{\min}) < 0$. Hence, all $\mu \in M$ have a 2-order Hausdorff phase transition at $q_{\min}$ and a 2-order packing phase transition at $q_{\min}$.
iv)-v) Similar to the proofs of ii) and iii). □

List of Notation

1_A	characteristic function for the set A.
$\prec$	26,137
$\|\cdot\|_\infty$	essential supremum.
$\|\cdot\|_{-\infty}$	essential infimum.
$[\cdot]$	26
$\lll$	188
$\lfloor$	denotes the restriction of a measure μ to a subset E, i.e. $(\mu\lfloor E)(A) := \mu(E \cap A)$. 52
$\vert\cdot\vert$	26,137
$\vee$	$\vee_{a\in A} a := \sup_{a\in A} a$ for $A \subseteq [-\infty, \infty]$.
$\wedge$	$\wedge_{a\in A} a := \inf_{a\in A} a$ for $A \subseteq [-\infty, \infty]$.
$\varnothing$	the empty set.
α	$\alpha := -\beta'$. 42
$\alpha_\mu(x)$	local dimension of the measure μ at x (if it exists). 1
$\underline{\alpha}_\mu(x)$	lower local dimension of the measure μ at x. 1
$\overline{\alpha}_\mu(x)$	upper local dimension of the measure μ at x. 1
$\underline{a}$	43
$\overline{a}$	43
$\underline{a}_n$	123
$\overline{a}_n$	123
$a_{\max}$	45
$a_{\min}$	45
$\underline{a}_\mu$	21
$\overline{a}_\mu$	21
$\underline{A}_\mu$	21
$\overline{A}_\mu$	21
$\underline{a}_u(\mu)$	212
$\overline{a}_u(\mu)$	212
$\underline{A}_u(\mu)$	212
$\overline{A}_u(\mu)$	212
$\mathcal{A}_{u,\Gamma}$	σ-algebra generated by the family $(\pi_{u,\alpha\|\|\alpha\|-1})_{\alpha\in\Gamma}$. 138

$\mathcal{A}_{u,n}$	σ-algebra generated by the family $(\pi_{u,\alpha\|\|\alpha\|-1})_{\alpha\in E_u^{(n)}}$. 138
$A(q)$	42
$A(q,t)$	40
$A_{u,v}(q)$	42
$A_{u,v}(q,t)$	40
β	40
β^*	the Legendre transform of β, i.e. $\beta^*(\alpha) := \inf_q(q\alpha + \beta(\alpha))$ for $\alpha \in \mathbb{R}$. 21
$\hat{\beta}$	54
b_μ	generalized multifractal Hausdorff dimension function. 20
B_μ	generalized multifractal packing dimension function. 20
$\mathcal{B}(X)$	Borel algebra on the topological space X.
$B(x,r)$	closed ball with center x and radius r.
c	constant in condition (IV). 39
$\mathbb{C}$	complex numbers.
$\mathcal{C}^s$	centered s-dimensional Hausdorff measure. 17
$\overline{\mathcal{C}}^s$	centered s-dimensional pre-Hausdorff measure. 17
$\overline{\mathcal{C}}_\delta^s$	δ-approximative centered s-dimensional pre-Hausdorff measure. 16
$C_u, C_u(\omega)$	random graph directed self-similar sets. 28
$C_{u,\gamma}$	28
$C_n^{\mathbf{K}}$	29
$C_{u,n}^{\mathbf{K}}$	28
card	cardinality of a set.
$\mathcal{C}(X)$	family of compact subsets of X.
$C(E)$	box dimension of E (if it exists). 23
$\underline{C}(E)$	lower box dimension of E. 23
$\overline{C}(E)$	upper box dimension of E. 23
$C_\mu^q(E)$	multifractal q-box dimension of E (if it exists). 24
$\underline{C}_\mu^q(E)$	lower multifractal q-box dimension of E. 24
$\overline{C}_\mu^q(E)$	upper multifractal q-box dimension of E. 24
$C^0(X)$	continuous functions on the topological space X.
Con(X)	family of contractions on the metric space X.
δ_a	Dirac measure concentrated at a.
Δ	constant in condition (I). 39
Δ_u	27
$\Delta_\mu(a)$	$\Delta_\mu(a) := \{x \in \operatorname{supp}\mu \mid \alpha_\mu(x) = a\} = \{x \in \operatorname{supp}\mu \mid \lim_{r\searrow 0} \frac{\log \mu B(x,r)}{\log r} = a\}$. 2
$\overline{\Delta}^a(\mu)$	$\overline{\Delta}^a(\mu) := \{x \in \operatorname{supp}\mu \mid \overline{\alpha}_\mu(x) \le a\} = \{x \in \operatorname{supp}\mu \mid \limsup_{r\searrow 0} \frac{\log \mu B(x,r)}{\log r} \le a\}$. 2

$\overline{\Delta}_a(\mu)$	$\overline{\Delta}_a(\mu) := \{x \in \operatorname{supp}\mu \mid a \le \overline{\alpha}_\mu(x)\} = \{x \in \operatorname{supp}\mu \mid a \le \limsup_{r\searrow 0} \frac{\log \mu B(x,r)}{\log r}\}$. 2
$\underline{\Delta}^a(\mu)$	$\underline{\Delta}^a(\mu) := \{x \in \operatorname{supp}\mu \mid \underline{\alpha}_\mu(x) \le a\} = \{x \in \operatorname{supp}\mu \mid \liminf_{r\searrow 0} \frac{\log \mu B(x,r)}{\log r} \le a\}$. 2
$\underline{\Delta}_a(\mu)$	$\underline{\Delta}_a(\mu) := \{x \in \operatorname{supp}\mu \mid a \le \underline{\alpha}_\mu(x)\} = \{x \in \operatorname{supp}\mu \mid a \le \liminf_{r\searrow 0} \frac{\log \mu B(x,r)}{\log r}\}$. 2
$\Delta(E)$	logarithmic index of E. 17
$\Delta^q_\mu(E)$	generalized multifractal logarithmic index of E. 19
$\overset{\mathrm{d}}{=}$	equality in distribution.
d^q_μ	lower q-Rényi dimension of the measure μ. 24
D	$D := \max_v \operatorname{diam}(X_v)$. 172
D_X	Hausdorff metric on $\mathcal{F}_b(X)$.
D^q_μ	upper q-Rényi dimension of the measure μ. 24
diam	diameter of a subset of a metric space.
$\dim(E)$	Hausdorff dimension of E. 17
$\operatorname{Dim}(E)$	packing dimension of E. 17
$\dim^q_\mu(E)$	generalized multifractal Hausdorff dimension of E. 19
$\operatorname{Dim}^q_\mu(E)$	generalized multifractal packing dimension of E. 19
ε_q	136
$\underline{e}$	45
$\overline{e}$	45
$E^{(n)}$	set of (finite) paths of length n. 26
$E^{(*)}$	set of (finite) paths. 26
E_u	set of edges with initial vertex u. 26
$E^{(n)}_u$	set of (finite) paths of length starting at u. 26
$E^{(*)}_u$	set of (finite) paths starting at u. 26
$E^{(\mathbb{N})}_u$	set of infinite paths starting at u. 26
E_{uv}	set of edges with initial vertex u and terminal vertex v. 26
$E^{(n)}_{uv}$	set of (finite) paths of length n from u to v. 26
$E^{(*)}_{uv}$	set of (finite) paths from u to v. 26
$e_\mu(q)$	63
$E_\mu(q)$	63
$\mathbf{E}_P[X]$	expectation of the random variable X w.r.t. the distribution P.
$\mathbf{E}_P[X\|\mathcal{G}]$	conditional expectation of the random variable X given the σ-algebra $\mathcal{G}$ w.r.t. the distribution P.
φ	function defined in Lemma 7.1.3. 172
φ_u	34
Φ_u	34
$\Phi(q,t)$	spectral radius of the matrix $A(q,t)$. 40

f_μ	Hausdorff multifractal spectrum function of the measure μ. 2
F_μ	packing multifractal spectrum function of the measure μ. 2
$\mathcal{F}(X)$	family of closed subsets of X.
$\mathcal{F}_b(X)$	family of closed and bounded subsets of X.
Γ_u	27
$h^q_\delta(\mu)$	generalized (q,δ)-Rényi entropy of the measure μ. 24
$\mathcal{H}^t$	t-dimensional Hausdorff measure. 16
$\mathcal{H}^t_\delta$	δ-approximative t-dimensional pre-Hausdorff measure. 16
$\mathcal{H}^{q,t}_\mu$	generalized multifractal Hausdorff measure. 18
$\overline{\mathcal{H}}^{q,t}_\mu$	generalized centered multifractal pre-Hausdorff measure. 18
$\overline{\mathcal{H}}^{q,t}_{\mu,\delta}$	generalized δ-approximative centered multifractal pre-Hausdorff measure. 18
$\iota(e)$	initial vertex of the edge e. 26
$I^q_{\mu,r}$	25
$\underline{I}^q_\mu$	25
$\overline{I}^q_\mu$	25
κ_γ	42
λ^d	d-dimensional Lebesgue measure in $\mathbb{R}^d$.
λ_u	34
Λ, Λ_u	34
$\Lambda_u, \Lambda_{\lambda,u}$	34
Λ^q_μ	generalized multifractal logarithmic index function. 20
ℓ_γ	43
L_X	metric on $\mathcal{P}(X)$ defined by $L(\mu,\nu) := \sup_{f\in \mathrm{Lip}_1(X)} \left\| \int f\,d\mu - \int f\,d\nu \right\|$. 100
$\mathrm{Lip}(f)$	Lipschitz constant of f. 27
$\mathrm{Lip}_1(X)$	family of maps $f : X \to \mathbb{R}$ with $\mathrm{Lip}(f) \leq 1$.
$\hat\mu$	32
$\bar\mu$	32
$\mu_u, \mu_u(\omega)$	random graph directed self-similar measures. 32
$\hat\mu_u, \hat\mu_u(\omega)$	31
$\bar\mu_u, \bar\mu_u(\omega)$	31
μ^γ_n	31
$\mu^\gamma_{u,n}$	31
$\mathcal{M}_{u,q}$	random multifractal construction measure. 157
$\hat{\mathcal{M}}_{u,q}$	random multifractal construction measure. 164

$\mathcal{M}(X)$	family of Radon measures on X.
$\mathbf{Max}(u)$	family of all finite maximal antichains in $E_u^{\mathbb{N}}$. 137
$\mathbb{N}$	positive integers.
$N_\delta(E)$	largest number of disjoint balls of radius δ with centers in E. 23
Ω	27
Ω_0	28
Ω_u	27
Ω'_u	156
$(\Omega_u)_0$	28
$\mathcal{O}(X,Y)$	compact open topology on $\mathrm{Con}(X,Y)$. 33
π_u	30
$\pi_{u,\alpha}$	27
p_e	27,28
$p_{\alpha e}$	27
P, P_λ	34
$P_u, P_{\lambda,u}$	34
$\mathcal{P}^t$	t-dimensional packing measure. 17
$\overline{\mathcal{P}}^t$	t-dimensional pre-packing measure. 17
$\overline{\mathcal{P}}^t_\delta$	δ-approximative t-dimensional pre-packing measure. 17
$\mathcal{P}^{q,t}_\mu$	generalized multifractal packing measure. 18
$\overline{\mathcal{P}}^{q,t}_\mu$	generalized multifractal pre-packing measure. 18
$\overline{\mathcal{P}}^{q,t}_{\mu,\delta}$	generalized δ-approximative multifractal pre-packing measure. 18
$\mathcal{P}(X)$	family of Borel probability measures on X.
$\mathcal{P}_0(X)$	22
$\mathcal{P}_1(X)$	22
$\mathcal{P}_0(X,E)$	22
$\mathcal{P}_1(X,E)$	22
$q_{\max}$	45,46,51
$q_{\min}$	45,46,51
$\mathbb{Q}$	rational numbers.
$Q_{u,q}$	165
$\rho_u, \rho_u(q)$	42
$\underline{\rho}(q)$	$\underline{\rho}(q) := \min_{u\in V} \rho_u(q)$. 42
$\overline{\rho}(q)$	$\overline{\rho}(q) := \max_{u\in V} \rho_u(q)$. 42
$\mathbb{R}$	real numbers.
$r(A)$	spectral radius of the matix A.
$\sigma_u, \sigma_u(q)$	42

References

Ar M. Arbeiter, *Random recursive construction of self-similar measures. The non-compact case*, Probab. Th. Rel. Fields **88** (1991), 497–520.

As R. Ash, *Real Analysis and Probability*, Academic Press, 1972.

Av V. Aversa & C. Bandt, *The Multifractal Spectrum of Discrete Measures*, Acta Universitatis Carolinae–Mathematica et Physica **31** (1990), 5–8.

Ban C. Bandt, *Self-Similar Sets I. Topological Markov Chains and Mixed Self-Similar Sets*, Math. Nachr. **142** (1989), 107–123.

BanG C. Bandt & S. Graf *Self-Similar Sets 7. A characterization of self-similar fractals with positive Hausdorff measure*, Proc. Amer. Math. Soc. **114** (1992), 995–1001.

Bar M. Barnsley, J. Elton & D. Hardin, *Recurrent Iterated Function Systems*, Constr. Approx. **5** (1989), 3–31.

Bed T. Bedford & M. Urbanski, *The box and Hausdorff dimension of self-affine sets*, Ergodic Theory Dyn. Sys. **10** (1990), 627–644.

Bi1 P. Billingsley, *Hausdorff dimension in probability theory*, Illinois J. Math. **4** (1960), 187–209.

Bi2 P. Billingsley, *Ergodic Theory and Information*, Wiley, 1965.

Bo T. Bohr & D. Rand, *The entropy function for characteristic exponents*, Physica**25D** (1987), 387–393.

Bow R. Bowen, *Equilibrium States and the Ergodic Theory of Anosov Diffeomorphism*, Lecture Notes in Mathematics **470**, Springer Verlag, Berlin, New York, 1975.

Br G. Brown, G Michon & J. Peyrière, *On the Multifractal Analysis of Measures*, Journal of Statistical Physics **66** (1992), 775–790.

Ca R. Cawley & R. D. Mauldin, *Multifractal Decomposition of Moran Fractals*, Advances in Mathematics **92** (1992), 196–236.

Co1 P. Collet, J. L. Lebowitz & A. Porzio, *The Dimension Spectrum of Some Dynamical Systems*, Journal of Statistical Physics **47** (1987), 609–644.

Co2 P. Collet, *Hausdorff Dimension of the Singularities for Invariant Measures of Expanding Dynamical Systems*, Proceedings, Dynamical Systems Valparaiso 1986, pp. 47–58, Lecture Notes in Mathematics, No. 1331, Springer Verlag, 1988.

Cu1 C. D. Cutler, *Connecting ergodicity and dimension in dynamical systems*, Ergod. Th. & Dynam. Sys. **10** (1990), 451–462.

Cu2 C. D. Cutler, *Measure disintegrations with respect to σ-stable monotone indices and pointwise representation of packing dimension*, Proceedings of the 1990 Measure Theory Conference at Oberwolfach. Supplemento Ai Rendiconti del Circolo Mathematico di Palermo, Ser. II, No 28 (1992), 319–340.

Cu3 C. D. Cutler, *Some Results on the Behavior and Estimation of the Fractal Dimensions of Distributions on Attractors*, Journal of Statistical Physics **62** (1991), 651–708.

Del A. Deliu, J. S. Geronimo, R. Shonkwiler & D. Hardin, *Dimension associated with recurrent self-similar sets*, Math. Proc. Camb. Phil. Soc. **110** (1991), 327–336.

Du1 L. E. Dubins & D. A. Freedman, *Random distribution functions*, Bull. Amer. Math. Soc. **69** (1963), 548–551.

Du2 L. E. Dubins & D. A. Freedman, *Random distribution functions*, Proceedings, Fifth Berkeley Symp. on Math. Statistics and Probability (editors L. M. LeCam & J. Neuman), pp. 183–214, Univ. of California Press, Berkeley/Los Angeles, 1967.

Edg G. A. Edgar, *Measure, topology, and fractal geometry*, Springer Verlag, New York, 1990.

Ed G. A. Edgar & R. D. Mauldin, *Multifractal Decompositions of Digraph Recursive Fractals*, Proc. London Math. Soc. **65** (1992), 604–628.

Fa1 K. J. Falconer, *The Geometry of Fractal Sets*, Cambridge Tracts in Mathematics, No 85, Cambridge Univ. Press, New York/London, 1985.

Fa2 K. J. Falconer, *Fractal Geometry-Mathematical Foundations and Applications*, John Wiley & Sons, 1990.

Fa3 K. J. Falconer, *Random fractals*, Math. Proc. Camb. Phil. Soc. **100** (1986), 559-582.

Fa4 K. J. Falconer, *The Multifractal Spectrum of Statistically Self-Similar Measures*, preprint (1993).

Fa5 K. J. Falconer, *The Hausdorff dimension of self-affine fractals*, Math. Proc. Camb. Phil. Soc. **103** (1988), 339–350.

Fa6 K. J. Falconer, *The dimension of self-affine fractals II*, Math. Proc. Camb. Phil. Soc. **111** (1992), 169–179.

Fed H. Federer, *Geometric measure theory*, Grundlehren der Math. Wiss. Band 153, Springer Verlag, 1969.

Fen S. Feng, *An Extension of Multifractal Decomposition of Generalized Moran Fractals*, preprint (1993).

Fr U. Frisch & G. Parisi, *On the singularity structure of fully developed turbulence*, appendix to U. Frisch, *Fully developed turbulence and intermittency*, Turbulence and Predictability in Geophysical Fluid Dynamics and Climate Dynamics, Proc. Int. Sch. Phys., "Enrico Fermi" Course LXXXVIII, pp.84–88, North Holland, Amsterdam, 1985.

Fro O. Frostman, *Potentiel d'équilibre et capacité des ensembles avec quelques applications à la théorie des fonctions*, Meddel. Lunds Univ. Mat. Sem.**3** (1935), 1–118.

Ge J. S. Geronimo & D. Hardin, *An exact formula for the measure dimension associated with a class of piecewise linear maps*, Constr. Approx. **5** (1989), 89–98.

Gra1 S. Graf, *Statistically Self-Similar Fractals*, Probab. Th. Rel. Fields **74** (1987), 357–394.

Gra2 S. Graf, R. D. Mauldin & S. C. Williams, *The exact Hausdorff dimension in random recursive constructions*, Mem. Am. Math. Soc. **71** no. 381 (1988).

Gra3 S. Graf, R. D. Mauldin & S. C. Williams, *Random Homeomorphisms*, Advances in Mathematics **60** (1987), 239–359.

Gr1 P. Grassberger & I. Procaccia, *Characterization of strange attractors*, Phys. Rev. Lett. **50** (1983), 346–349.

Gr2 P. Grassberger, *Generalized dimensions of strange attractors*, Phys. Lett. **97A** (1983), 227–230.

Haa1 H. Haase, *On the dimension of product measures*, Mathematika **37** (1990), 316–323.

Haa2 H. Haase, *Dimension of Measures*, Acta Universitatis Carolinae–Mathematica et Physica **31** (1990), 29–34.

Haa3 H. Haase, *A survey of the dimensions of measures*, Proceedings of conference "Topology and Measure VI", Warnemünde, Germany, August 1991 (editors C. Bandt, J. Flachsmeyer & H. Haase), pp. 66–75, in Mathematical Research, Vol. 66, Topology, Measures, and Fractals, Akademie Verlag, 1992.

Ha T. C. Halsey, M. H. Jensen, L. P. Kadanoff, I. Procaccia & B. J. Shraiman, *Fractal measures and their singularities: The characterization of strange sets*, Phys. Rev. A **33** (1986), 1141–1151.

He H. Hentschel & I. Procaccia, *The infinite number of generalized dimensions of fractals and strange attractors*, Physica **8D** (1983), 435–444.

Ho R. Holley & E. C. Waymire, *Multifractal Dimension and Scaling Exponents for Strongly Bounded Random Cascades*, The Annals of Applied Probability **2** (1992), 819–845.

Hu J. Hutchinson, *Fractals and self-similarity*, Indiana Univ. Math. J. **30** (1981), 713–747.

Kah1 J.-P. Kahane, *Sur le modèle de turbulence de Benoit Mandelbrot*, C. R. Acad. Sci. Paris **278** (1974), 621–623.

Kah2 J.-P. Kahane, *Sur le Chaos Multiplicatif*, Ann. Sc. Math. Québec **9** (1985), 105–150.

Kah3 J.-P. Kahane, *Positive Martingales and Random Measures*, Chinese Ann. Math. **8b** (1987), 1–12.

Kah4 J.-P. Kahane, *Multiplications aléatoires et dimensions de Hausdorff*, Ann. Inst. Poincare **23** (1987), 289–296.

Kah5 J.-P. Kahane, *Random multilplications, random coverings, and multiplicative chaos*, Analysis at Urbana, Proceedings of the Special Year in Modern Analysis at the University of Illinois, 1986–87 (editors E. Berkson, N. Peck and J. Uhl), pp. 196–255, London Mathematical Society Lecture Note Series, Vol. 137, Cambridge Univ. Press, Cambridge, 1989.

Kah6 J.-P. Kahane, *Produits de poids aléatoires indépendants et applications*, Proceedings of the NATO Advanced Study Institute and Séminaire de mathématiques supérieures on Fractal Geometry and Analysis, Montréal, Canada, July 3–21, 1989 (editors J. Bélair & S. Dubuc), pp. 277–324, NATO ASI Series, Series C: Mathematical and Physical Sciences, Vol. 346, Kluwer Academic Press, 1991.

Kah7 J.-P. Kahane & J. Peyrière, *Sur Certaines Martingales de Benoit Mandelbrot*, Advances in Mathematics **22** (1976), 131–145.

Kal O. Kallenberg, *Random Measures*, Akademie Verlag, Berlin, 1983.

Ke R. Kenyon & Y. Peres, *Hausdorff Dimensions of Affine-Invariant Sets and Sierpinski Sponges*, preprint (1992).

Ki1 J. King, *The Singularity Spectrum for General Sierpinski Carpets*, preprint (1992).

Ki2 J. King & J. S. Geronimo, *Singularity spectrum for recurrent IFS attractors*, Nonlinearity **6** (1992), 337–348.

Kin J. R. Kinney & T. S. Pitcher, *The Dimension of the Support of a Random Distribution Function*, Bull. Amer. Math. Soc. **70** (1964), 161–164.

Le1 J. Lee, P. Alstrøm & H. E. Stanley, *Is there a phase transition in the multifractal spectrum of DLA*, Fractals' Physical Origin ond Properties, Proceedings of the Special Seminar on Fractals at the Ettore Majorana Centre for Scientific Culture, Erice (Trapani), Italy, October, 1988 (editor L. Pietronero), pp. 217–226, Plenum Press, New York, 1989.

Le2 J. Lee, P. Alstrøm & H. E. Stanley, *Phase Transition on DLA*, Proceedings of the NATO Advanced Study Institute on Random Fluctuations and Pattern Growth: Experiments and Models, Cargése, Corsica, France, July 18–31, 1988 (editors H. E. Stanley & N. Ostrowsky), p. 311, NATO ASI Series, Series E: Applied Sciences, Vol. 157, Kluwer Academic Press, 1988.

Lo1 A. O. Lopes, *The Dimension Spectrum of the Maximal Measure*, SIAM J. Math. Anal. **20** (1989), 1243–1254.

Lo2 A. O. Lopes, *Dimension Spectra and a Mathematical Model for Phase Transition*, Advances in Applied Mathematics **11** (1990), 475–502.

Lo3 A. O. Lopes, *Entropy and large deviation*, Nonlinearity **3** (1990), 527–546.

Lo4 A. O. Lopes, *A First-Order Level-2 Phase Transition*, in Thermodynamic Formalism Journal of Statistical Physics **60** (1990), 395–411.

Lo5 A. O. Lopes, *The zeta function, non-differentiability of pressure and the critical exponent of transition*, Advances in Mathematics **101** (1993), 133–165.

Man1 B. Mandelbrot, *Intermittent turbulence in self similar cascades: Divergence of high moments and dimension of the carrier*, J. Fluid Mech. **62** (1974), 331–358.

Man2 B. Mandelbrot, *Multiplications aléatoires itérées et distributions invariantes par moyenne pondérée aléatoire*, C. R. Acad. Sci. Paris **278** (1974), 289–292.

Man3 B. Mandelbrot, *Multiplications aléatoires itérées et distributions invariantes par moyenne pondérée aléatoire: quelqeus extensions*, C. R. Acad. Sci. Paris **278** (1974), 355–358.

Mat P. Mattila, *Differentiation of Measures on Uniform Spaces*, Proceedings Measure Theory Oberwolfach 1979, pp. 261–282, Lecture Notes in Mathematics, No. 794, Springer Verlag, 1980.

Mau1 R. D. Mauldin & S. C. Williams, *Random Recursive Constructions: Asymptotic Geometric and Topological Properties*, Trans. Am. Math. Soc. **295** (1986), 325–346.

Mau2 R. D. Mauldin & S. C. Williams, *Hausdorff Dimension in Graph Directed Constructions*, Transactions of the American Math. Soc. **309** (1988), 811–829.

Mc C. McMullen, *The Hausdorff dimension of general Sierpinski Carpets*, Nagoya Math. J. **96** (1984), 1–9.

Me M. Mendes-France & G. Tenenbaum, *A one-dimensional model with phase transition*, preprint (1992).

Min H. Minc, *Nonnegative matrices*, Wiley–Interscience series in discrete mathematics and optimization, Wiley, New York, 1988.

Mo P. A. P. Moran, *Additive functions of intervals and Hausdorff measure*, Proceedings of the Cambridge Philosophical Society **42** (1946), 15–23.

Ol L. Olsen, *A Multifractal Formalism*, Advances in Mathematics, to appear.

Pa G. Paladin & A. Vulpiani, *Anomalous scaling laws in multifractal objects*, Physics Reports **156** (1987), 147–225.

Pat N. Patzschke & U. Zähle, *Self-Similar random measures IV — The Recursive Construction Model of Falconer, Graf and Mauldin and Williams*, Math. Nachr. **149** (1990), 285–302.

Per Y. Peres, *The self-affine carpets of McMullen and Bedford have infinite Hausdorff measure*, preprint (1992).

Pes Y. Pesin & H. Weiss, *On the dimension of a general class of geometrically defined deterministic and random Cantor-like sets*, preprint (1993).

Pe1 Ya. Pesin, *Dimension type characteristics for invariant sets of dynamical systems*, Russian Math. Surveys **43** (1988), 111–151.

Pe2 Ya. Pesin, *Generalized spectrum for the dimension: the approach based on Carathéodory's construction*, Constantin Carathéodory: an international tribute, pp. 1108–1119, World Sci. Publishing, Teaneck, 1991.

Pe3 Ya. Pesin, *On Rigorous Mathematical Definitions of Correlation Dimension and Generalized Spectrum for Dimensions*, Journal of Statistical Physics **71** (1993), 529–547.

Pey1 J. Peyrière, *Turbulence et dimension de Hausdorff*, C. R. Acad. Sci. Paris **278** (1974), 567–569.

Pey2 J. Peyrière, *Calculs de Dimensions de Hausdorff*, Duke Mathemetical Journal **44** (1977), 591–601.

Pey3 J. Peyrière, *Multifractal Measuures*, Proceedings of the NATO Advanced Study Institute on Probabilistic and Stochastic Methods in Analysis with Applications, Il Ciocco, pp. 175–186, NATO ASI Series, Series C: Mathematical and Physical Sciences, Vol 372, Kluwer Academic Press, Dordrecht, 1992.

Ra D. Rand, *The singularity spectrum $f(\alpha)$ for cookie-cutters*, Ergod. Th. & Dynam. Sys. **9** (1989), 527–541.

Ray X. S. Raymond & C. Tricot, *Packing regularity of sets in n-space*, Math. Proc. Camb. Phil. Soc. **103** (1988), 133–145.

Re1 A. Rényi, *Some Fundamental Questions of Information Theory*, Magyar Tud. Akad. Mat. Fiz. Oszt. Közl **10** (1960), 251–282.

Re2 A. Rényi, *On Measures of Entropy and Information*, Proceedings 4th Berkeley Symposium on Mathematical Statistics and Probability 1960, pp. 547–561, Univ. of California

Press, Berkeley, 1961.
Rob A. Roberts & D. Varberg, *Convex Functions*, Academic Press, 1973.
Ru D. Ruelle, *Thermodynamic Formalism*, Addison-Wesley, Reading, MA, 1978.
Sa H. Sato, *Global Density Theorem for a Federer Measure*, Tôhoku Math. J. **44** (1992), 581–595.
Sc A. Schief, *Separation properties of self-similar sets*, Proc. Amer. Math. Soc., to appear.
Se E. Seneta, *Non-Negative Matrices and Markov Chains, 2. ed.*, Springer Verlag, New York, 1981.
Sh M. Shereshevsky, *A complement to Young's theorem on measure dimension: the difference between lower and upper pointwise dimensions*, Nonlinearity **4** (1991), 15–25.
Ste S. Stella, *On Hausdorff Dimension of Recurrent Net Fractals*, Proc. Am. Math. Soc. **116** (1992), 389–400.
Str R. S. Strichartz, *Self-Similar Measures and their Fourier Transforms III*, preprint (1992).
Ta S. J. Taylor & C. Tricot, *Packing Measure, and its Evaluation for a Brownian Path*, Trans. Amer. Math. Soc. **288** (1985), 679–699.
Tay1 S. J. Taylor, *The measure theory of random fractals*, Math. Proc. Camb. Phil. Soc. **100** (1986), 383–406.
Tay2 S. J. Taylor, *A measure theory definition of fractals*, Proceedings of the Measure Theory Conference at Oberwolfach. Supplemento Ai Rendiconti del Circolo Mathematico di Palermo, Ser. II, No 28 (1992), 371–378.
Te1 T. Tel, *Fractals, multifractals and thermodynamics*, Zschr. f. Naturf. A **43** (1988), 1154–1174.
Te2 T. Tel, *Dynamical spectrum and thermodynamic functions of strange sets from an eigenvalue problem*, Phys. Rev. **36A** (1987), 2507–2510.
To H. Tohoki & Y. Tsujii, *A remark on random fractals*, Hiroshima Math. J. **19** (1989), 563–566.
Tr C. Tricot, *Two definitions of fractional dimension*, Math. Proc. Camb. Phil. Soc. **91** (1982), 57–74.
Tsu1 Y. Tsujii, *Generalized random ergodic theorems and Hausdorff*, measures of random fractals Hiroshima Math. J. **19** (1989), 363–377.
Tsu2 Y. Tsujii, *Markov-self-similar sets*, Hiroshima Math. J. **21** (1991),491–519.
Wa P. Walters, *An Introduction to Ergodic Theory*, Springer Verlag, New York, 1982.
Way1 E. Waymire & S. C. Williams, *Markov cascades*, preprint (1993).
Way2 E. Waymire & S. C. Williams, *Multipoicative cascades: Dimension spectra and dependence*, preprint presented at: Colloque en l'honneur de J.-P. Kahane, Université de Paris-Sud, Orsay, France, 1993.
Wi D. Williams, *Probability with Martingales*, Cambridge University Press, Cambridge, 1991.
Yo L.-S. Young, *Dimension, entropy and Lyapunov Exponents*, Ergod. Th. & Dynam. Sys. **2** (1982), 109–124.
Zä1 U. Zähle, *Self-similar random measures. I — Notion, carrying Hausdorff dimension, and hyperbolic distribution*, Prob. Th. Rel. Fields **80** (1988), 79–100.
Zä2 U. Zähle, *Self-similar random measures. II — Generalization to self-affine measures*, Math. Nachr. **146** (1990), 85–98.
Zä3 U. Zähle, *The Fractal character of localizable measure value processes. III — Fractal carrying sets of branching diffusions*, Math. Nachr. **138** (1988), 293–311.

Printed and bound by CPI Group (UK) Ltd, Croydon, CR0 4YY
23/10/2024
01778230-0009